COMPUTATIONAL CHEMISTRY METHODOLOGY IN STRUCTURAL BIOLOGY AND MATERIALS SCIENCES

COMPUTATIONAL CHEMISTRY METHODOLOGY IN STRUCTURAL BIOLOGY AND MATERIALS SCIENCES

Edited by

Tanmoy Chakraborty, PhD

Prabhat Ranjan, BE, MTech

Anand Pandey, PhD

AAP | APPLE ACADEMIC PRESS

Apple Academic Press Inc.
3333 Mistwell Crescent
Oakville, ON L6L 0A2 Canada

Apple Academic Press Inc.
9 Spinnaker Way
Waretown, NJ 08758 USA

Printed in the United States of America on acid-free paper
International Standard Book Number-13: 978-1-77188-568-3 (Hardcover)
International Standard Book Number-13: 978-1-315-20754-4 (eBook)

Library and Archives Canada Cataloguing in Publication

Computational chemistry methodology in structural biology and materials sciences / edited by Tanmoy Chakraborty, PhD, Prabhat Ranjan, BE, MTech, Anand Pandey, PhD.

Includes bibliographical references and index.
Issued in print and electronic formats.
ISBN 978-1-77188-568-3 (hardcover).--ISBN 978-1-315-20754-4 (PDF)

1. Chemistry, Physical and theoretical--Data processing. 2. Chemistry, Physical and theoretical--Methodology. 3. Biology. 4. Materials science. I. Chakraborty, Tanmoy, editor II. Ranjan, Prabhat (Mechatronics professor), editor III. Pandey, Anand, editor

QD455.3.E4C66 2017 541.0285 C2017-903269-0 C2017-903270-4

Library of Congress Cataloging-in-Publication Data

Names: Chakraborty, Tanmoy, editor. | Ranjan, Prabhat, (Mechatronics professor) editor. | Pandey, Anand, editor.
Title: Computational chemistry methodology in structural biology and materials sciences / editors, Tanmoy Chakraborty, PhD, Prabhat Ranjan, BE, MTech, Anand Pandey, PhD.
Description: Toronto; New Jersey: Apple Academic Press, 2017. | Includes bibliographical references and index.
Identifiers: LCCN 2017021264 (print) | LCCN 2017028825 (ebook) | ISBN 9781315207544 (ebook) | ISBN 9781771885683 (hardcover : alk. paper)
Subjects: LCSH: Proteins--Structure--Mathematical models. | Materials science--Mathematical models. | Chemistry--Electronic data processing.
Classification: LCC QP551 (ebook) | LCC QP551 .C7124 2017 (print) | DDC 572/.633--dc23
LC record available at https://lccn.loc.gov/2017021264

Apple Academic Press also publishes its books in a variety of electronic formats. Some content that appears in print may not be available in electronic format. For information about Apple Academic Press products, visit our website at **www.appleacademicpress.com** and the CRC Press website at **www.crcpress.com**

CONTENTS

LIST OF CONTRIBUTORS

A. K. Bandyopadhyay
Government College of Engineering and Ceramic Technology, W. B. University of Technology, 73, A. C. Banerjee Lane, Kolkata–700010, India, E-mail: asisbanerjee1000@gmail.com

Vincenzo Barone
Piazza dei Cavalieri 7, 56126 Pisa, Italy, E-mail: vincenzo.barone@sns.it

Pakiza Begum
Department of Chemical Sciences, Tezpur University, Tezpur, Napaam, 784 028, Assam, India

Andrea Brogni
Piazza dei Cavalieri 7, 56126 Pisa, Italy

Tanmoy Chakraborty
Department of Chemistry, Manipal University Jaipur, Jaipur, Rajasthan, India–303007, E-mail: tanmoy.chakraborty@jaipur.manipal.edu; tanmoychem@gmail.com

Santanu Das
Department Materials Science and Engineering, University of North Texas, Denton, TX 76207, USA

Ramesh C. Deka
Department of Chemical Sciences, Tezpur University, Tezpur, Napaam, 784 028, Assam, India, E-mail: ramesh@tezu.ernet.in

Andrea Fratalocchi
Primalight, Faculty of Electrical Engineering; Applied, Mathematics and Computational Science, King Abdullah University of Science and Technology (KAUST), Thuwal 23955-6900, Saudi Arabia

Aniruddha Ghosal
Institute of Radiophysics and Electronics, Calcutta University, Calcutta, India

Juan Sebastian Totero Gongora
Primalight, Faculty of Electrical Engineering; Applied, Mathematics and Computational Science, King Abdullah University of Science and Technology (KAUST), Thuwal 23955-6900, Saudi Arabia

John C. Hackett
Institute for Structural Biology and Drug Discovery, Virginia Commonwealth University, 800 East Leigh Street, Richmond, Virginia 23219, USA

Christopher M. Hadad
Department of Chemistry, The Ohio State University, 100 W. 18th Ave., Columbus, Ohio 43210, USA

Carrigan J. Hayes
Department of Chemistry, Otterbein University, 1 South Grove Street, Westerville, Ohio 43081, USA

M. A. Jaseela
Department of Chemistry, University of Calicut, Malappuram, Kerala, 673635, India

Rita Kakkar
Computational Chemistry Group, Department of Chemistry, University of Delhi, Delhi–110007, India, Tel.: +91-11-27666313; E-mail: rkakkar@chemistry.du.ac.in

Heena Khanchandani
Department of Metallurgical and Materials Engineering, MNIT, Jaipur, India

Ju Young Kim
Institute for Cell Engineering, Johns Hopkins University School of Medicine, 733 N Broadway, Baltimore, Maryland 21205, USA

Ajay Kumar
Department of Mechatronics, Manipal University Jaipur, Jaipur, Rajasthan, India–303007

Nitish Kumar
Centre for Nanotechnology, Central University of Jharkhand, Ranchi–835205, India

Rupesh Kumar
Department of Metallurgical and Materials Engineering, MNIT, Jaipur, India

Vinod Kumar
Assistant Professor, Department of Metallurgical and Materials Engineering, MNIT, Jaipur–302017, India, Tel.: +91 141 2713457, E-mail: vkt.meta@mnit.ac.in

Moklesa Laskar
Institute of Radiophysics and Electronics, Calcutta University, Calcutta, India

Daniele Licari
Piazza dei Cavalieri 7, 56126 Pisa, Italy

Giordano Mancini
Piazza dei Cavalieri 7, 56126 Pisa, Italy

Ornov Maulik
Department of Metallurgical and Materials Engineering, MNIT, Jaipur, India

Narges Mohammadi
Molecular Model Discovery Laboratory, Department of Chemistry and Biotechnology, School of Science, Faculty of Science, Engineering and Technology, Swinburne University of Technology, Hawthorn, Melbourne, Victoria, 3122, Australia

K. Muraleedharan
Department of Chemistry, University of Calicut, Malappuram – 673635, India; Tel.: +91-494-2407413; Fax: +91-494-2400269; E-mail: kmuralika@gmail.com

Vijisha K. Rajan
Department of Chemistry, University of Calicut, Malappuram – 673635, India; Tel.: +91-494-2407413; Fax: +91-494-2400269

Prabhat Ranjan
Department of Mechatronics, Manipal University Jaipur, Jaipur, Rajasthan, India–303007

David Saffen
Department of Cellular and Genetic Medicine, School of Basic Medical Sciences, Fudan University, 130 Dongan Rd, Shanghai 200032, P.R. China

Andrea Salvadori
Piazza dei Cavalieri 7, 56126 Pisa, Italy

G. Saranya
Department of Physics, Bharathiar University, Coimbatore–641046, India

Babusona Sarkar
Department of Materials Science, Indian Association for the Cultivation of Science, Jadavpur, Kolkata–700032, India

K. Senthilkumar
Department of Physics, Bharathiar University, Coimbatore–641046, India, Fax: +91-422-2422387, E-mail: ksenthil@buc.edu.in

Priyanka Sharma
Department of Metallurgical and Materials Engineering, MNIT, Jaipur, India

Jaibeer Singh
Department of Metallurgical and Materials Engineering, MNIT, Jaipur, India

T. M. Suhara
Department of Chemistry, MES Ponnani College, University of Calicut, Kerala, 679577, India

Peng Tao
Department of Chemistry, Southern Methodist University, 3215 Daniel Avenue, Dallas, Texas 75275–0314, USA, E-mail: ptao@smu.edu

Feng Wang
Molecular Model Discovery Laboratory, Department of Chemistry and Biotechnology, School of Science, Faculty of Science, Engineering and Technology, Swinburne University of Technology, Hawthorn, Melbourne, Victoria, 3122, Australia, Tel.: +61 3 9214 5065; Fax: +61-3-9214-5921; E-mail: fwang@swin.edu.au

LIST OF ABBREVIATIONS

4hC	4-hydroxy coumarin
4mE	4-methyl esculetin
7h4mC	7-hydroxy-4-methyl coumarin
7hC	7-hydroxy coumarin
ADME	absorption, distribution, metabolism and excretion
BLYP	Becke-Lee-Yang-Parr
BSSE	basis set superposition error
C	coumarin
CFs	core functions
CI	cyberinfrastructures
CIS	configuration interaction singles
COs	core orbitals
CP	critical point
CPCM	conductor-like polarizable continuum model
CT	charge transfer
DA	dissociative adsorption
DBs	discrete breathers
DC	3,4-dihydrocoumarin
DCA	dichloroacetate
DFT	density functional theory
DOS	density of states
DSSC	dye sensitized solar cells
E	esculetin
EA	electron affinity
EGI	European grid infrastructure
EM	electromagnetic
EMM	molecular mechanics energies
Evdw	van der Waals interactions
FDTD	finite difference-time domain
FFs	force fields
FWHM	full width at half-widths maximum

GGA	generalized gradient approximation
HOMO	highest occupied molecular orbital
HPC	high-performance computer
IC	internal coordinates
ILM	intrinsic localized modes
IP	ionization potential
KB	Kleinman and Bylander
LCPO	linear combination of pairwise overlaps
LIA	linear interaction approximation
LOB	large object datatypes
LP	lone pair
LUMO	lowest unoccupied molecular orbitals
MB	Maxwell-Bloch
MC	Monte Carlo
MD	molecular dynamics
MI	magneto-inductive
MM	molecular mechanics
MM	metamaterial
NBO	natural bond orbital
NCPPs	norm-conserving pseudo potentials
NIR	near infra-red
NLKG	nonlinear Klein-Gordon
NLO	nonlinear optical
NLSE	nonlinear Schrodinger equation
NMODE	normal mode
NPA	natural population analysis
NSF	National Science Foundation
OFETs	organic field effect transistors
OLEDs	organic light emitting diodes
OPVs	organic photovoltaic cells
OSC	organic solar cell concentrators
P	polarization
PAW	projected augmented wave
PBC	periodic boundary conditions
PBS	portable batch system
PCM	polarizable continuum model

PDC pyruvate dehydrogenase complex
PDE partial differential equations
PDHK pyruvate dehydrogenase kinase
PDOS partial density of states
PEN pentacene
PP pseudopotential
PSA polar surface area
pWT phosphoSer768 wild-type
QBs quantum breathers
QM quantum mechanics
QNPC quadratic nonlinear photonic crystal
QPM quasi-phase matching
RCSB Research Collaboratory for Structural Bioinformatics
RIC redundant internal coordinates
RLC resistor-inductor-capacitor
RMSD root-mean square deviation
RR Resonance Raman
S Softness
SBH Schottky barrier height
SCF similar convergence
SCF self-consistent field
SFOs symmetrized fragment orbitals
SGB surface generalized born
SHG second harmonic generation
SRR split-ring-resonators
TD-DFT time dependent density functional theory
TET targeted energy transfer
TM transmembrane
TPA triphenylamine
TPBS two-phonon bound state
TPSA topological PSA
TRPC transient receptor potential-canonical
TRPC6 transient receptor potential-canonical-6
TS transition state
UPML uniaxial perfectly matched layer
vdW van der Waals

VPAC	Victorian Partnership for Advanced Computing
VRM	virtual reality modeling
XP	extra precision
XSEDE	Extreme Science and Engineering Discovery Environment

PREFACE

This book, *Computational Chemistry Methodology in Structural Biology and Material Sciences*, provides a survey of research problems in theoretical and experimental chemistry. The subject matter covered in the book varies from materials science to biological activity. Part 1 of the book emphasizes new developments in the domain of theoretical and computational chemistry and its applications to bio-active molecules, whereas in Part 2 the study of materials science has been depicted vividly.

In Chapter 1, the authors have computed the pK_a value of a number of alkylamines using the density functional theory (DFT) methodology. Considering versatility and importance of amines in different domain, this particular study is very useful and relevant. It will help to explain the mechanistic feature of CO_2 capturing processes by amines. A close agreement is observed between experimental parameters with the computed data.

Keeping in view the wide biological importance of coumarins, this report is very useful. The study on the effects of unsaturation of chemical reactivity of coumarins has been reported in Chapter 2. Invoking DFT-based descriptors, the authors have shown the reactivity variations by substitution. Site selectivity has been also predicted by using local DFT-based descriptors.

In Chapter 3, molecular dynamics simulations have been utilized to study the interaction between FKBP12 (FK506 binding protein-12 kDa) and transient receptor potential-canonical 6 (TRPC6). The computed data have identified thermodynamically favorable binding affinity with FKBP12. The study reveals the formation of specific binding pockets for the recognition and interaction of FKBP12 with the TRPC6 intracellular domain.

In Chapter 4, the author has worked on finding inhibitors of the pyruvate dehydrogenase kinase (PDHK). He has explored the interaction within dichloroacetate (DCA) and PDHK2. The results of virtual screening are in similar line with the experimental findings. A search for more potent inhibitors is discussed.

The evolution of computational chemistry is mapped in this report. Two parallel approaches of computational chemistry viz. quantum mechanics and molecular mechanics have been discussed. The importance of two approaches, different computational techniques, and latest development has been noted in Chapter 5.

The application of computational chemistry to design new materials is nicely reflected in Chapter 6. Designing of photoactive materials is an active field of research. In this report, computational processes for designing and modeling of photoactive compounds having application in the solar cells are reported. The unique features of dye-sensitized solar cells have been studied in terms of computational processes, and predictions have been done toward new photovoltaic materials in terms of modeling.

In Chapter 7, the authors have tried to predict stable adsorption geometry of organic molecule on metal surface invoking using theoretical technique. Some of electronic properties have been considered to characterize charge transfer properties of organic molecules.

In Chapter 8, the conversion of methane to liquid fuels in terms of DFT has been reported. C-H bond activation of methane promoted by Pt and Pd sub-nanoclusters have been investigated and reported. The study reveals the efficacy of Pt clusters in breaking of C-H bond in methane.

In Chapter 9, the electronic, magnetic and optical properties of copper-silver nano alloy clusters have been studied in terms of DFT-based descriptors in this analysis. Computed DFT descriptors nicely correlate the optical properties of instant compound. Theoretical parameters show a hand-in-hand trend with experimental data.

In Chapter 10, the authors have derived nonlinear Klein Gordon-equation for metamaterials in terms of the model behavior of split-ring resonators for the application in antenna. The variation principle has been applied to reach mathematical equations. Multisolitone behavior of metamaterial system has also been explained by this work.

In Chapter 11, the model has been developed for dispersive active materials in finite-difference time-domain (FDTD) framework. The dispersive materials having optical properties have been modeled in terms of relative dielectric permittivity. A second-order optimized algorithm has been used to deal with the dispersion. In addition, there is a discussion of the Maxwell-Bloch (MB) formalism and solution.

In Chapter 12, a synthesizing technique of nanocrystalline AlMgFeCu-CrNi has been discussed. Experimental technique successfully predicts the behaviors of the synthesize compounds. The phase transfer observation is also discussed.

Overall, this book, *Computational Chemistry Methodology in Structural Biology and Material Sciences*, is a collection of chapters that cover a wide range of subject matter regarding the application of theoretical and experimental chemistry, materials science, and biological domain. The research present in this book is very important in the context of contemporary research problems.

ABOUT THE EDITORS

Tanmoy Chakraborty, PhD
Associate Professor, Department of Chemistry, Manipal University, Jaipur, India

Tanmoy Chakraborty, PhD, is Associate Professor in the Department of Chemistry at Manipal University Jaipur, India. He has been working in the challenging field of computational and theoretical chemistry for the last six years. He has completed his PhD from the University of Kalyani, West-Bengal, India, in the field of application of QSAR/QSPR methodology in the bioactive molecules. He has published many international research papers in peer-reviewed international journals with high impact factors. Dr. Chakraborty is serving as an international editorial board member of the *International Journal of Chemoinformatics and Chemical Engineering*. He is also reviewer of the *World Journal of Condensed Matter Physics* (WJCMP). Dr. Tanmoy Chakraborty is the recipient of prestigious Paromeswar Mallik Smawarak Padak, from Hooghly Mohsin College, Chinsurah (University of Burdwan), in 2002.

Prabhat Ranjan, BE, MTech
Assistant Professor, Department of Mechatronics, Manipal University, Jaipur, India

Prabhat Ranjan is now working as Assistant Professor in the Department of Mechatronics Engineering at Manipal University Jaipur, India. He has been working in the area of computational nanomaterials for the last four years. He has published high-quality research papers in peer-reviewed international journals and has also participated in various conferences and workshops at the national and international level. He was awarded the Manipal University Jaipur-President Award in the year 2015 for his contribution toward the development of the university and also received a Materials Design Scholarships 2014 for his contribution in the area of materials modeling. He has completed his Bachelor of Engineering in Electronics and Communication and Master of Technology in Control System Engineering from Manipal University, Manipal.

Anand Pandey, PhD
Associate Professor, Department of Mechanical Engineering, Manipal University, Jaipur, India

Anand Pandey, PhD, is now working as Associate Professor in Department of Mechanical Engineering, Manipal University Jaipur, India. Dr. Pandey's research area includes macro- and micro-machining of aerospace materials, design of experiments, hybrid machining, and fabrication of metal-composites. He has published in 28 research articles published in leading international journals and conferences. He has completed his doctoral degree from the Sant Longowal Institute of Engineering & Technology (Deemed University) in Longowal, India.

PART I

COMPUTATIONAL CHEMISTRY METHODOLOGY IN BIOLOGICAL ACTIVITY

CHAPTER 1

STUDY OF pKa VALUES OF ALKYLAMINES BASED ON DENSITY FUNCTIONAL THEORY

VIJISHA. K. RAJAN and K. MURALEEDHARAN

Department of Chemistry, University of Calicut, Malappuram – 673635, India; Tel.: +91-494-2407413; Fax: +91-494-2400269; E-mail: kmuralika@gmail.com

CONTENTS

ABSTRACT

A computational investigation based on the density functional theory (DFT) has been performed for predicting the pKa value of alkylamines. Amines are versatile and widely studied class of organic compounds. Their importance in biochemistry, natural product chemistry, industrial purposes, etc., make them highly frequent subject in physical as well as in (bio) chemical studies. The pKa studies are important in explaining the acid-base properties of amines. In the study of macroligands with a large number of ionisable sites such as dendrimers, we must calculate the pKa values of their chemical building blocks, i.e., amines and for that we have to get validate and accurate method. Another important and developing property of amines is that they can serve as the novel candidates for CO_2 capture and sequestration processes. Here also the values of pKa are important. The pKa values are very important in explaining the mechanism (not included in this work) of CO_2 capture process and their capturing capacity. Reasonable prediction of pKa values would therefore be of great value in screening for new candidates for CO_2 capture process. In the present work, we have employed a DFT/B3LYP level of theory with 6–311++G(d,p) basis set for the amines and the computational works were done through Gaussian 03 version. We have computed the gas phase basicity, gas phase proton affinity, free energy of solvation, electronic and non-electronic contributions to free energy and pKa values of alkylamines. The free energy of solvation was computed using IEF-PCM models with water as the solvent. The gas phase basicity, gas phase proton affinity and the electronic and non-electronic contributions to free energy were found to increase with the number of carbon atoms. The pKa values were calculated using different types of basis sets; 6–311++G(d,p) has been found to be the best. The pKa values of all the amines studied were computed using the basis set 6–311++G(d,p) and are optimized both in gas and solution phase (water). The alkylamines studied include primary, secondary and tertiary amines. It has been found that the computed values are in good agreement with the experimental results which shows that the DFT (B3LYP) based analysis with 6–311++G(d,p) basis set is a reliable and easy method for the computation of pKa values.

1.1 INTRODUCTION

Amines are a versatile class of compounds, which have importance in physical chemistry as well as in biochemistry. They have a lot of applications in industries too. They are present in biological tissues, proteins, natural products, etc., and can serve as the building or preceding materials for a number of important compounds. Amines may serve as prototypes for side-chain groups of proteins and also becomes a suitable candidate in the area of macroligands like dendrimers [14]. Another application of amines is in the absorption of gases like CO_2, CH_4, SO_2, etc. Global warming, a major concern today is due to the increased release of green house gases [11]. Carbon dioxide is the major component of green house gases, which partakes the 60% of global warming. Removal of these kinds of gases is the immediate requirement of our nature and this is an interesting area in research. Though it is a well studied and proven technology called carbon dioxide capture and storage (CCS) [12], efforts are ongoing to get simpler and environmentally friendly technologies. It is proved that amine based solvents [5, 13] can serve as a very good candidate for carbon dioxide removal and hence the studies on amines are very pivotal.

The acid dissociation constant, i.e., pKa, is an important variable in explaining reaction mechanisms. Most of the reactions start with the protonation or deprotonation of reactant molecules and to know, where the protonation or deprotonation occurs, a knowledge of the pKa values is a must. The sharing of protons depends on the differences in the basic strengths and this in turn will depend on their pKa values [3]. Many experimental studies for the determination of pKa values of organic compounds are known. In the present work, the prediction/evaluation of pKa values of different amines using DFT method is described.

Most of the works reported in literature explaining the computational models to predict the pKa values are based on the free energy calculations. Charif et al. [2] perform free energy calculations using density functional theory-B3LYP (DFT-B3LYP) method in the solvent media. DFT-B3LYP level of theory have been used by Yun et al. [17] for the determination of pKa values of guanine in water and employed the electrostatic and non-electrostatic contributions to free energy in the presence of solvent. B3LYP density functional theory, IEF-PCM solvation modeling with a modified UFF cavity, and Boltzmann weighting of tautomers have been performed by Vincenzo et al. [15] to compute the pKa values of nucleobases and the guanine oxidation products. The use of quantum chemical descriptor to estimate pKa values of about 57 amines (primary, secondary and tertiary) has been done by Ivan Juranic [9] who employed the parameterization method 6 (PM6) and conductor-like screening model (COSMO) implemented in molecular orbital package (MOPAC) and an index variable to distinguish between primary, secondary and tertiary amines for his study.

The prediction of pKa values of amines are widely done for CCS technologies. The goal is to reduce energy requirement, the environmental impact and the capture cost. Discovering a solvent with high capture capacity, less economic, fast and uniform absorption capacity, high equilibrium sensitivity and low enthalpy of absorption would definitely be a solution for this problem [9]. The pKa values are very important in explaining the mechanism of CO_2 capture process and their capturing capacity. Reasonable prediction of pKa values would therefore be of great value in screening new candidates for CO_2 capture process. The gas phase basicities, deprotonation energies and proton affinities are very important variables in predicting the gas phase and aqueous phase Brönsted acidities [2].

The present work employed DFT-B3LYP level of theory coupled with IEF-PCM model of solvation to estimate the pKa values. In the present work, we studied 15 different alkylamines including primary, secondary and tertiary for the analysis of their gas phase basicity, proton affinity and pKa values. The pKa values were calculated from free energy measurements. The computed results are in agreement with the experimental results. All the calculations were carried out by using Gaussian 03 software [7].

1.1.1 THEORY

Just like the *ab initio* and semi-empirical calculations, DFT is also based on the solution of Schrodinger wave equation. DFT method directly estimates the electron distribution or density, instead of calculating the wave function and is faster than the *ab initio* but are slower than the semi-empirical calculations. DFT method becomes a tool for predicting a number of physical and chemical properties because of its high predicting power [8]. Most of the pKa value calculation studies are carried out by using DFT method. The level of theory adopted was B3LYP, which consists of Becke's exchange functional [1] in conjunction with Lee-Yang–Parr correlational functional [10] and the basis set used is 6–311++G(d,p).

The solution phase reactions are usually carried out by using the continuum models. In the Polarization Continuum Models (PCM), the solvent is described by a dielectric medium and a cavity is defined inside this dielectric medium [11]. This cavity is formulated to insert the molecule in the solvent phase. These models are basically parameterized to calculate the free energy of solvation as given by Eq. (1):

$$\Delta G_{solv} = \Delta G_{es} + \Delta G_{vdw} + \Delta G_{cav} \tag{1}$$

where ΔG_{es} is the electrostatic, ΔG_{vdw} is the Van der Waals and ΔG_{cav} is the activation energy contributions to free energy of solvation which are obtained from the output of the Gaussian file. There are different PCM models available which are designed to improve the computational performance of the method. One example is the IEF-PCM model which have been used in this work, is an Integral Equation Formalism for solving relevant Self-Consistent Reaction Field (SCRF) equations which facilitates computation of gradients and molecular response properties an extension to permit application to infinite periodic systems in one and two dimensions, and an extension to liquid/liquid and liquid/vapor interfaces [4].

The pKa value, given by Eq. (2), shows a linear relationship with the free energy. The free energy can be calculated with the help of the thermodynamic cycle shown in Figure 1.1 [11].

The necessary equations for the calculation of various parameters are:

$$pKa = \Delta G_{ps}/(2.303RT) \tag{2}$$

$$\Delta G_{ps} = \Delta G_{pg} + \Delta G_{s} \tag{3}$$

$$\Delta G_{pg} = G_{g}(B) + G_{g}(H^{+}) - G_{g}(BH^{+}) \tag{4}$$

$$\Delta G_{s} = \Delta G_{s}(B) - \Delta G_{s}(BH^{+}) \tag{5}$$

where, ΔG_{ps} is the change in free energy of protonation. Here the quantities such as free energy of proton in gas phase (–6.29 kcal/mol) and that in

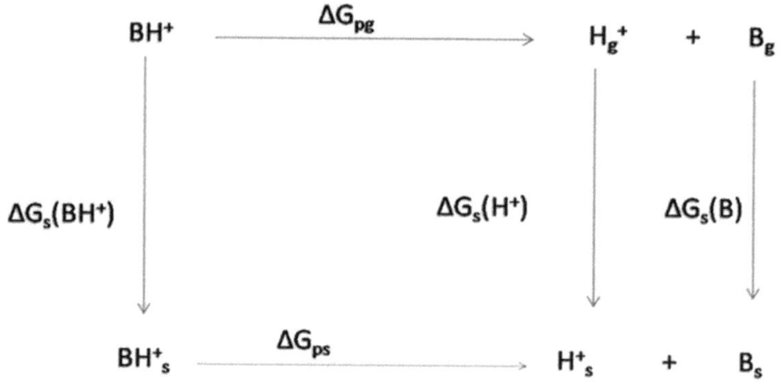

FIGURE 1.1 Thermodynamic cycle employed for the calculation of pKa value.

solution (–263.977 kcal/mol) used for the calculation are as given in the literature [11] and all others are computed using Gaussian 03 software.

The gas phase basicity can be calculated from Eq. (4). The proton transfer reactions and ionization reactions plays an important role in atmospheric chemistry and biochemistry [16]. The gas phase proton affinity is the change in enthalpy of protonation in gas phase and is calculated by Eq. (6):

$$\Delta H_{pg} = H_g (B) + H_g (H^+) - H_g (BH^+) \tag{6}$$

where, the gas phase enthalpy of proton (1.48 kcal/mol) is as given in the literature [17].

The output files of Gaussian 03 in solution phase give the quantities like ΔG_{el} and $\Delta G_{non\text{-}el}$, the electrostatic and non-electrostatic contributions to free energy in the presence of a solvent. The non-electrostatic contributions include cavitation, repulsion and dispersion energy to solvation energy [6].

$$\Delta G_{non\text{-}el} = \Delta G_{cav} + \Delta G_{rep} + \Delta G_{dis} \tag{7}$$

The gas phase basicity, proton affinity, pKa values, electrostatic and non-electrostatic contributions of free energy (in the solution phase) of alkylamines has been computed using Eq. (7). All the calculations are done at a temperature of 298 K and at a pressure of 1 atmosphere.

1.2 COMPUTATIONAL DETAILS

The method employed in the present work is a computational approach to predict the pKa values of amines and focuses to reproduce the trend in pKa values. The necessary steps involved in the study are:

1. Selection of the basis set: Methylamine is optimized with basis sets such as 6–31G, 6–31G(d), 6–31G(d,p), 6–31++G, 6–31++G(d) and 6–311++G(d,p) with DFT-B3LYP level of theory. The pKa value is calculated with the results obtained from the optimization by all these basis sets and compared with the experimental pKa value of methylamine. The basis set 6–311++G(d,p), which agree well with the experimental value (Table 1.1), is chosen for further study.
2. All the amines are optimized in gas phase using 6–311++G(d,p) and computed the properties.
3. Optimization of amines in solvent (aqueous) phase: The optimized geometry from the gas phase is used as the input for solution phase calculation. Though wide variety of PCM models are available for solution phase analysis, there is however no clear guidelines to say, whether a specific model is more suitable or reliable. In the present work we have chosen the IEF-PCM model with its default settings in Gaussian 03 for solution phase analysis.

The optimized gas phase geometries of amines (neutral and protonated) studied is shown in Figures 1.2a, 1.2b and 1.2c.

TABLE 1.1 pKa Values of Methylamine Obtained by Using Different Basis Sets

Basis set	**pKa values at 298 K**		**Relative deviation (%)**
	Calculated	**Experimental**	
6–31G	16.34	10.50	55
6–31G(d)	14.15	10.50	38
6–31G(d,p)	14.90	10.50	41
6–31++G	12.21	10.50	16
6–31++G(d)	9.99	10.50	4
6–311++G(d,p)	10.45	10.50	**0.4**

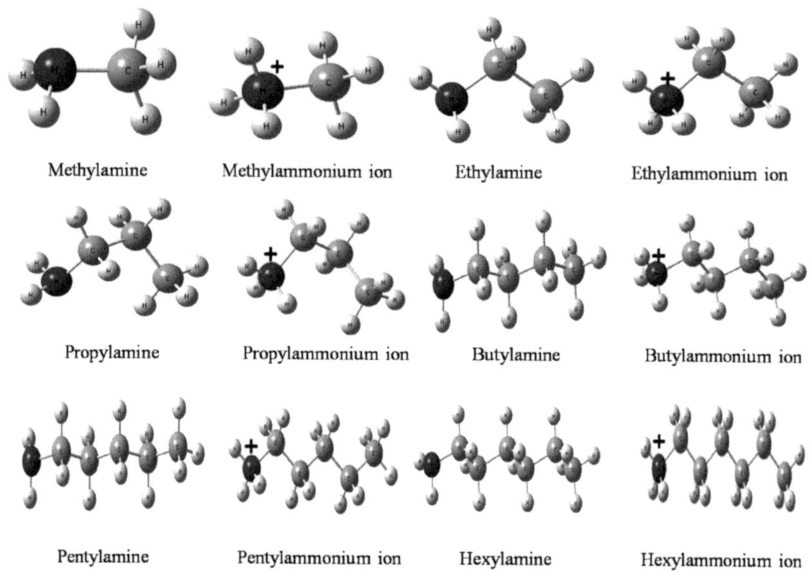

FIGURE 1.2A Optimized geometries of neutral and protonated primary amines.

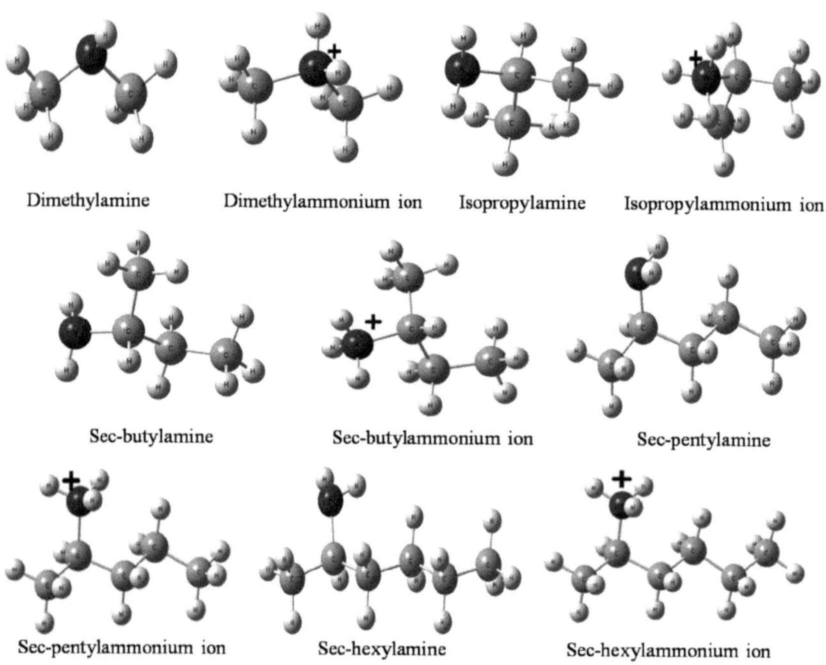

FIGURE 1.2B Optimized geometries of neutral and protonated secondary amines.

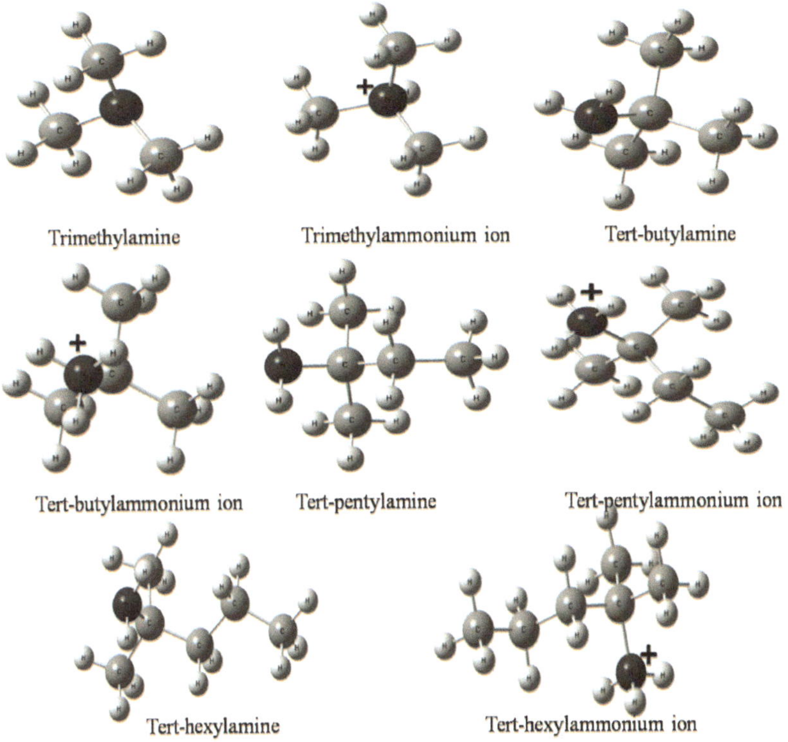

FIGURE 1.2C Optimized geometries of neutral and protonated secondary amines.

1.3 RESULTS AND DISCUSSION

1.3.1 PREDICTION OF GAS PHASE BASICITY

The first step for the calculation of gas phase basicity is the calculation of Gibbs free energy changes associated with the protonation reaction of amines. All the amines and their respective protonated cations are optimized in gas phase using DFT (B3LYP) level of theory with 6–311++G(d,p) as basis set. The Gibbs free energy values are directly obtained from the thermochemistry part of the output file. The change in free energy is computed with the help of the thermodynamic cycle (see Figure 1.1) and Eq. (4). The calculated gas phase basicity of all the amines studied is given in Table 1.2 (only available experimental data is given).

TABLE 1.2 Gas Phase Basicity Values of Different Alkylamines

Number	Amine	Gas phase basicity at 298 K	
		Calculated	**Experimental**
1	Methylamine	206.47	206.60
2	Dimethylamine	214.07	214.30
3	Trimethylamine	218.34	219.40
4	Ethylamine	210.59	210.00
5	Propylamine	212.17	211.30
6	Isopropylamine	213.73	212.50
7	Butylamine	212.63	-
8	Sec-butylamine	214.95	214.10
9	Tret-butylamine	216.37	215.10
10	Pentylamine	213.06	212.60
11	Sec-pentylamine	216.03	-
12	Tert-pentylamine	217.80	215.98
13	Hexylamine	213.27	213.50
14	Sec-hexylamine	216.43	-
15	Tert-hexylamine	218.52	-

From Table 1.2 it is clear that the computed values show good agreement with the experimental results. There is only slight deviation from the experimental values. The graphical representation of calculated versus experimental gas phase basicities of different amines are given in Figure 1.3a. As expected from the experimental results, the gas phase basicities increases with the number of carbon atoms in the alkylamines and also from primary to secondary and to tertiary amines (See Figures 1.3b and 1.3c). In Figure 1.3c, the amines which are numbered from 1–6 are primary, 7–10 are secondary and 11–13 are tertiary.

1.3.2 PREDICTION OF GAS PHASE PROTON AFFINITY

Similar to the calculation of gas phase basicity, the gas phase proton affinity (PA) can also be calculated with the help of the thermodynamic cycle shown in Figure 1.1. The enthalpy values of both neutral and protonated amines are computed, and then with the help of Eq. (6), PA can be

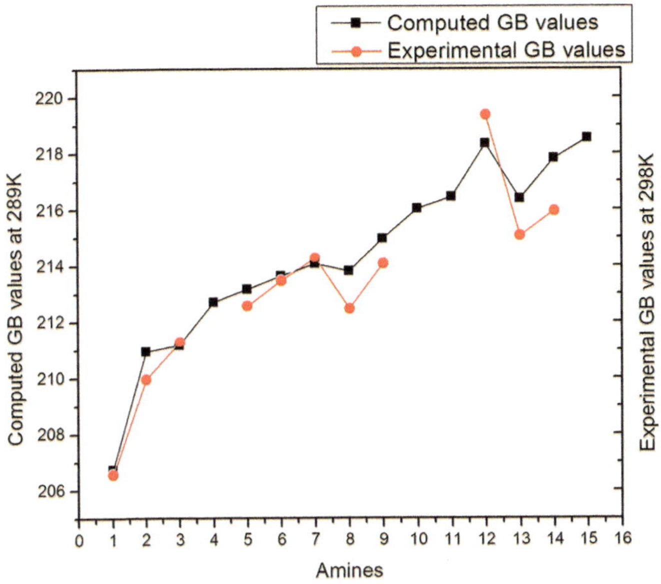

FIGURE 1.3A Calculated and experimental gas phase basicity values of different alkylamines at 298K.

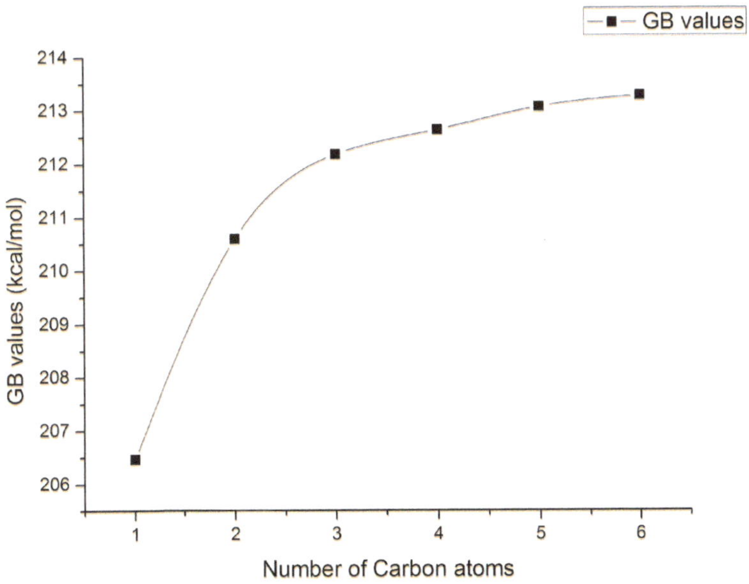

FIGURE 1.3B The dependence of gas phase basicity values on the number of carbon atoms of different alkylamines.

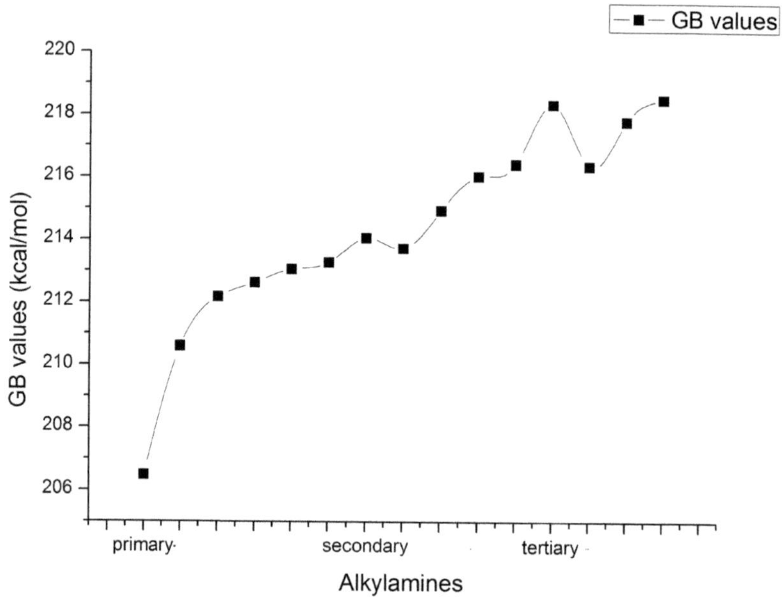

FIGURE 1.3C Dependence of gas phase basicity values of different alkylamines on the nature of carbon atom.

computed. All the calculations are done using B3LYP/6–311++G(d,p) in Gaussian 03. The calculated and the experimental gas phase proton affinities of alkylamines studied are listed in Table 1.3 (only available experimental values are given).

The PA increases with the number of carbon atoms but it is not general. The calculated values are in good agreement with the experimental results (see Figure 1.4).

1.3.3 CALCULATION OF pKa VALUES

The pKa value is a measure of basic strength of amines as it increases with the basic strength. pKa value gives an idea about how much acidic is the hydrogen in a molecule [3]. The basicity of amines is due to the presence of non-bonded pair of electrons on nitrogen atom and relatively low electronegativity of this atom. The pKa values of alkylamines are calculated

TABLE 1.3 Calculated and Experimental Gas Phase Proton Affinity Values of Different Alkylamines

Number	Amine	Gas phase proton affinity values at 298 K	
		Calculated	Experimental
1	Methylamine	214.71	214.90
2	Dimethylamine	221.62	222.20
3	Trimethylamine	225.84	226.40
4	Ethylamine	217.82	218.00
5	Propylamine	221.71	222.80
6	Isopropylamine	221.23	220.80
7	Butylamine	220.13	223.30
8	Sec-butylamine	222.46	222.19
9	Tret-butylamine	223.78	223.30
10	Pentylamine	220.53	220.70
11	Sec-pentylamine	223.53	–
12	Tert-pentylamine	225.11	224.13
13	Hexylamine	220.76	221.60
14	Sec-hexylamine	224.03	–
15	Tert-hexylamine	222.85	–

by using Eq. (1) and the thermodynamic cycle shown in Figure 1.1, and are in good agreement with the experimental results. The computed and the experimental pKa values of amines are tabulated in Table 1.4 (only available experimental values are given).

From Table 1.4, it is clear that there is no periodic increase or decrease in the pKa values of different amines. The pKa values of primary, secondary and tertiary amines in solution varies as secondary > primary > tertiary. The electron releasing groups such as alkyl groups when attached to the nitrogen atom of amines increases the stability of its positively charged conjugate acid and they enhance the basicity by (+) I mechanism, i.e., the dimethylamine (pKa = 10.74) is more basic than the methylamine (pKa = 10.44) and which in turn is more basic than the trimethylamine (pKa = 10.01). Steric factor also plays an important role in the prediction of pKa values. When there is large

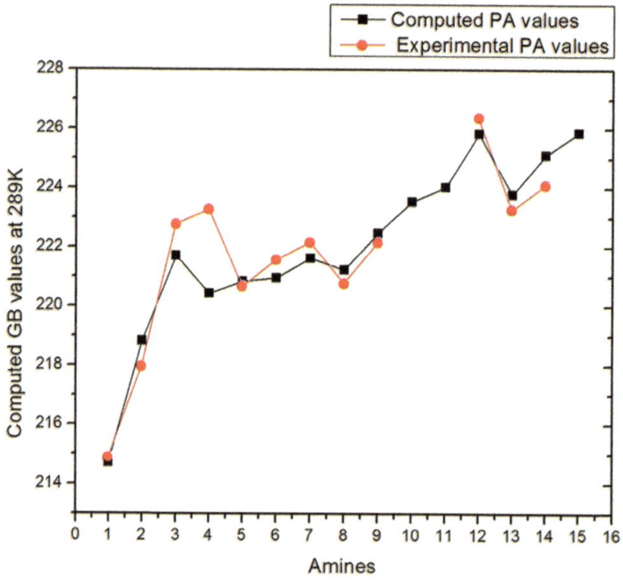

FIGURE 1.4 Calculated and experimental gas phase proton affinity values of different alkylamines.

TABLE 1.4 pKa Values (Calculated and Experimental) of Different Alkylamines

Number	Amine	pKa values at 298 K	
		Calculated	Experimental
1	Methylamine	10.44	10.50
2	Dimethylamine	10.74	10.73
3	Trimethylamine	10.10	09.80
4	Ethylamine	10.42	10.60
5	Propylamine	10.55	10.67
6	Isopropylamine	10.61	10.63
7	Butylamine	10.66	10.68
8	Sec-butylamine	10.89	10.80
9	Tret-butylamine	10.44	10.43
10	Pentylamine	10.79	10.63
11	Sec-pentylamine	10.83	–
12	Tert-pentylamine	11.01	10.85
13	Hexylamine	10.72	10.56
14	Sec-hexylamine	10.95	–
15	Tert-hexylamine	10.36	–

or multiple substituents they hinders the bond formation with proton and reduces the basicity. So the tertiary amines are less basic than the secondary amines. The results obtained from the theoretical study also leads to the same conclusion. The graphical representation of calculated and the experimental pKa values of amines are shown in Figure 1.5.

1.3.4 CALCULATION OF ELECTROSTATIC AND NON-ELECTROSTATIC CONTRIBUTIONS TO FREE ENERGY

The electrostatic and non-electrostatic contributions to free energy are directly obtained from the output files of Gaussian 03. The electrostatic contribution is the same as that of the free energy of solvation of amines which are given in Table 1.5. The non-electrostatic contributions include the cavitation, repulsion and dispersion energies to the free energy in the presence of a solvent.

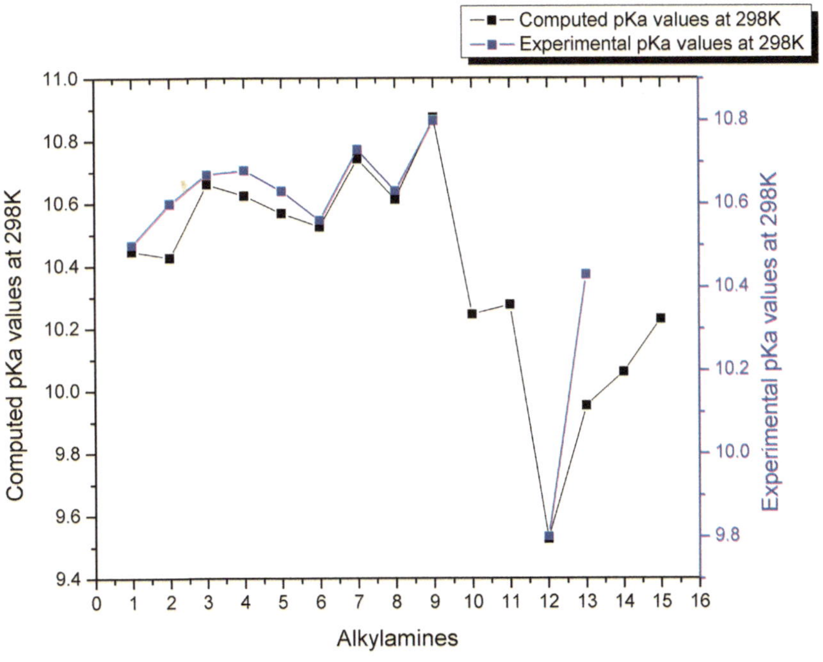

FIGURE 1.5 Calculated and experimental pKa values of different alkylamines at 298K.

TABLE 1.5 Electrostatic and Non-Electrostatic Free Energy Values of Different Amines

Number	Amine	ΔG_{ele}	$\Delta G_{non\text{-}ele}$
1	Methylamine	–5.16	4.55
2	Methylammonium ion	–70.55	4.33
3	Dimethylamine	–4.08	6.46
4	Dimethylammonium ion	–64.66	6.20
5	Trimethylamine	–1.75	7.89
6	Trimethylammonium ion	–57.15	7.76
7	Ethylamine	–5.18	4.98
8	Ethylammonium ion	–69.05	4.81
9	Propylamine	–5.01	5.60
10	Propylammonium ion	–67.76	5.45
11	Isopropylamine	–5.24	6.47
12	Isopropylammonium ion	–66.87	6.24
13	Butylamine	–5.66	6.39
14	Butylammonium ion	–68.78	6.14
15	Sec-butylamine	–5.39	7.08
16	Sec-butylammonium ion	–66.71	6.86
17	Tret-butylamine	–4.99	7.76
18	Tret-butylammonium ion	–64.20	7.56
19	Pentylamine	–5.74	7.11
20	Pentylammonium ion	–68.78	6.85
21	Sec-pentylamine	–5.24	8.14
22	Sec-pentylammonium ion	–65.64	7.91
23	Tert-pentylamine	–5.01	8.33
24	Tert-pentylammonium ion	–63.88	8.14
25	Hexylamine	–5.84	7.84
26	Hexylammonium ion	–68.80	7.58
27	Sec-hexylamine	–5.32	8.89
28	Sec-hexylammonium ion	–65.60	8.65
29	Tert-hexylamine	–4.88	9.52
30	Tert-hexylammonium ion	–61.44	9.22

The non-electrostatic contributions to free energy increases for higher amines. The free energy values are more for the protonated ions than their corresponding neutral amines. The electrostatic contribution (free energy of solvation) of neutral and protonated amines is also obtained during the calculation of pKa values. It has been observed that the values so obtained are similar to that given in the output file which confirms that the calculations are true. A comparison of the calculated free energy values of solvation and that obtained in the output file are given in Table 1.6.

1.4 CONCLUSION

A computational study has been performed to calculate the pKa values of alkylamines. The gas phase basicity and gas phase proton affinity values has also been computed. All the calculations were performed

TABLE 1.6 Free Energy Values of Solvation of Neutral Alkylamines

Amine	ΔG_{ele} calculated	ΔG_{ele} from output
Methylamine	–4.85	–5.16
Dimethylamine	–3.71	–4.08
Trimethylamine	–1.57	–1.75
Ethylamine	–4.91	–5.18
Propylamine	–4.67	–5.01
Isopropylamine	–5.01	–5.24
Butylamine	–5.56	–5.66
Sec-butylamine	–526	–5.39
Tret-butylamine	–4.77	–4.99
Pentylamine	–5.81	–5.74
Sec-pentylamine	–5.14	–5.24
Tert-pentylamine	–4.78	–5.01
Hexylamine	–6.12	–5.84
Sec-hexylamine	–5.54	–5.32
Tert-hexylamine	–4.88	–5.01

TABLE 1.7 Free Energy Values of Solvation of Protonated Alkylammonium Ions

Protonated ammonium ion	ΔG_{ele} calculated	ΔG_{ele} from output
Methylammonium ion	–70.32	–70.55
Dimethylammonium ion	–63.91	–64.66
Trimethylammonium ion	–56.49	–57.15
Ethylammonium ion	–68.11	–69.05
Propylammonium ion	–66.31	–67.76
Isopropylammonium ion	–65.32	–66.87
Butylammonium ion	–67.05	–68.78
Sec-butylammonium ion	–64.75	–66.71
Tert-butylammonium ion	–62.22	–64.20
Pentylammonium ion	–67.07	–68.78
Sec-pentylammonium ion	–63.47	–65.64
Tert-pentylammonium ion	–61.59	–63.88
Hexylammonium ion	–67.07	–68.80
Sec-hexylammonium ion	–63.59	–65.60
Tert-hexylammonium ion	–61.44	–63.71

using DFT-B3LYP level of theory in combination with IEF-PCM solvation model and 6–311++G(d,p) basis set, in Gaussian 03 software package. The computed results are in good agreement with the experimental results. It has been found that the pKa values vary linearly with the free energy values of protonation. The gas phase basicity and proton affinity increases with the number of carbon atom as given in the literature. The basicity values in solution (water) decreases in the order secondary > primary > tertiary as it is clear from the pKa values (pKa increases with basic strength). The pKa values of primary amines are approximately the same, i.e., near to 10.7. The pKa values of secondary amines are slightly higher and those of tertiary amine were slightly lower than the primary amines. The method adopted is simple and easy to perform. In future, this study can be extended to various fields such as screening of solvents for CO_2 capture, macromolecular ligand studies, acid-base solubility studies, etc.

KEYWORDS

- **alkylamines**
- **density functional theory**
- **free energy of solvation**
- **gas phase basicity**
- **gas phase proton affinity**
- **pKa values**

REFERENCES

1. Becke, A. D. (1993). *J. Chem. Phys. 98*, 5648–5652.
2. Charif, I. E., Mekelleche, S. M., Villemin, D., & Mora-Diez, N. (2007). *Journal of Molecular Structure. 818*, 1–6.
3. Clayden, Greeves; & Warren, Wothers. *Organic Chemistry*. Oxford University Press: Oxford, 2001.
4. Cramer, J. C. *Essentials of Computational Chemistry: Theories and Models*; Second Edition: John Wiley & Sons Ltd: USA, 2004.
5. Eirik Falck da Silva. Computational Chemistry Study of Solvents for Carbon Dioxide Absorption. Norwegian University of Science and Technology. 2005.
6. Farhad, K., Amr, H., & Allan, L. L. (2009). *East. J. Mol. Struc-Theochem. 916*, 1–9.
7. Gaussian 03, Revision C.02, Frisch, M. J., Trucks, G. W., Schlegel, H. B., Scuseria, G. E., Robb, M. A., Cheeseman, J. R., et al., Gaussian, Inc., Wallingford CT, 2004.
8. Gotthard, S., & Jan-Ole, J. (2012). *WIREs Comput Mol Sci. 2*, 456–465.
9. Juranić, I. (2014). *Croat. Chem. Acta. 87*(4), 343–347.
10. Lee,C., Yang, W., & Parr, R. G. (1988). *Phys. Rev. B. 37*, 785–789.
11. Mayuri, G., Eirik Falck da Silva; & Hallvard, F. S. (2012). *Energy Procedia. 23*, 140–150.
12. Peter, F. *CRS Report*. Congressional Research Service, 2014.
13. Prachi, S. Amine Based Solvent for CO_2 Absorption from Molecular Structure to Process. University of Twente, 2011.
14. Vyacheslav, S. B., Mamadou, S. D., & William, A. G. J. (2007). *Phys. Chem. A. 111*, 4422–4430.
15. Vincenzo, V., Roberto, C., Barbara, H. M., & Bernhard, S. H. (2008). *J. Phys. Chem. B. 112*, 16860–16873.
16. Younes, V., Hossein, F., & Mahmoud, T. (2014). *J. Chem. Sci. 126*(4), 1209–1215.
17. Yun, H. J., William, A. G. III., Katherine, T. N., Lawrence, C. S., Sungu, H., & Doo, S. C. (2003). *J. Phys. Chem. B. 107*, 344–357.

ADDITIONAL READING

1. Douglas, C. C., Robert, D., & Damian, M. (2002). *J. Org. Chem. 67*, 5098–5105.
2. Guohui, L., & Qiang, C. (2003). *J. Phys. Chem. B. 107*, 14521–14528.
3. Hunter, E. P., & Lias, S. G. (1998). *J. Phys. Chem. Ref. Data. 27*(3), 413–656.
4. Kristin, S. A., & George, C. S. (2010). *Annu. Rep. Comput. Chem. 6*, 113–137.
5. Paul, G. S. (2008). *Int. J. Quantum Chem. 108*, 2849–2855.
6. Washburn, E. W. (1929). *ICT*. New York, McGraw-Hill, *6*, 261.
7. Yanbo, D., Dan, N., Karsten, K. J., & Ronald, M. L. (1995). *J. Phys. Chem. 99*, 11575–11583.

CHAPTER 2

A DFT INVESTIGATION OF THE INFLUENCE OF α,β UNSATURATION IN CHEMICAL REACTIVITY OF COUMARIN AND SOME HYDROXY COUMARINS

M. A. JASEELA,[1] T. M. SUHARA,[2] and K. MURALEEDHARAN[1]

[1]*Department of Chemistry, University of Calicut, Malappuram, Kerala, 673635, India*

[2]*Department of Chemistry, MES Ponnani College, University of Calicut, Kerala, 679577, India, Tel.: +91-494-2407413; Fax: +91-494-2400269; E-mail: kmuralika@gmail.com*

CONTENTS

ABSTRACT

Coumarins are of great interest in the field of research because of their tremendous biological properties. It makes these compounds so attractive for further backbone derivatisation and screening to meet specific applications. The present study depicts the correlation between structural variations of Coumarin (C), its metabolite 3,4-dihydrocoumarin (DC) and their respective substituted compounds with chemical reactivity which in turn correlates to the influence of α,β unsaturation on structures. All Quantum mechanical computational calculations are done in density functional theory (DFT) method using the Gaussian 09W simulation package. The geometries and energies of these molecules have been computed at B3LYP/6–311G (2d, 2p) level of theory. Natural Bond Orbital (NBO) analysis of these molecules provides a quantitative evaluation of electron density distribution and which is used to make a qualitative conclusion of donor-acceptor properties of substituents with parent C and DC scaffolds. It helps to predict the reactive centers and nature of feasible reactions in the presence of ring activating and deactivating groups. Time dependent density functional theory (TD-DFT) based computation is adopted for the evaluation of absorption (UV-Vis) spectra of the molecules under consideration by including one set of diffuse function in basis set, which is utilized to analyze the shifting of absorption maxima as the consequence of different types of substituents on third position of the parent scaffold. The calculation of dipole moment, polarization and hyper polarization are showing the extent of the application of these compounds in the field of nonlinear optical organic materials. Global reactive descriptors like ionization potential, electron affinity, electronegativity, electrophilicity, softness, hardness, etc., are calculated in order to analyze the extent of reactivity of considering molecules. The local reactivity descriptors such as Fukui functions and relative nucleophilicity were also calculated for the better understanding of the nature of reactive site and attempted to predict the type of reactions that are preferable for the considering chemical systems. In addition to Quantum mechanical calculation, druggability of some important hydroxy coumarin systems were also examined using Molinspiration property calculator available online.

2.1 INTRODUCTION

Coumarins comprise a large class of compounds having different biological and chemical activities mainly found in the plant kingdom. These plants are widely distributed in tropical rain forests where several species are used in folk medicines [19]. Although this class of compounds is distributed in whole part of the plant, highest level is in fruits (e.g., Bilberry, Cloudberry), followed by the roots, stems and leaves. The occurrence of these compounds in diverse parts of the plant is dependent on environmental conditions and seasonal changes. However the well-known source of Coumarins is the higher plants with richest sources being in the family of *Rutaceae, Fabiaceae, Umbelliferae* [22], interestingly some of the important members in Coumarins have been discovered from microbial sources also, e.g., *Novobiocin* and *Coumermycin* from *Streptomycets*, and *Aflatoxins* from *Aspergillus* species [9, 10].

In chemical aspects, Coumarin is a member of benzopyrone family, all of which are consisting of benzene ring joined with α-pyrone ring. Benzopyrones are further classified into two categories of benzo-α-pyrone to which Coumarins belong and benzo-γ-pyrone of which flavanoids are the principal candidates. The Coumarins are of great interest due to their tremendous pharmacological properties and characteristic conjugated molecular architecture [25]. The basic structure of Coumarin is as given in Figure 2.1.

By recognizing the importance of the Coumarin backbone structure, its various natural and synthetic derivatives are utilized to meet potential

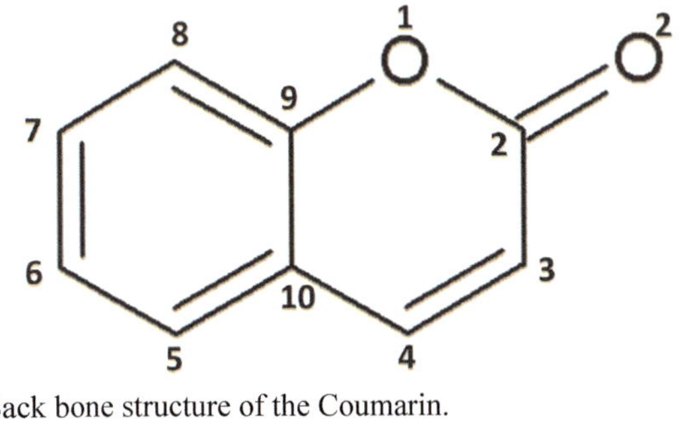

FIGURE 2.1 Back bone structure of the Coumarin.

applications in various fields. Many of them are already find their role due to their important pharmacological effects, including analgesic, anti-arthritis, anti-inflammatory, anti-pyretic, anti-bacterial, anti-viral and anti-cancer [6] agents. Among Coumarins, hydroxy coumarins and their derivatives have been effectively used as anticoagulants for the treatment of disorders due to the excessive undesirable clotting, such as thrombophlebitis and certain cardiac conditions. There are a number of comparative pharmacological investigations of the hydroxyl coumarin derivatives; e.g., 4-hydroxy coumarins have shown a good anticoagulant activity with low side effects and little toxicity [29]. The major subtypes of Coumarins are simple Coumarins, Furanocoumarins, Pyranocoumarins and Pyrone substituted Coumarins. The simple Coumarins are hydroxylated, alkylated, alkoxylated or glycozylated coumarins like 7-hdroxy coumarin and substitution always comes to the benzene ring not on pyrone ring. Furanocoumarins consist of five membered furan ring attached with Coumarin backbone and which are further categorized to linear or angular based on substitution at one or both side of benzene moiety, Pyranocoumarins are analogous to Furanocoumarins but consists of a six membered ring and 4-hydroxy coumarin is an example for pyrone substituted Coumarin which indicate that substitutions are present at pyrone ring and not to benzene moiety [25]. Many synthetic compounds are also exists in this division. By virtue of structural simplicity and easy of synthesis, 7 hydroxy coumarin is regarded as the parent of complex Coumarins. One of the important application of hydroxy coumarins is their role in designing chemo sensors to detect anions especially cyanide. It is noted that the π system bearing an X-H moiety (where X is an electronegative atom such as N or O) are considered as favorable candidates for the selective anion recognition as chemo sensors. Among various chemo sensors, fluorescent chemo sensors have its own advantages such as high selectivity, low cost, ease of detection and suitability as diagnostic tools for biological concerns.

The present study reports the electronic properties and chemical reactivity of Coumarin structure by comparing with its metabolite 3,4-dihydro coumarin and also with their substituted structures. The chemical reactivity of selected hydroxy coumarins are explained using density functional theory (DFT) which entails a better understanding of their role and importance of basic structure of Coumarin in biological world.

On the basis of previous description, it is possible to establish that the chemical and biological activity of a system is directly depends on its ability to interact covalently as well as non-covalently in a specific way. The computational chemistry is a useful tool for these kinds of investigations, studies of chemical and biological properties with specificity, rapidity and low cost [15]. In addition, the study of chemical reactivity descriptors along with substitutions at same level of theory and *in silico* druggability tests of some hydroxy coumarins are also done in order to understand their potential applications.

2.1.1 THEORY

Computational Chemistry simulates chemical structures and reactions numerically based on fundamental laws of physics. It opens new frontiers for the chemist to study chemical phenomena by running calculations on computers rather than by examining reactions of compounds experimentally. The recent impact of DFT methods, similar to ab-initio method in many ways, by the development of Quantum chemical calculation is very considerable. The main idea of DFT is to describe an interacting system of fermions via its density and not via its many body wave function [36].

2.1.1.1 NBO Analysis

The mutual influences of the electronic substituents on conjugated rings are still at the interest area of research because of the variation in reactivity of the chemical system as the cause of substitution. With the help of Quantum-Chemical calculations, it is now possible to differentiate the contributions from orbital interactions and predict the effect of a substituent on reaction center. NBO analysis is based on an approach of transforming multi electron wave functions of molecules into the localized form that corresponds to single-center [lone pair (LP)] and two centered [natural bond and antibonding orbitals (BD and BD*, respectively)] elements. It gives a deep insight into the intra and intermolecular orbital interactions in molecules between filled donor and empty acceptor NBOs, which enables us to give a quantitative evaluation and thereby results a qualitative conclusion of donor-acceptor properties of substituents [54]. For each donor (i) and acceptor (j) interactions are quantitatively expressed by means of second-order perturbation interaction energy $E^{(2)}$ associated with delocalization $i \rightarrow j$ is estimated as

$$E^{(2)} = q_i \frac{F^2(i,j)}{\varepsilon_j - \varepsilon_i} \qquad (1)$$

where q_i is the electronic occupancy in donor orbital, F(i, j) is the off diagonal NBO Fock matrix element and ε_i and ε_j are diagonal elements in orbital energies. By analyzing the interactions between occupied Lewis NBO (bond pair or lone pair) as donor and an unoccupied non-Lewis NBO (anti-bonding or Rydberg) as acceptor with their second order perturbation energy will give clear information regarding the origin of stabilization of that molecule. If the stabilization energy $E^{(2)}$ associated with an interaction is large, more will be the extent of stabilization. The NBO analysis provides an efficient method for studying inter and intra molecular bonding interactions and also extent a convenient basis for investigating charge transfer or conjugative interactions in the molecular system [16, 42].

2.1.1.2 Global Reactive Descriptors

The utility of global reactive descriptors is of great importance to answer some fundamental questions in chemistry such as reactive feasibility (whether a reaction will take place or not) or intra molecular selectivity. The basic relationships within the conceptual framework of DFT are very useful to predict chemical potential, electronegativity, chemical hardness and chemical softness with respect to the number of electrons present in considering chemical system. The chemical hardness fundamentally signifies the reluctance towards the deformation or polarization of the electron cloud of the atoms, ions or molecules under small perturbation encountered during chemical processes [18]. Chemical Softness is the measure of the capacity of a molecule to receive electrons; more precisely it is related with the groups or atoms present in that molecule and inversely proportional to chemical hardness [36]. The chemical potential in DFT, measures escaping tendency of an electron from equilibrium, is accounted by the first derivative of energy with respect to the number of electrons [57] and which is also the negative of electronegativity, is a measure of the tendency to attract electrons in a chemical bond [36].

Global reactive descriptors can be calculated by two different methods, one is based on the difference in total electronic energy of neutral molecule and its corresponding anion and cation, obtained from the geometry of the neutral molecule in order to keep the external potential as constant, and usually call it as 'energy vertical' [46].

$$IP_E = E_{cation} - E_{neutral} \tag{2}$$

$$EA_E = E_{neutral} - E_{anion} \tag{3}$$

The other one is based on the approximation suggested by Koopman's theorem [23], by considering the difference between highest occupied molecular orbital (HOMO) and lowest unoccupied molecular orbitals (LUMO) energies of the neutral molecule under study and call it as 'orbital vertical' [53]. By making use this approximation the global reactive descriptors can be written as given below.

$$\text{Ionisation Potenial (IP)} \approx - E_{HOMO} \tag{4}$$

$$\text{Electron Affinity (EA)} \approx - E_{LUMO} \tag{5}$$

where E_{HOMO} is the energy of HOMO and E_{LUMO} is the energy of LUMO.

$$\text{Hardness } (\eta) \approx (IP - EA)/2 \tag{6}$$

$$\text{Electronegativity } (\chi) \approx (IP + EA)/2 \tag{7}$$

$$\text{Softness (S)} \approx 1/2n \tag{8}$$

$$\text{Chemical potential } (\mu) \approx -\chi \tag{9}$$

$$\text{Electrophilicity index } (\omega) \approx \mu^2/2n \tag{10}$$

The global electrophilicity index (ω) is a measure of lowering in energy due to maximal electron flow between donor and acceptor. The electrophilicity index is built up from the electronic structure of molecules, independent from its nucleophilic counterpart, is replaced by an unspecified environment viewed as sea of electrons [35]. The molecule having high HOMO energy shows more reactivity with electrophiles while low LUMO energy is good for molecular reaction with nucleophile [41]. Since hardness corresponds to the gap between HOMO and LUMO energy level, high energy gap results higher hardness and lower reactivity and it also can be correlated with aromaticity by defining that larger HOMO-LUMO gap has been associated with higher aromatic nature. Hence the absolute hardness is considered as a criterion with high significance for determining stability and reactivity of a chemical system [3, 8].

2.1.1.3 Local Reactive Descriptors

Beside the global reactive descriptors, it is possible to define its local (regional) counterpart condensed to specific atoms of a molecule. The local reactivity parameters can be analyzed through the evaluation of Fukui indices. Fukui function is an electron density based reactivity index and giving information regarding which atom has higher tendency to accept or loose electrons than others in a particular molecule, means which are more electrophilic or nucleophilic in nature. Based on this, reactive sites of a molecule can be classified as sites for electrophilic, nucleophilic or radical attack.

Fukui function is given by the equation

$$f(r)=\left(\frac{\delta\rho(r)}{\delta N}\right)_{v(r)}=\left(\frac{\delta\mu}{\delta v(r)}\right)_{N} \qquad (11)$$

where $\rho(r)$ is the electron density, N is the number of electron and $v(r)$ is the external potential exerted by the nucleus, μ is electronic chemical potential.

The calculations of condensed Fukui functions are based on the finite difference approximation and partitioning of the electron density $\rho(r)$ between atoms in molecular systems.

For reaction with electrophiles

$$f_j^- = q_j(N) - q_j(N-1) \qquad (12)$$

For reaction with nucleophiles

$$f_j^+ = q_j(N+1) - q_j(N) \tag{13}$$

For free radical reaction

$$f_j^0 = \frac{1}{2}\left[q_j(N+1) - q_j(N-1)\right] \tag{14}$$

Here is the charge on the jth atomic site calculated from Mullikan population or natural population on anion ($q_j(N+1)$), cation ($q_j(N-1)$) and $q_j(N)$ neutral species [36]. Population analysis is a mathematical way of partitioning the wave function or electronic density to obtain bond orders, nuclear charges, etc. But the natural population is more accepted than Mullikan population for Fukui calculation because of its improved numerical stability and which will help to give a better description for electronic distribution of chemical systems having high ionic character [43].

The local softness indices, s_k, are closely related to condensed Fukui functions and global softness and it can be used for comparing reactivity of atomic centers between different molecules.

The local softness [56] is defined as

$$S(r) = \left(\frac{\partial n(r)}{\partial \mu}\right)_{v(r)} \tag{15}$$

Upon integration, yields global softness

$$S = \int S(r)\partial r = \left(\frac{\partial N}{\partial \mu}\right)_{v(r)} \tag{16}$$

The relationship between the local softness and Fukui function is given below

$$S(r) = Sf(r) = \left(\frac{\delta \rho(r)}{\delta N}\right)_{v(r)} \left(\frac{\partial N}{\partial \mu}\right)_{v(r)} \tag{17}$$

It reflects that these quantities, condensed Fukui functions and local softness, are giving the same information about the relative site of reactivity in a molecule. The relative nucleophilicity indices, S_k^-/S_k^+ represent the nucleophilicity of an atomic center compared with its own electrophilicity and defined as quotient of the condensed local softness indices of the molecular system. The atomic center having the highest S_k^-/S_k^+ value is considered as the most reactive site for electrophilic attack [45].

2.1.1.4 Polarisability and Hyper Polarisability

Polarizabilities and hyperpolarizabilities determine the strength of molecular interactions, cross section of different scattering, collision processes and also non-linear optical properties. The development of organic non-linear optical (NLO) materials is motivated by the prominence of their cost effective applications in telecommunication, computing, signal transmission, embedded network sensing and even in the medical field for the development of sensors. The NLO property of a molecule is originated from the non centrosymmetric alignment of NLO chromophores while interacting with electromagnetic waves leads to the production of new fields altered in phase, frequency, amplitude or other propagation characteristics from the incident fields. It is well known that impinging a beam of light on a material leads to the oscillation of charges on each atom. The amount of charge displacement in linear materials is proportional to their instantaneous magnitude of electric field with the same frequency of the incident light. The displacement of charge from the equilibrium, means polarization (P), for small fields is directly proportional to the applied field, E,

$$P = \alpha E \tag{18}$$

where α is the linear polarizability.

For a NLO material, the displacement of charges from its equilibrium position should be represented using a nonlinear function of electric field. Likewise, materials which subjected to a high electric field will cause for the deviation of their polarisabilility from the linear regime (Gunter, 2000).

For nonlinear polarization,

$$P = P_0 + \chi^1 E + \chi^2 E^2 + \chi^3 E^3 + \ldots \tag{19}$$

It is also noted that the total dipole moment of a chemical system arises while applying electric field should be calculated by accounting four components present in the given equation

$$\mu = \mu_0 + \alpha\, F + \tfrac{1}{2}\, \beta\, F^2 + 1/6\, \gamma^3 + \ldots \tag{20}$$

where μ_0 is permanent dipolemoment, α is polarisability, β is first order polarisability and γ is the second order polarisability. The mean polarizability 'α' can be represented in terms of x, y, z components as follows [59].

$$\alpha = (\alpha_{xx} + \alpha_{yy} + \alpha_{zz})/3 \tag{21}$$

The microscopic first order hyper polarisability (β) is a third rank tensor having 27 tensor components. But for a molecule with no symmetry, these 27 elements are reduced to 10 according to Kleinman symmetry [24, 58]. So the calculations for finding out the average hyperpolarisability based on the Eq. (22) is composed of these 10 irreducible tensor components of β_{xxx}, β_{xyy}, β_{xzz}, β_{yyy}, β_{xxy}, β_{yzz}, β_{zzz}, β_{xxz} and β_{yyz} [51].

$$\beta_{tot} = (\beta^2_x + \beta^2_y + \beta^2_z)^{1/2} \tag{22}$$

where

$$\beta_x = (\beta_{xxx} + \beta_{xyy} + \beta_{xzz}) \tag{23}$$

$$\beta_y = (\beta_{yyy} + \beta_{xxy} + \beta_{yzz}) \tag{24}$$

$$\beta_z = (\beta_{zzz} + \beta_{xxz} + \beta_{yyz}) \tag{25}$$

2.1.1.5 In Silico Druggability Test

Lipnski's Rule of Five [28] is generally used for analyzing the extent of oral bioavailability properties of considering chemical systems. This rule is based on certain observations, an empirical rule, states that a molecule is likely to be orally active if it satisfies the following conditions: (a) the molecular weight should be less than 500; (b) logP value (calculated from octanol/water partition coefficient) should not be greater than 5; (c) number of hydrogen bond donors (–OH and –NH groups) should not be greater than 5; and (d) number of hydrogen bond acceptors (mainly N or O atoms) should be at most 10. Noticeably the limits of all condition of this rule are 5 or its multiple, hence popularly called as 'Rule of Five.' The rule describes the molecular properties, which are important for a drug's pharmacokinetics in the human body, including their absorption, distribution, metabolism and excretion (ADME). However, the rule does not predict whether a molecule is pharmacologically active or not. A compound is likely to be orally active as long as not more than one rule is violated. The extension to the Lipinski's rule of five states that polar surface area (PSA) should be less than or equal to 140 Å^2 and number of rotatable bonds should be within 10 [52]. PSA is formed from polar atoms in a molecule and calculated by using the method, termed topological PSA (TPSA), based on the summation of the tabulated surface contribution of polar fragments [14] usually from oxygen, nitrogen and attached hydrogen.

2.2 COMPUTATIONAL DETAILS

From the theoretical point of view, Quantum mechanical studies related with reactivity and physical properties of some Coumarin derivatives are demonstrated that DFT/B3LYP/6–311G (2d, 2p) is a reliable method and basis set for the calculation of geometries and related properties of them [39]. The gas phase structures of molecules under study in the ground state are optimized by performing DFT with Beck's three parameters [2] for exchange interaction and Lee-Yang-Parr [26] is used to consider correlation functional (B3LYP) with 6–311G (2d, 2p) basis set. The vibration analysis is also performed in order to ensure that the optimized geometries are corresponding to a true minimum not a transition state. All computational chemistry calculations are done using Gaussian 09W simulation package [17]. Natural Bond Orbital (NBO) analysis implemented in GAUSSIAN 09W package is carried out at the same level of theory to get a good understanding of contributions from each atom in the formation of different molecular orbitals and various second order interactions exist between the subsystems. In order to study the extent of NLO property the polarisbility and hyper polarisability are calculated. The local (or regional) reactivity descriptors, condensed Fukui functions, are evaluated from single point calculation in terms of the molecular coefficients and the overlap matrix of the system. The optimized molecular geometries are utilized for the single point energy calculations with the same level of theory for anions and cations of the considering molecules in the ground state with doublet multiplicity. The individual atomic charges are calculated by natural population analysis (NPA) from NBO results, are used to calculate the Fukui functions of corresponding reactive sites. TD-DFT is used for the evaluation of absorption (UV-Vis) spectra of molecules by including one set of diffuse function in the basis set.

The molecules selected for computational chemistry calculations of this present work are Coumarin(C), 3,4 dihydrocoumarin (DC) and with substitution of nitro, amino and trifluoro methyl groups at third position of parent ring which represents like $C.NO_2$, $C.NH_2$, $C.CF_3$ and $DC.NO_2$, $DC.NH_2$ and $DC.CF_3$ (Figure 2.2).

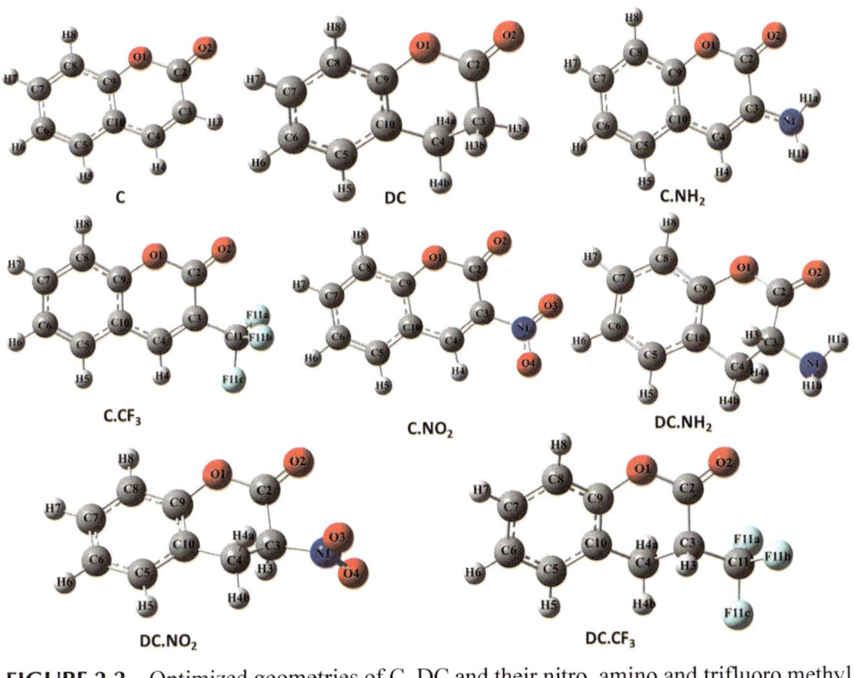

FIGURE 2.2 Optimized geometries of C, DC and their nitro, amino and trifluoro methyl substituted structures using B3LYP/6–311G (2d, 2p).

In order to recognize the importance of chemical and biological properties of Coumarins, we have calculated Quantum mechanical reactivity descriptors as well as In silico druggability parameters of some hydroxy coumarins such as 7-hydroxy coumarin (7hC), 4-hydroxy coumarin (4hC), 7-hydroxy–4-methyl coumarin (7h4mC), Esculetin (E), 4-methyl esculetin (4mE) (see Figure 2.3).

Biological studies related to druggability of considering molecules are predicted by calculating Lipnski's Rule of Five [28], explains the extent of oral bio availability properties with the help of free on-line Cheminformatics tool namely Molinspiration property explorer. The SMILES [55] notations of considering structures are generated using Open Babel [33] and then fed into the Molinspiration software to calculate molecular properties such as logP, topological polar surface area, number of hydrogen bond donors and acceptors, molecular weight, volume and number of rotatable bonds.

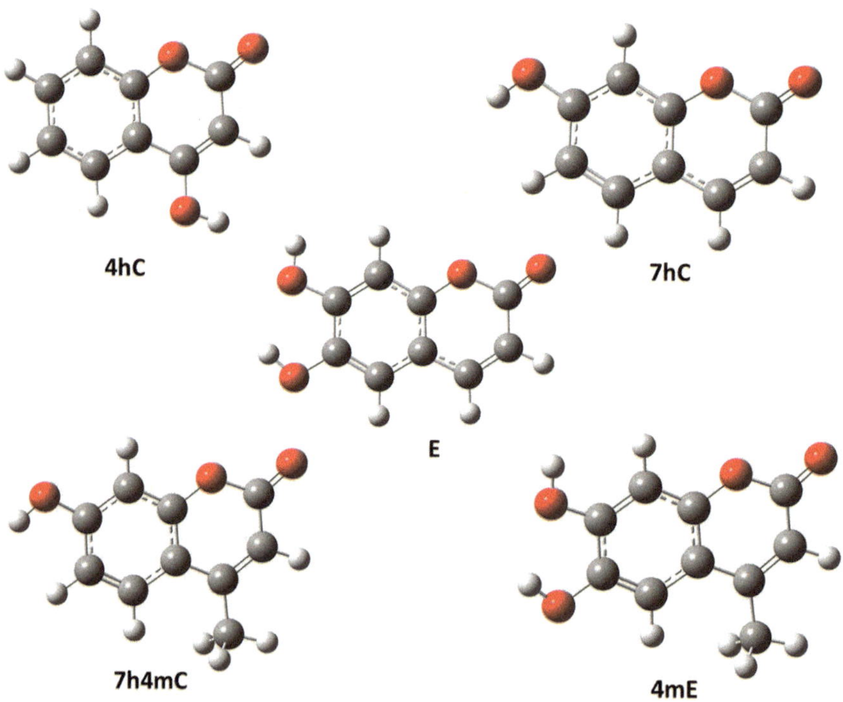

FIGURE 2.3 Optimized geometries of hydroxy coumarins studied using B3LYP/6–311G (2d, 2p).

2.3 RESULTS AND DISCUSSION

2.3.1 DETERMINATION OF CHARACTERISTICS OF STRUCTURES FROM NBO ANALYSIS

It has been generally accepted that the presence of soft π electron density is the characteristics of C=C double bond in unsubstituted alkene. The more electronegative character of sp^2 hybrid with respect to the sp^3 is considered as one of the reasons for the better stabilization of a negative charge in alkene. Therefore, the expected reactivity of the C=C functionality is the attack of electrophiles. However, this behavior can be interestingly modified by inserting suitable substitutions. The presence of an electron withdrawing substituents at the adjacent place of C=C double bond facilitates the attack of nucleophile on this reactive site. The nucleophile activation of α,β-unsaturated carbonyl compounds are the well recognized examples for this type of modifications [7]. The peculiar properties of α-benzapyrone are also correlated with α,β unsaturated ester group present in the parent ring. In this context, studies regarding the structure of coumarin (C) and dihydrocoumarin (DC) by considering the possible intra molecular orbital interactions are certainly important. The experimentally observed aromatic nature of compound C and DC are accurately described with the help of computational thermo chemical studies by designing isodesmic reactions, suggests that DC is more stable than C even both are aromatic in nature [32]. On accounting the structural variations, the present study consists of the prediction of relative aromaticity of the considering systems by using NBO calculations. The differences in the thermodynamic and kinetic stability of these compounds also can be explained in terms of aromaticity and delocalization of electron density through π molecular orbitals.

Table 2.1 displays the calculated natural hybrids of atoms present in cyclic ring, the σ (C6–C5) bond is formed from the combined influence of hybrid $sp^{1.76}$ of C6 atom (which is formed from the mixture of 36.16% s, 63.74% p and 0.10% d atomic orbital's) and $sp^{1.72}$ of C5 (which is from 36.72% s, 63.19% p and 0.09% d atomic orbital's). It means that the above described molecular orbital is formed by the contribution of C6 and C5 atoms in that proportion and the total occupancy of electrons in σ C6–C5 bond is 1.9778. The C–C single bond is formed by the linear combination

TABLE 2.1 Occupancy of NBOs and Hybrids of C and DC Calculated by B3LYP/6–311G (2d, 2p)

NBOs	Occupancy	Hybrid	AO
Coumarin			
σ C6–C5	1.97781	C6 $sp^{1.76}$	s (36.16%) p (63.74%) d (0.10%)
		C5 $sp^{1.72}$	s (36.72%) p (63.19%) d(0.09%)
π C6–C5	1.69633	C6 s^0p^1	s (0.00%) p (99.96%) d (0.04%)
		C5 s^0p^1	s (0.00%) p (99.96%) d (0.04%)
σ C6–C7	1.97835	C6 $sp^{1.8}$	s (35.62%) p (64.28%) d (0.10%)
		C7 $sp^{1.77}$	s (36.03%) p (63.86%) d (0.10%)
σ C5–C10	1.96935	C5 $sp^{1.82}$	s (35.44%) p (64.46%) d (0.09%)
		C10 $sp^{1.83}$	s (35.26%) p (64.68%) d (0.06%)
σ C10–C9	1.97091	C10 $sp^{2.1}$	s (32.19%) p (67.71%) d (0.10%)
		C9 $sp^{1.63}$	s (37.95%) p (62.00%) d (0.06%)
π C10–C9	1.59162	C10 s^0p^1	s (0.00%) p (99.98%) d (0.02%)
		C9 s^0p^1	s (0.00%) p (99.96%) d (0.04%)
σ C10–C9	1.97117	C10 $sp^{2.09}$	s (32.38%) p (67.55%) d (0.06%)
		C9 $sp^{1.90}$	s (34.42%) p (65.48%) d (0.10%)
σ C9–C8	1.97331	C9 $sp^{1.65}$	s (37.65%) p (62.30%) d (0.05%)
		C8 $sp^{1.87}$	s (34.77%) p (65.10%) d (0.12%)
σ C9–O1	1.98981	C9 $sp^{3.15}$	s (24.08%) p (75.79%) d (0.13%)
		O1 $sp^{1.87}$	s (35.20%) p (64.62%) d (0.18%
σ C8–C7	1.97546	C8 $sp^{1.73}$	s (36.60%) p (63.31%) d (0.09%)
		C7 $sp^{1.78}$	s (35.96%) p (63.93%) d (0.11%)
π C8–C7	1.68262	C8 s^0p^1	s (0.00%) p (99.95%) d (0.05%)
		C7 s^0p^1	s (0.00%) p (99.96%) d (0.04%)
σ C4–C3	1.98109	C4 $sp^{1.66}$	s (37.57%) p (62.33%) d (0.10%)
		C3 $sp^{1.56}$	s (39.02%) p (60.87%) d (0.11%)
π C4–C3	1.82984	C4 s^0p^1	s (0.00%) p (99.93%) d (0.07%)
		C3 s^0p^1	s (0.00%) p (99.90%) d (0.10%)
σ C2–C3	1.98374	C2 $sp^{1.48}$	s (40.31%) p (59.65%) d (0.04%)
		C3 $sp^{2.17}$	s (31.50%) p (68.36%) d (0.14%)
σ C2–O2	1.99079	C2 $sp^{2.98}$	s (25.06%) p (74.81%) d (0.12%)
		O2 $sp^{2.25}$	s (30.69%) p (69.10%) d (0.21%)

TABLE 2.1 (Continued)

NBOs	Occupancy	Hybrid	AO
σ C2–O1	1.99440	C2 $sp^{1.91}$	s (34.37%) p (65.55%) d (0.09%)
		O1 $sp^{1.36}$	s (42.17%) p (57.32%) d (0.50%)
π C2–O2	1.97940	C2 s^0p^1	s (0.00%) p (99.84%) d (0.16%)
		O2 s^0p^1	s (0.00%) p (99.62%) d (0.38%)
Dihydro coumarin			
σ C4–C3	1.97378	C4 $sp^{2.70}$	s (27.04%) p (72.87%) d (0.09%)
		C3 $sp^{2.45}$	s (28.97%) p (70.93%) d (0.10%)
σ C2–C3	1.98632	C2 $sp^{1.54}$	s (39.40%) p (60.56%) d (0.04%)
		C3 $sp^{2.86}$	s (25.89%) p (73.95%) d (0.16%)

of two sp^3 hybridized carbon atoms while sp^2 hybridized carbon atoms take part in the formation of C=C bond and C≡C is derived from two sp hybridized atoms with the maximum occupancy of electrons in a chemical bond is considered as to be 2 [50]. On analyzing the composition of π orbitals of other carbon centers of both C and DC reveals that only p orbitals take part in the formation of π orbitals. For example π C6–C5 bond is formed by the combined influence of C6 (mixture of 0.00% s, 99.96% p and 0.04% of d orbital's) and C5 (mixture of 0.00% s, 99.96% p and 0.04% of d orbital's) with the occupancy of 1.6933. Likewise, it has been observed that p orbitals of six carbon atoms in benzene ring take part together in the formation of π orbital and their occupancy is nearly to 1.5, indicates the electrons are delocalizing among these π orbitals of the benzene ring in both C and DC. Even though the p orbitals are the source of π bonds in C3=C4 and C2=O2 of C but some slight variations are seen in their occupancy of 1.829 and 1.979 respectively. It gives an indication for the existence of some sort of delocalization even which is not so strong enough to the same extent that seen in benzene moiety of the C. The C3–C4 and C2=O2 bond in DC with occupancy of 1.97378 and 1.9889, respectively, indicate the lack of involvement of π orbital electrons in delocalization as a conjugated system due to the absence of bridged C3=C4 double bond which extends the conjugation from benzene ring to pyrone ring. From the close observation of hybrids of each atom in C reveals that, the carbon centers in benzene ring are nearly in sp^2 hybridization and extended to carbon centers C10, C4 and C3 of C10–C4 and C3=C4 bonds which are the part of α-pyrone ring. The hybridization of C2 is observed as $sp^{1.91}$, nearly to sp^2 type, indicates the possibility of the extension of the delocalization of π electrons through C3=C4 and C2=O2. However, the electron occupancy of these bonds raises the questions regarding the strength of delocalization of these π electrons as the continuation of benzene ring. The shape of sp^2 hybridized orbitals of carbon centers are planar with p atomic orbital is placed right angle to the plane of the nucleus [15] and all carbon centers in C are nearly 120° with slight variations are seen for angles C9–C10–C4 and C3–C2–O1 with value of 117.4° and 115.9°, respectively. In contrast to α-pyrone moiety, all carbon centers in benzene moiety are showing bond angle of 120°. The optimized geometry of DC consists of a planar benzene ring and a heterocyclic ring with large distortion in planarity as compared with C. The effect of angular strain within the sp^3 hybridized carbon atoms and repulsion between contiguous $-CH_2-$ groups results the staggered conformation of hydrogen atoms which in turn distort the planarity of pyrone ring in DC. Contrasting with the above behavior, C consists of sp^2 hybridized fragment and is found to be adopted a completely planar structure at its most stable geometry. Furthermore, this conclusion can be verified by the analysis of bond order of each C–C bond in both C and DC which are given in Table 2.2.

If a system having alternated double bonds and there exists an aromatic π electron delocalization then every bond which involved in conjugation should be with the bond order nearly of 1.5 [7]. The bond order for C2–C3 of C, which is placed between C3=C4 and C2=O2, is observed as 1.0962 and likewise bond order of C4–C10, present between bridged benzene ring and C3=C4, is found to be 1.1560. These above mentioned bonds in DC are found to be with the bond order of 0.9912 and 1.0125, respectively. The benzene moiety present in both C and DC show the bond order nearly of 1.5, gives a positive correlation to the aromaticity and the bond order of considering systems obtained from NBO analysis (see Table 2.2). Even though the strong aromatic conjugation exists in the benzene moiety of both compounds, the extent of delocalization is varied in α-pyrone moiety. It can be reasoned out for DC by considering the absence of alternate double bond in α-pyrone moiety for the delocalization. The bond orders of C enlighten the existence of delocalization even which is not as strong

TABLE 2.2 Bond Orders for Each Bond of C and DC from NBO Analysis

Bonds	C	DC
C6–C7	1.3904	1.4273
C6–C5	1.4784	1.4363
C5–C10	1.3295	1.4006
C10–C9	1.2905	1.3542
C9–C8	1.3671	1.3852
C8–C7	1.4598	1.4406
C10–C4	1.1560	1.0125
C4–C3	1.7220	1.0067
C3–C2	1.0962	0.9912
C2–O2	1.7380	1.7961
C2–O1	0.9062	0.9540
C9–O1	1.0008	0.9345

as aromatic delocalization seen in benzene moiety. Therefore, it can be concluded by accounting electronic occupancies and bond orders that, the α-pyrone moiety present in DC has no any aromatic nature while in C there exists some delocalization towards C2=O2 bond which may cause for the reduction of aromatic nature of C, and hence DC is thermodynamically more stable than C. Table 2.3 displays the second order perturbation energies of donor-acceptor interactions between NBO's of C and DC due to resonance effect.

The higher stabilization energy values indicate the existence of strong interactions between $\pi \rightarrow \pi^*$ orbitals in the benzene moiety of both molecules even a slight higher value for DC than C. The interactions between π C7–C8 orbital to π*C5–C6 and π*C9–C10 orbitals with stabilization energy of 17.74 and 22.88 kcal/mol, respectively, is an evidence for delocalization of one π orbital with two adjacent π* orbitals. Similar type of interaction, $\pi \rightarrow \pi^*$, also exist between other carbon centers in the benzene ring. Besides the strength of perturbation it gives additional information regarding the preferable direction of π electron flow, so that tracking of these orbital interactions will help us to predict the possible reactive centers and mechanism of electrophilic or nucleophilic substitution on

TABLE 2.3 Second Order Perturbation Energy Between NBOs for Explaining Resonance Effect in C and DC

Donor NBOs	Acceptor NBOs	$E^{(2)}$ kcal/mol	
		C	DC
	π* C9–C10	17.73	20.31
π C6–C7	π* C7–C8	21.10	21.19
	π* C6–C7	19.67	20.46
π C9–C10	π* C7–C8	16.43	19.02
	π* C3–C4	15.51	0.00
	π* C6–C7	17.74	19.34
π C7–C8	π* C9–C10	22.88	22.81
	π* C9–C10	10.97	0.00
π C3–C4	π* C2–O2	22.65	0.00
π C2–O2	π* C3–C4	5.58	0.00
	σ* C9–C10	7.19	6.33
LP(1) O1	σ* C2–C3	4.80	5.53
	π* C9–C10	30.30	22.44
LP(2) O1	π* C2–O2	35.97	38.25

Note: $E^{(2)}$ second order perturbation energy.

both ring. The observed stabilizing interaction of πC9–C10→π*C3-C4 with 15.51 kcal/mol second order perturbation energy in C also explains the interaction between benzene and pyrone moiety even which is lower than average interaction energy among aromatic orbitals of 22 kcal/mol. But such type of interaction is not present in compound DC because of the absence of C3=C4 bond.

A very strong interaction also has been observed in molecule C between the p-type orbital containing lone electron pair of O1 and the neighboring π*C9–C10, π*C2–O2 antibonding orbitals with perturbation energy of 30.30 and 35.97 kcal/mol respectively, while in the case of DC, this perturbation energy varied as 22.44 kcal/mol and 38.75 kcal/mol for the same π* molecular orbitals. It reveals that the lone electron pair of O1 in C is more interacting with the benzene ring than adjacent C2=O2 bond. Likewise, the other interactions present in C such as π C9–C10→π*C3–C4 and π C3–C4→π*C9–C10 with perturbation energy of 15.51 and 10.94 kcal/mol respectively also can be correlated to the lowering of bond order of C9–C10, 1.29, than other bonds in aromatic moiety, giving a strong support to the lowering of aromatic nature of C as the consequence of delocalization of lone electron pair in O1 to benzene ring. The bond order of C9–C10 in DC, nearly at 1.35, ensures the strong delocalization with in the benzene ring than in C which in turn leads to the higher aromatic nature of DC.

The other interactions which have high contribution in the reactivity of a molecule is π*→π* [11]. These types of interactions are π*C9–C10→π*C3–C4 and π*C2–O2→π*C3–C4 with perturbation energy of 185.50 and 105.41 kcal/mol respectively present in C, but such interactions between antibonding orbitals with high amount is absent in DC.

2.3.2 ELECTRONIC EFFECTS OF SUBSTITUENTS FROM NBO ANALYSIS

This portion of the work is devoted to a comparative estimation of the electronic effect of functional groups like NO_2, CF_3 and NH_2 on third position (α-pyrone moiety) of C and DC structures. The inductive and resonance effects are the basis for the classification of substituents. Nitro and trifluoromethyl groups are associated with the withdrawal of electrons from the parent ring while amino group make available its lone electron pair to the π conjugated system [7].

It is generally accepted that during the interaction between a substituent and its parent ring, the manifestation of two main electronic effects is possible. One is inductive effect, also known as polar effect; arise from the covalent single bond between unlike atoms. The electron pair forming the σ bond between unlike atoms is never shared absolutely equally for the two atoms, but always tends to be attracted towards the more electronegative atoms among the two [50]. It attenuates in proportion to the distance from a carbon atom bonded to the substituent. All inductive effects are permanent polarizations in the ground state of a molecule, and are therefore manifested in its physical properties, such as dipole moment and their polarisability. In addition, inductive effects operating through the bonds of a chemical system, an essentially analogous effect can operate either through the space surrounding to the molecule or in solution, via molecules of solvent surround it. As all the computational calculations of these considering structures are performed in gas phase, the solvent effect on reactivity can't be predicted from the results.

The other electronic effect is conjugative effect, also known as resonance or mesomeric effect. The present system deals with the electronic distributions occur in unsaturated, especially in conjugated, systems via their π orbitals. It has been observed for almost all cases, along with mesomeric effect there will also be an inductive effect with much smaller in amount than mesomeric effect as σ electrons are much less polarizable and hence less readily shifted than π electrons. The difference between this transmission of electrons via a conjugated system and the inductive effect in a saturated system is that the mesomeric effect suffers much less diminution by its transmission, and the polarity at adjacent carbon atoms alternates. The mesomeric, like inductive, the effects are permanent polarizations in the ground state of a molecule and so reflect in their physical properties.

The essential difference between inductive and mesomeric effects is reflected in their way of action, the inductive effects can operate in both saturated and unsaturated compounds while mesomeric effects can operate only in unsaturated, especially in conjugated, chemical systems. The former involve the electrons in σ bonds while the latter deals with π bonds and orbitals. Inductive effects transmit over only quite short distances in saturated chains before dying away, whereas mesomeric effects may be transmitted from one end to the other of quite large molecules provided that conjugation (delocalized π orbital) is present, through which they can proceed.

The energy of the donor-acceptor interaction between the orbitals of a substituent with the benzopyrone ring can be utilized for the evaluation of the impact of an electronic effect. In accordance with the generally accepted designations, –I and –C effects characterize the electron acceptor properties of substituents while +I and +C designations for electron donor properties. The orbital interactions determined within NBO analysis by using B3LYP/6–311G (2d, 2p) level of theory is characterized according to the scheme in Table 2.4, presents the total energies of the donor-acceptor interaction of orbitals associated with various electronic effects.

TABLE 2.4 Energy Contributions (kcal/mol) From Orbital Interactions According to the Inductive and Mesomeric Effect for Nitro, Amino and Trifluoromethyl Substituted C and DC

	Coumarin (C)				Dihydrocoumarin (DC)			
	Interaction energy (kcal/mol)							
NO_2	+I	+8.61	$\Sigma E_I = -1.81$	$\Sigma E_i = -16.15$	+I	+2.15	$\Sigma E_I = -4.07$	$\Sigma E_i = -4.07$
	–I	–10.42			–I	–6.22		
	+C	+3.89	$\Sigma E_C = -14.34$		+C	0	$\Sigma E_C = 0$	
	–C	–18.23			–C	0		
CF_3	+I	+1.56	$\Sigma E_I = -7.82$	$\Sigma E_i = -7.82$	+I	+2.48	$\Sigma E_I = -2.52$	$\Sigma E_i = -2.52$
	–I	–9.24			–I	–5.00		
	+C	0	$\Sigma E_C = 0$		+C	0	$\Sigma E_C = 0$	
	–C	0			–C	0		
NH_2	+I	+17.14	$\Sigma E_I = +7.06$	$\Sigma E_i = +41.85$	+I	+14.1	$\Sigma E_I = +7.35$	$\Sigma E_i = +7.35$
	–I	–10.08			–I	–6.75		
	+C	+34.79	$\Sigma E_C = +34.79$		+C	0	$\Sigma E_C = 0$	
	–C	0			–C	0		

Note: E is interaction energy; $\Sigma E_i = \Sigma E_I + \Sigma E_C$.

The energies related to individual components of the +I, –I, +C and –C effects, the total energies ΣE_I and ΣEc for the inductive and resonance effects respectively, the energy $\Sigma E_i = \Sigma E_I + \Sigma Ec$ corresponds to the total effect of substituent. If $\Sigma E_i < 0$, the substituent shows acceptor properties with parent ring and partial transfer of electron density from the orbitals of ring to substituent. If $\Sigma E_i > 0$, the substituent has donor properties in prominent. The extent of interaction of substituents on the parent ring of C and DC are shown in Figures 2.4a and 2.4b, respectively from NBO analysis.

The transfer of electron density from the natural σ-bonding orbital of benzopyrone ring to natural σ* antibonding orbital of substituent indicates the electron accepting inductive effect (which refer to as –I effect). The interaction between orbitals σ (C–C) of bezopyrone ring to σ (C–N) in C.NO_2 molecule models the partial transfer of electron density from benzopyrone ring to NO_2 group with 10.42 kcal/mol energy is an example for –I effect, hence give a negative sign and characterized the value as −10.42 kcal/mol. The same interaction in DC.NO_2 is giving the stabilization energy of –6.22 kcal/mol only.

The transfer of electron density from the σ orbitals of a substituent to the σ* antibonding orbitals of benzopyrone ring is designated an electron donating +I inductive effect, so the energy of interaction is indicated with a positive sign. In the case of C.NO_2, interaction of σ (C–N) → σ* (C–C) takes +8.61 kcal/mol stabilization energy while for DC.NO_2, +2.14 kcal/mol are attributed to the +I effect. Electron donating mesomeric +C effect is characterized by the transfer of electron density from natural bonding π orbitals of substituents to π* orbitals of benzopyrone ring. The +C effect of C.NO_2 due to the interaction of π (O–N) → π* (C–C) is giving the energy of +3.89 kcal/mol while the energy of accepting interaction of π (C–C) → π*(O–N) with stabilization energy of −18.23 kcal/mol corresponds to –C effect. All donor-acceptor interactions of which their threshold interaction energy exceeded by the value of 0.5 kcal/mol is taken for the total electronic effect of particular functional group with benzopyrone and dihydrobenzopyron ring for determining the effect of substituents. It means that the characterizing energy of the +I effect of –NO_2 group in C.NO_2 molecule is determined from the contributions of the five components and the –I effect of the same from the contributions of four components.

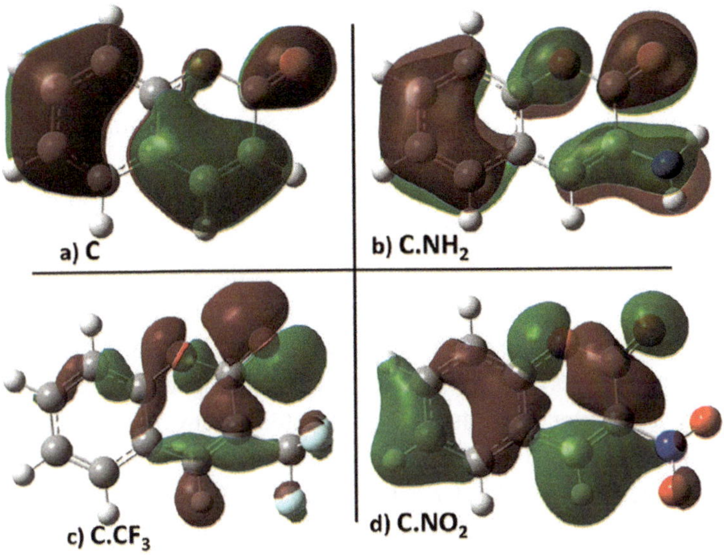

FIGURE 2.4A Molecular orbital diagram showing the interaction of substituents with C.

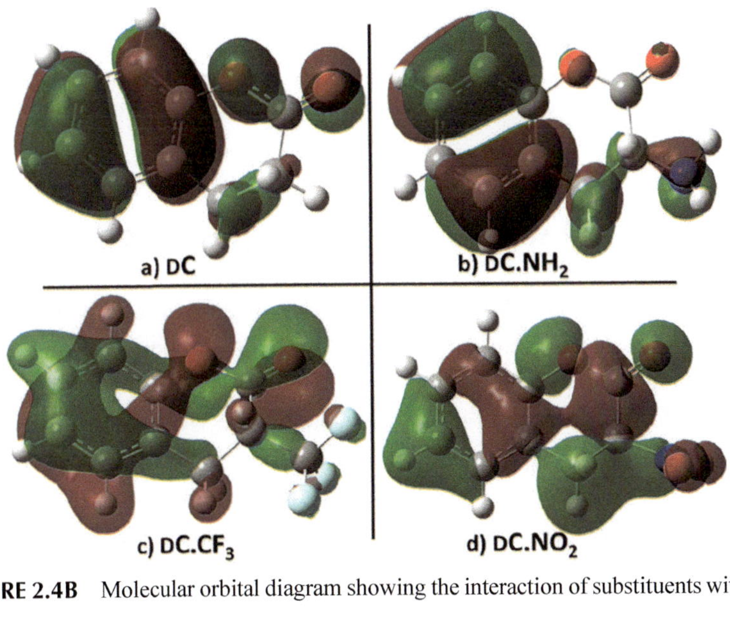

FIGURE 2.4B Molecular orbital diagram showing the interaction of substituents with DC.

The negative values of the total energies (ΣE_I) of $C.NO_2$ and $DC.NO_2$ ensures the general manifestation of electron-acceptor properties of NO_2 functional group. Even the NO_2 group in $C.NO_2$ is showing +I and –I as well as +C and –C effects, the overall effect indicate that NO_2 is substantially stronger π acceptor (with greater negative value of ΣEc) than σ-acceptor (with lower negative value of ΣE_I). It has been observed that the competition exists between +I and –I effects for $DC.NO_2$ because of the absence of C3=C4 double bond where the substituent is attached which in turn reduces the possibility for the conjugative effect of $–NO_2$ group. On comparing the electronic effect of $–CF_3$ in C and DC, –I effect is prominent as the presence of highly polarized three σ C-F bonds of $–CF_3$ group. By analyzing the influence of selected electron withdrawing substituents ($–CF_3$ and $–NO_2$) on third position of benzopyrone ring, NO_2 acts as strong π acceptor while $–CF_3$ shows mild σ electron accepting nature. The electron donating substituent, $–NH_2$, activate the ring by exhibiting a strong +C effect because of the delocalization of its lone electron pair on nitrogen to π*C3–C4 orbital with interaction energy of +34.79 kcal/mol. Even though the conjugative effect is prominent for $–NH_2$ group in C, there exists some sort of +I and –I effect because of the polarization of C–N bond towards the nitrogen atom but comparatively lower in its value (see Table 2.4). The value of $\Sigma E_I > 0$ for both $C.NH_2$ and $DC.NH_2$ enlighten the donor property of $–NH_2$ group even conjugative effect is absent in $DC.NH_2$.

2.3.3 FRONTIER MOLECULAR ORBITAL ANALYSIS

The frontier orbital (HOMO and LUMO) are of great importance in defining the reactivity of chemical species. The HOMO energy of a molecule represents its ability to donate electrons, means that it is the highest orbital in which electrons are residing in the ground state while LUMO, which is the electron orbital just above the HOMO, considered as electron accepting level and its energy represents the extent of electron accepting ability of that molecule. The reactivity of a molecule mainly depends on these frontier orbitals. The fully occupied or unoccupied orbital energy can be lowered or raised by electronic perturbation. An electron-withdrawing group (electron acceptor) lowers the orbital energy to the square of the atomic orbital coefficient at the center where the substituent is attached, while an electron donor substituent is interested in raising the energy of molecular orbitals [12]. The HOMO-LUMO energy separation of DC and C are 5.984 and 4.625 eV respectively, emphasizes the higher stability of DC as compared with C in support of the results obtained from NBO analysis. Even though the aromatic nature of DC and C arise from the fused benzene ring present in their structures but lower than the stability of benzene itself with HOMO-LUMO separation of 6.737 eV. The variation of HOMO-LUMO gap of C and DC as the result of substitutions are shown in Table 2.5.

The conjugated molecules are generally characterized by a small HOMO-LUMO energy gap, as the result of significant intramolecular charge transfer (CT) from electron donor groups to efficient electron acceptor through π conjugated path. Even though the electron donating groups destabilize (raise the energy) the molecular orbitals, affect much more in HOMO than LUMO energy as seen in Table 2.6, displays the deviation of HOMO and LUMO energy of substituted structures from its unsubstituted parent molecules.

Comparing the deviation values, the destabilization is more affected in HOMO than LUMO energy while substituting amino group, but a slight variation is seen for DC. It is clearly evident that DC shows more aromatic nature with high-energy gap than C, because of the absence of α,β

TABLE 2.5 Quantum Chemical Parameters like HOMO, LUMO Energy, Eg and ENERGY of C, DC and Their Substituted Structures

Molecule	HOMO (eV)	LUMO (eV)	E gap (eV)	Energy (a.u)
C	–6.699	–2.074	4.625	–497.1663
C.NO_2	–7.307	–3.165	4.142	–701.723
C.CF_3	–7.111	–2.559	4.552	–834.309
C.NH_2	–5.799	–1.578	4.211	–552.547
DC	–6.707	–0.723	5.984	–498.382
DC.NO_2	–7.228	–2.225	5.003	–702.932
DC.CF_3	–7.017	–1.058	5.958	–835.524
DC.NH_2	–6.728	–0.769	5.959	–553.747

Note: Eg – Energy gap between HOMO-LUMO energy.

TABLE 2.6 Deviation of HOMO, LUMO and Eg Values on Substitution with Parent Ring

Molecule	d(HOMO)	d(LUMO)	d(Eg)
C	0	0	0
C.NO_2	–0.608	–1.091	–0.483
C.CF_3	–0.412	–0.485	–0.073
C.NH_2	+0.900	+0.496	–0.414
DC	0	0	0
DC.NO_2	–0.521	–1.502	–0.981
DC.CF_3	–0.310	–0.335	–0.026
DC.NH_2	–0.021	–0.046	–0.025

Note: d: deviation; –ve sign: stabilization; +ve sign: destabilization; –ve Eg: lowering of Eg.

unsaturated C3=C4 bond for delocalization from benzene ring to pyrone ring. Amino group envisages its electron donating ability by facilitating its lone electron pair to delocalize through the conjugation of attached chemical system which in turn leads to the destabilizatrion of HOMO of parent ring.

Two types of electron withdrawing substituents are selected on the basis of prominence of conjugative and inductive effect while they are functioning. The electron withdrawing nature of nitro group is predominantly by its electron accepting conjugative effect compared to electron accepting inductive effect whereas electron withdrawing nature of trifluoromethyl group is because of its electron accepting inductive effect due to the presence of three fluorine atoms attached on carbon. It has been observed that these electron-withdrawing substituents are stabilizing the LUMO level than HOMO level of both C and DC, but substitution of nitro group is more effective than that of trifluoromethyl group. The lowering of HOMO-LUMO gap is the consequence of the large stabilization of the LUMO due to the electron accepting ability of the electron withdrawing groups from ring system [47]. Even though the electron accepting conjugative effect is predominant in nitro group there exist some sort of electron accepting inductive effect due to the presence of two strong resonating N–O bonds. Hence, the LUMO level of DC is also stabilized by nitro and trifluoro methyl groups even in the absence of α,β unsaturated C=C double bond.

The effect of substitution on frontier orbitals gives a general trend, even the HOMO and LUMO energies are dependent on the nature of the molecule. The substitution of electron donating group leads to the destabilization of molecular orbitals especially HOMO energy while electron withdrawing group stabilizes much more the LUMO level in a conjugated chemical system. This trend on substitution can be effectively utilized for the designing of better pericyclic reactions, concerted in nature, purely depending on energies and nature of HOMO, LUMO and reaction condition (thermal energy or light energy) only. The effect of substituents in HOMO-LUMO gap of C and DC are shown in Figures 2.5a and 2.5b, respectively.

2.3.4 GLOBAL REACTIVITY DESCRIPTORS

The computed energy using B3LYP/6–311G (2d, 2p) theory indicate that DC is more stable compared to C by the amount of 763.04 kcal/mol.

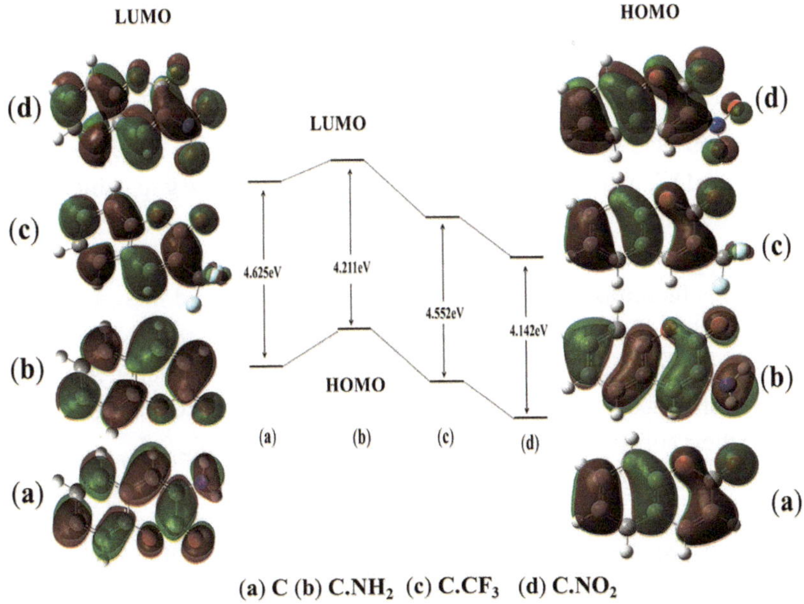

FIGURE 2.5A Frontier MO and energy gap for C and its substituted structures.

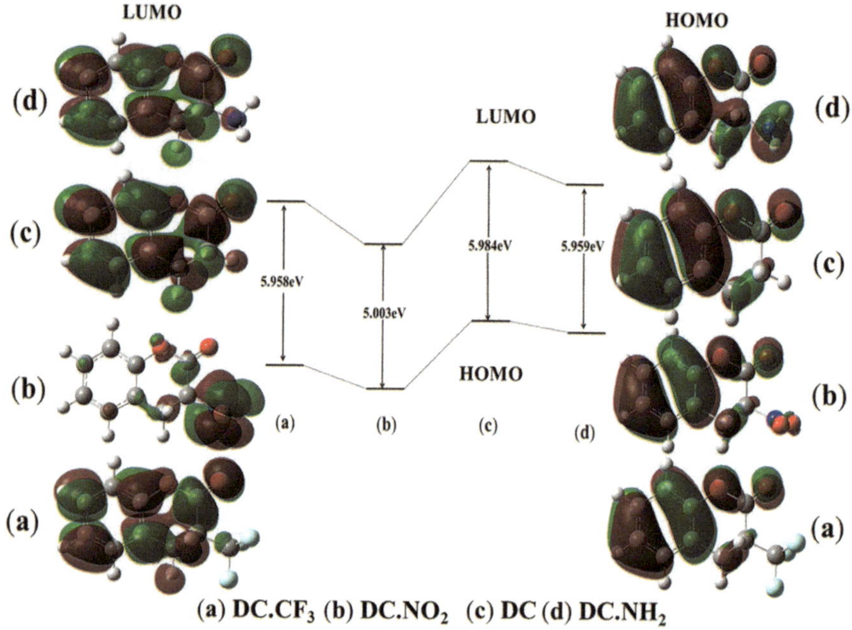

FIGURE 2.5B Frontier MO and energy gap for DC and its substituted structures.

Even though the conjugation is higher in C than DC, NBO reveals that the higher aromatic nature of DC is giving the extra stability to the system. This also reflects in HOMO-LUMO energy gap of both the compounds, i.e., C and DC, where later is showing higher energy gap which leads to the lowering of chemical reactivity. The global reactivity descriptors calculated by the orbital vertical method requires HOMO-LUMO energy gap ensures its importance in determining the kinetic and thermodynamic stability of the chemical system. The calculated global reactivity descriptors such as ionization potential, electron affinity, global hardness, global softness, chemical potential, electronegativity and electrophilicity of C, DC and their substituted structures are tabulated in Table 2.7.

The studies regarding the reactivity of aromatic aldehyde towards acid catalyzed aromatic exchange reactions with DFT-based theory interpret the influence of aromatic ring on reactivity of molecules through delocalization of π electrons conclude that the aromatic nature is positively correlated with hardness of a chemical system and so increase in aromaticity leads to the increase in hardness which in turn leads to the decrease in

TABLE 2.7 Global Reactivity Descriptors (eV) Calculated for C, DC and Their Substituted Structures Through Orbital Vertical Method Calculated by B3LYP/6–311G (2d, 2p)

Molecule	IP	EA	μ	H	ω	χ	S
C	6.699	2.074	4.386	2.312	4.163	–4.386	0.216
$C.NO_2$	7.307	3.165	5.236	2.071	6.619	–5.236	0.241
$C.CF_3$	7.111	2.559	4.835	2.276	5.135	–4.835	0.219
$C.NH_2$	5.799	1.578	3.688	2.111	3.223	–3.688	0.237
DC	6.707	0.723	5.984	3.715	2.992	–3.715	0.134
$DC.NO_2$	7.228	2.225	4.726	2.502	4.465	–4.726	0.199
$DC.CF_3$	7.017	1.058	4.037	2.977	2.735	–4.037	0.167
$DC.NH_2$	6.728	0.769	3.748	2.979	2.358	–3.748	0.168

reactivity [44]. The particular results of C and DC for hardness are 2.312 and 3.715 eV respectively confirms the above mentioned interpretation that higher aromatic nature and hardness correspond to higher stability and lower reactivity for particular aromatic system. The influence of electronic substituents on global reactivity descriptors of present systems are studied by inserting electron withdrawing ($–NO_2$ and $–CF_3$) and donating ($–NH_2$) groups at C3 position on the ring and which in turn has correlated with their chemical reactivity. The more changes in reactivity on substitution observes for the system having lower energy gap of its unsubstituted form whereas greater separation causes smaller change in reactivity by accounting the nature of substituents. Domingo et al. states that the more reactive nucleophile is characterized by lower **ω**, **μ** values and inversely a good electrophile is characterized by high value of **ω** and **μ** (2002). The calculated values for global electrophilicity index (**ω**) are showing the nucleophilic power of considering heterocycles. The compound DC and its substituted forms are showing lower value for **ω** and hence they are more nucleophilic whereas C and its substituted compounds are electrophilic in nature.

2.3.5 *LOCAL REACTIVITY DESCRIPTORS*

Besides the global reactivity descriptors, it is possible to define its regional counterpart condensed to atoms from the projection of the global quantity into any atomic centers in the molecule by using Fukui functions. Reactivity of a chemical system is influenced by the factors related to the availability of electrons (electron density) in particular bonds, or at particular atoms. A position with low electron availability is likely to be attacked by electron rich reagent (nucleophile) while position with more electron availability is easily attacked by electron deficient reagent (electrophile). In order to study the local reactivity descriptors, the optimized molecular geometries are utilized in single-point energy calculations which have been performed at the DFT/UB3LYP level of theory for the anions and cations of title molecules using ground state with doublet multiplicity. The individual atomic charges are obtained by NPA have been used to calculate the Fukui functions. Since this NPA exhibit improved numerical stability which results a better description of electronic distribution of compounds having high ionic character, such as those consist of metallic atoms or carbon structures with polarizing groups as seen in present molecules [42].

The values of Fukui functions, f^+ is giving the extent of nucleophilic attack while f^- is for electrophilic attack, of C and DC with their substituted structures are summarized in Table 2.8 reveal that the most reactive site for nucleophilic attack in C and its substituted structures present in α-pyrone moiety, more specifically C4 position, while DC and its substituted forms are showing the same in benzene moiety. The most favorable site for electrophilic attack in C occurs on C3 position but the substitutions make a displacement of this reactive site from α-pyrone to benzene moiety. The most feasible site for electrophilic attack in DC and its substituted structures present in benzene moiety. The inclusion of electron releasing amino group on α-pyrone ring slightly decreases the electrophilic nature of C4 position with value of 0.121 while the electron withdrawing groups increase the reactivity. It can be reasoned out by explaining the conjugation of lone electron pair of nitrogen atom in amino group facilitates higher electron density in C3═C4 bond even which is attached with cyclic ester group.

It has been observed that the conjugative effect is responsible for the change in reactive site for electrophilic attack at C7 position in $C.NH_2$ compared to $C.NO_2$ and $C.CF_3$ by facilitating more orbital interactions between π C9–C10→π* C8–C7 and π C5–C6→π* C8–C7 with perturbation energy

TABLE 2.8 Site Specific Reactivity Order Obtained for C, DC and Their Substituted Structures from the Calculation of Fukui Indices

Molecule	Fukui function	O1	C2	C3	C4	C5	C6	C7	C8	C9	C10
C	f+	0.046	0.032	0.141	**0.154**	0.087	0.013	0.124	0.047	0.032	–0.005
	f–	0.065	–0.022	**0.137**	0.015	–0.006	0.133	0.093	0.012	0.087	0.102
$C.NO_2$	f+	0.040	0.002	0.064	**0.172**	0.072	0.013	0.111	0.026	0.046	–0.025
	f–	0.062	–0.016	0.111	–0.004	–0.009	**0.138**	0.079	0.016	0.083	0.099
$C.NH_2$	f+	0.047	0.077	0.088	**0.121**	0.083	0.016	0.107	0.056	0.010	0.001
	f–	0.027	–0.007	0.056	0.122	0.049	0.035	**0.132**	0.006	0.058	0.025
$C.CF_3$	f+	0.046	0.018	0.129	**0.168**	0.083	0.013	0.122	0.039	0.041	–0.013
	f–	0.066	–0.019	0.128	–0.002	0.011	**0.145**	0.082	0.016	0.089	0.106
DC	f+	0.013	0.105	–0.006	–0.014	0.0436	0.054	**0.162**	0.029	0.026	0.142
	f–	0.105	–0.017	–0.003	–0.020	0.001	**0.202**	0.054	0.045	0.132	0.108
$DC.NO_2$	f+	0.023	0.022	0.007	–0.006	0.052	0.019	**0.102**	0.040	–0.002	0.056
	f–	0.088	–0.0142	–0.008	–0.022	–0.002	**0.199**	0.064	0.033	0.139	0.109
$DC.NH_2$	f+	0.014	0.128	0.001	–0.017	0.0245	0.068	0.141	**0.166**	0.032	0.122
	f–	0.085	–0.016	–0.015	–0.013	0.001	**0.158**	0.056	0.031	0.104	0.081
$DC.CF_3$	f+	0.014	0.136	–0.004	–0.014	0.025	0.071	**0.143**	0.015	0.035	0.125
	f–	0.099	–0.014	–0.001	–0.021	0.003	**0.206**	0.054	0.043	0.138	0.106

Note: f^+: for nucleophilic attack; f^-: electrophilic attack.

of 18.62 and 19.15 kcal/mol respectively. Likewise, DC and related structures exhibit the similar trend for nucleophilic attack with most reactive site of DC.NH_2 is C8 while DC, DC.NO_2 and DC.CF_3 exhibit the reactive site as C7 position. The most favorable site for electrophilic attack in DC and its substituted structures are at C6 with variation are seen in the extent of reactivity in the order of DC.CF_3 > DC > DC.NO_2 > DC.NH_2.

Relative nucleophilicity s_k^-/s_k^+ indices indicate the nucleophilicity of an atomic center compared with its own electrophilicity and defined as quotient of the condensed local softness indices of the molecular system. Atomic center having highest s_k^-/s_k^+ value is the most reactive nucleophilic site. Relative nucleophilicity has been successfully applied on the site selectivity studies on organic molecules in order to understand the reaction pathway.

It is found to be that, s_k^-/s_k^+ values are less subjected to the errors due to basis set implementation and correlation effects. On analyzing the relative nucleophilicity of carbon centers in C and DC summarized in Table 2.9 indicate that the highest value of relative nucleophilicity occurs at C6 position present in benzene moiety for both compounds.

TABLE 2.9 Relative Nucleophilicity (s_k^-/s_k^+) Indices for Atoms in Ring System of C and DC

Centers	C			DC		
	s_k^-	s_k^+	s_k^-/s_k^+	s_k^-	s_k^+	s_k^-/s_k^+
O1	0.0134	0.0099	1.3539	0.0175	0.0022	**7.9727**
C2	–0.0034	0.0069	–0.4927	–0.0028	0.0175	–0.1596
C3	0.0239	0.0305	0.7836	–0.0005	–0.0010	0.5000
C4	–0.0008	0.0333	–0.0240	–0.0033	–0.0023	1.4348
C5	–0.0019	0.0188	–0.1011	0.0002	0.0073	0.0219
C6	0.0298	0.0031	**9.6129**	0.0337	0.0091	**3.7155**
C7	0.0171	0.0268	0.0638	0.0090	0.0270	0.3331
C8	0.0034	0.0102	0.3333	0.0075	0.0048	1.5625
C9	0.0179	0.0071	2.5212	0.0022	0.0043	0.5232
C10	0.0214	–0.0011	–19.454	0.0018	0.0237	0.0761

2.3.6 UV-VISIBLE ABSORPTION SPECTRA

Referring to the comparative studies regarding the efficiency of TD-DFT and configuration interaction singles (CIS) approach, Jacquemin et al. [21] prescribe the TD-DFT-based calculation for excitation studies. TD-DFT gives a fast and often accurate estimation of transition energies between the energy levels of modeled system by using mono-determinant wave function [4]. An important advantage of TD-DFT compared to other excitation approaches is the inclusion of solvent contributions in computational calculations. Since the reactive medium has strong influence on the reactivity and excitation of chemical species, the inclusion of this solvent effect on computational calculation will give better theoretical prediction [38]. Even though the excitation energies are strongly over estimated in CIS, there exist some major chemical features that can be explained using this method. On comparing the efficiency of these two methods for explaining solvatochromism and symmetry of the excited state geometries, CIS method is not always good enough to reproduce the result consistently. Hence TD-DFT method is selected in order to achieve the information regarding vertical ground-to-excited state electronic transition in gas phase of the molecules under study which is directly related to absorption phenomena and can be computed with the only knowledge of the ground state geometry.

When a sample of molecule is exposed to light having enough energy to facilitate a transition with in the molecule, the required quanta of energy will be absorbed and there by occurs the excitation from lower to higher energy levels. In contrast to σ electrons, which are characterized by rotational symmetry of their wave function with respect to the bond direction, π electrons are characterized by a wave function with a node at the nucleus and its rotational symmetry is along a line through the nucleus. Normally π bonds are weaker than σ bonds because of their lower extent of overlapping between the atomic orbitals involved in the π bond formation due to their parallel orientations. This leads to the greater possibility for the excitation of π electrons by absorbing photons with lower energy [48]. If the observed absorption maxima is higher than 200 nm, the possible transitions are $n \rightarrow \pi^*$ and $\pi \rightarrow \pi^*$. Most of this transition will be occurred normally from singlet state HOMO to LUMO, and resulting species is also in singlet excited state. So, if the potential energy of excited state is lower, higher will be the excitation possibility and it can be achieved by increasing the size of conjugation. Since the NBO analysis of C and DC reveal the inevitable role of π molecular orbitals in determining the conjugative effect of substituents on the reactivity, the absorption bands are mainly derived from $\pi \rightarrow \pi^*$ electronic transitions. As considering conjugation, the chemical structure of C consists of an additional conjugation in α-pyrone ring fused with aromatic benzene while in DC the conjugation present only in benzene ring. All the electronic absorptions of C, DC and their substituted structures are summarized in Table 2.10, displays the transition from the ground state to the first excited state and are mainly described by one electron excitation from the HOMO to lowest unoccupied molecular orbital. Since the theoretical spectra are obtained from the gas phase calculations without considering the solvent effects, some difference has been observed in absorption wavelength while comparing with experimental result.

On comparing the absorption wavelength of C and DC, 300 and 245 nm, respectively, reveals the influence of extended conjugation in C which facilitates a red-shift in absorption (so called 'bathochromic shift') of 50 nm to it. Soon and Gordon [49] states that absorption maxima at

TABLE 2.10 Computed Energies (E), Absorption Wavelengths (λ), and Oscillator Strengths (f) of C and DC with Their Substituted Molecules Calculated by B3LYP/6–311+G (2d, 2p)

Molecules	E (eV)	Λ (nm)	f
C	4.129	300	0.1146
$C.NO_2$	3.482	356	0.1102
$C.CF_3$	4.006	309	0.0997
$C.NH_2$	3.946	314	0.3522
DC	5.053	245	0.0134
$DC.NO_2$	4.175	297	0.0100
$DC.CF_3$	5.067	244	0.0142
$DC.NH_2$	5.052	246	0.0114

324 nm of Coumarin derivative is because of the presence of α-pyrone ring. The absorption wavelength of C.NO_2, C.CF_3 and C.NH_2 are 356, 309 and 314 nm respectively whereas DC.NO_2, DC.CF_3 and DC.NH_2 are shown the absorptions of 294, 244 and 246 nm, respectively. All these values reveal that the substitution at C3 position, whatever be the electron donating or withdrawing group, is giving a bathochromic shift of varying extent as per the nature of substituent and the shift follows the order of $-NO_2 > -NH_2 > -CF_3$. From NBO analysis it has been established that $-NO_2$ is showing higher electron withdrawing conjugative effect through its two resonance double bonds and $-NH_2$ has electron releasing conjugative effect by making use of its lone electron pair whereas $-CF_3$ has electron withdrawing inductive effect as the presence of three fluorine atoms. From a comparative study of some modeled conjugated structures, done by Carlos and Barry in order to explain the effect of the backbone conjugation on the potential energy surface of the ground and excited state intra molecular hydrogen transfer reaction concluded that over all conjugation of the system is larger for excited state than the ground state. It validates the conjugative effect of substituents are the reason for higher bathochromic shift of C.NO_2 and C.NH_2.

2.3.7 POLARISATION AND HYPERPOLARIZATION

Polarisability and hyperpolarisability values obtained from Gaussian 09W calculation is converted to its electrostatic unit (esu) α: 1 a.u. = 0.148×10^{-24} esu; β: 1 a.u. = 8.639×10^{-33} esu). The calculated dipole moment and polarization values for C and DC are 4.67 Debye, 15.461×10^{-24} esu and 3.88 Debye, 14.798×10^{-24} esu, respectively. The hyperpolarization values for these molecules are 3.871×10^{-31} and 2.877×10^{-31} esu, respectively are summarized in Table 2.11.

Higher dipole moment of C indicates its relatively higher polar nature and thereby shows good solubility in water than DC. Dipole moment is considered as an important descriptor used for the analysis of biological properties. The values of polarization and hyperpolarization are showing extent of the ability of these referred molecules to polarize other molecules. It has been observed that the substitution of electron withdrawing groups at C3 position of both C and DC increase the values of these three

TABLE 2.11 Polarizability, Hyperpolarizability and Dipole Moment Calculated for C, DC and Their Substituted Molecules

Molecules	Polarisability × 10^{-24} (esu)	Hyperpolarisability × 10^{-31} (esu)	Dipole moment (Debye)
C	15.461	3.871	4.67
$C.NO_2$	18.402	6.709	7.45
$C.CF_3$	17.486	1.291	6.21
$C.NH_2$	18.037	3.472	3.61
DC	14.798	2.877	3.88
$DC.NO_2$	16.909	4.927	6.16
$DC.CF_3$	16.4904	3.427	4.82
$DC.NH_2$	16.044	2.321	3.62

descriptors, while the electron donating substituent decreases the values of these descriptors.

Urea is considered as one of the standard compounds used as a threshold for comparing NLO properties of molecules. Dipole moment, polarization and hyper polarization values for urea are 1.5256 Debye, 5.048 × 10^{-24} esu and 7.803 × 10^{-31} esu, respectively [40]. While comparing with these threshold values of urea, the dipole moment and polarization values are higher for all considering molecules but hyper polarization values are lower in nature.

2.3.8 CHEMICAL REACTIVITY OF HYDROXY COUMARINS

Lin et al. [27] has investigated the effect of suppression of reactive oxygen species (ROS) of eight selected Coumarin derivatives (Coumarin, dihydrocoumarin, 7-hydroxy coumarin, Esculetin, Scopoletin, 7-hydroxy-4-methyl coumarin, 4-methyl esculetin and 4-hydroxy coumarin) under oxidative condition which results the highest activity is obtained for Esculetin than other seven. Here, these Coumarins (see Figure 2.3) are subjected to computational calculations for obtaining global reactivity descriptors, which are summarized in Table 2.12, and as well as druggability parameters in order to ensure the importance of computational tools in

TABLE 2.12 Global Reactivity Descriptors (eV) Calculated for considering Hydroxy Coumarins Through Orbital Vertical Method Calculated by B3LYP/6–311G (2d, 2p)

Molecules	IP	EA	E_g	μ	η	ω	χ	S
4hC	6.608	1.795	4.813	4.201	2.406	3.667	–4.201	0.208
7hC	6.352	1.874	4.478	4.113	2.239	3.778	–4.113	0.223
7h4mC	6.291	1.757	4.534	4.024	2.267	3.571	–4.024	0.221
E	6.025	1.865	4.159	3.945	2.080	3.741	–3.945	0.240
4mE	5.954	1.739	4.215	3.846	2.107	3.510	–3.846	0.237

predicting the reactivity of chemical systems and also for recognizing the idea about the importance of Coumarin scaffold.

Among these considering hydroxy coumarins E shows high reactivity with HOMO-LUMO energy gap of 4.159 eV and hardness of 2.080 eV, lower compared to other considered molecules. The ionization potential energy of E is found to be as 6.025 eV, describes the donor properties of a molecule and the electron affinity is 1.865 eV. The global softness is also higher for E with the value of 0.24 eV. The nucleophilicity is lower for 4 mE with value of 3.51 eV, which indicate that 4 mE is more reactive nucleophile among these hydroxy coumarins.

2.3.9 *IN SILICO* DRUGGABILITY TEST

The in silico druggability test is carried out with the objective of screening out the above mentioned hydroxy coumarins based on the bioactivity and druggability with the help of Molinspiration property calculator, an on-line service for the calculation of important molecular properties (logP, polar surface area, number of hydrogen bond donors and acceptors and others), as well as prediction of bioactivity score for the most important drug targets (GPCR ligands, kinase inhibitors, ion channel modulators and nuclear receptors).

In the context of pharmacokinetics, the partition coefficient has a strong influence on ADME properties (absorption, distribution, metabolism and excretion). The hydrophobicity of a compound (as measured by its partition coefficient) is considered as a major determinant of how drug-like it is. More specifically, in order to absorb a drug orally, it has to pass through the lipid bilayers present in intestinal epithelium (a process known as transcellular transport) at first. For the efficient transport, the drug must be hydrophobic enough to partition into the lipid bilayer, but not so hydrophobic. When it once crosses the lipid bilayers it will not partition out again. Hydrophobicity plays a major role in determining where drugs are distributed with in the body after absorption and as a consequence how rapidly they are metabolized and excreted [30]. Typically, a low solubility is a reason for bad absorption, and therefore, the general aim is to avoid poorly soluble compounds.

On analyzing each criteria for satisfying Lipinski rule of five, molecular weight of all studied molecules is less than 500. The number of H-bond donors and acceptors never cross the limit. The detailed result of analysis of Lipinski's rule of five for all molecules under study could be seen from Table 2.13.

Since logP represents the octanol/water ratio, the low hydrophilicity leads to high logP value may cause poor absorption or permeation. It is essential that the logP value must not be greater than five for molecule to have a reasonable probability of being well absorbed. Based on this, all the molecules in the present work are in acceptable limit and they show the values in between 1.00–2.01 range. The higher value of logP, 2.01, is shown by C and other hydroxy coumarins are with values lower than 2.00 but greater than 1.00. The lowest logP value (1.02) is shown by the molecule E, a kind of dihydroxy coumarin, due to the presence of two

TABLE 2.13 Compounds Following Parameters of Lipinski's Rule for Drug Likeness

Molecule	Lipinski's Parameters					Extensions	
	logP	MW	HBA	HBD	Violations	TPSA	nrot
C	2	146.15	2	0	0	30.21	0
DC	1.79	148.16	2	0	0	26.30	0
4hC	1.72	162.14	3	1	0	50.44	0
7hC	1.51	162.14	3	1	0	50.44	0
4m7hC	1.89	176.17	3	1	0	50.44	0
E	1.02	178.14	4	2	0	70.67	0
4 mE	1.40	192.17	4	2	0	70.67	0

Note: MW: Molecular weight, logP: Octanol-water partition coefficient, HBA: Hydrogen bond acceptors, HBD: Hydrogen bond donors, TPSA: Topological polar surface area.

hydroxyl groups, which make a commendable variation in biological properties among E and C. As hydroxyl group is polar in nature, its water solubility is generally become higher, but the conjugated hydrocarbon ring present in the compound brings some sort of lipophilic nature also. This is the reason for the logP value of E falls in between 1 to 2, otherwise strong hydrophilic nature would be observed even negative for logP values. It is also clearly evident from the comparison of logP values of molecule E and 4 mE (which contain additional methyl group at C4 position of C compared to E) are 1.02 and 1.40, respectively, this difference is only due to the presence of an additional hydrophobic methyl group on parent Coumarin ring.

The bioavailability of a molecule can be assessed through its TPSA analysis. This descriptor has shown a correlation with passive molecular transport through membranes and therefore allows a prediction of transport properties of drugs and has been linked to drug bioavailability. Generally, it has been recognized that passively absorbed molecules with a PSA > 140 Å^2 are showing low oral bioavailability [52]. TPSA for popular drugs diclofenac is 49.33 Å^2 and for ibuprofen is 37.33 Å^2 indicate that even the limit is up to 140 Å^2, the value below 100 Å^2 is more favorable for good transportation through membrane [31]. It has been observed that all the molecules have PSA value within the prescribed limit and more favorably within 100 Å^2. All considering molecules are without rotatable bond hence so concluded that they should not be interested in conformational flexibility.

2.4 CONCLUSION

A computational study has been performed in order to investigate the influence of α,β unsaturation in chemical reactivity of Coumarin and some hydroxy coumarins. The gas phase structure of molecules in the ground state are optimized by performing calculations at B3LYP/6–311G (2d, 2p) level of theory with the help of Gaussian 09W simulation package. The NBO analysis shows the aromatic nature of Coumarin (C) and its metabolite 3,4 dihydro coumarin (DC) are originated from their benzene moiety; the latter shows higher aromaticity than former. The presence of extension of conjugation from benzene to α-pyrone moiety in C is found to be the reason for the reduction of aromatic nature (thermodynamic stability) and thereby increases the chemical reactivity. The observed correlation of thermodynamic stability with HOMO-LUMO energy gap extends a general conclusion that lower aromatic nature is reflected by the lowering of gap between HOMO-LUMO energy levels. The HOMO-LUMO gap for C and DC are 4.625 and 5.984 eV, respectively. The influence of electronic effects in both C and DC are also studied by inserting substitutions like $–NO_2$, $–NH_2$ and $–CF_3$ on third position of molecular structure which comes in α-pyrone ring, the results emphasis that the electron withdrawing groups stabilize the molecular orbitals while electron donating groups destabilize them. This variation is also reflected in HOMO-LUMO gap and thereby in chemical reactivity also. The calculation of dipole moment, polarization and hyper polarization suggest that the considering systems are polar in nature even variations are seen on substitution, which extends the application of these compounds in the field of NLO organic materials. TD-DFT method is used for the evaluation of absorption (UV-Vis) spectra of molecule by including one set of diffuse function in basis set. The substitutions, whatever be the electron donating or withdrawing group, is giving a bathochromic shift of varying extent as per the nature of substituent and the shift follows the order of $–NO_2 > –NH_2 > –CF_3$. Fukui functions are showing the site-specific electrophilic and nucleophilic nature of the compounds. The site for nucleophilic attack in C and its substituted structures is at C4 position in α-pyrone moiety while for DC and its substituted structures are showing the same in benzene moiety. Global reactivity descriptors like electronegativity, electrophilicity, hardness, softness, etc., are calculated using orbital vertical method and the variations are seen in the values of these descriptors on substitution. All considering hydroxy coumarins satisfy Lipinski rule of five, were done using Molinspiration property explorer to predict the bioactivity.

KEYWORDS

- **absorption spectra**
- **Coumarins**
- **density functional theory**

- **electronegativity**
- **Fukui function**
- **natural bond orbital**

REFERENCES

1. Baiz, C. R., Dunietz, B. D. (2007). *J. Phys. Chem. A. 111*, 10139–10143.
2. Becke, A. D. (1993). *J. Chem. Phys. 98*, 5648–5652.
3. Budzelaar, P. H. M., Cremer, D., Wallasch, M., Wurthwein, E. U., & Schleier, P. V. R. (1987). *J. Am. Chem. Soc. 21,* 6290–6299.
4. Burke, K., & Werschnik, J., Gross, E. K. U. (2005). J. *Chem. Phys. 123*, 062206–9.
5. Chen, S. G., & Chen, D. Z. J. (2004). *Photochem. Photobiol.* A: *Chem. 162*, 407–414.
6. Chohan, Z. H., Shaikh, A. U., Rauf, A., & Supuran, C. T. (2006). *J. Enz. Inhib. Med. Chem. 21*, 741–748.
7. Clayden, Greeves, & Warren, Wothers. Organic Chemistry; Oxford University Press: Oxford, 2001.
8. Cohen, Y., Klein, J., & Rabinovitz, M. (1986). *J. Chem. Soc. Chem. Commun.* 1071–1073.
9. Cooke, D., Fitzpatrick, B., O'Kennedy, R., McCormack, T., & Egan, D. Coumarin Biochemical Profile and Recent Developments. John Wiley & Sons, 1997.
10. Cooke, D. Recent Advances on Coumarin and Its Synthetic Derivatives; Dublin City University: Dublin, 1999.
11. Deepha, V., Praveena, R., & Sadasivam, K. (2015). *J. Mol. Struct. 1082*, 131–142.
12. Dewar, M. J. S. The Molecular Orbital Theory of Organic Chemistry. McGraw-Hill: New York, 1969.
13. Domingo, L. R., Aurell, M., Contreras, M., & Perez, P. (2002). *J. Phys. Chem. A. 106*, 6871–6875.
14. Ertl, P., Rohde, B., & Selzer, P. (2000). *J. Med. Chem. 43*, 3714–3717.
15. Foresman, J. B., & Frisch, A. E. Exploring Chemistry with Electronic Structure Methods: 2nd ed. Gaussian, Pittsburgh, 1996.
16. Foster, J. P., & Weinhold, F. (1980). *J. Am. Chem. Soc. 102*, 7211–7218.
17. Frisch, M. J., Trucks, G. W., Schlegel, H. B., Scuseria, G. E., Robb, M. A., Cheeseman, J. R. et al., (2009). Gaussian 09, Revision D.01, Gaussian, Inc., Wallingford CT.
18. Gilman, J. (1997). *J. Mater. Res. Innov. 1*, 71–76.
19. Guilet, D., Seraphin, D., Rondeau, D., Richomme, P., Bruneton, J. (2001). *Phytochemistry J. 58*, 559–571.
20. Günter, P. Nonlinear Optical Effects and Materials, Springer Series in Optical Sciences: Springer Verlag, 2000.
21. Jacquemin, D., Perpète, E. A., Assfeld, X., Scalmani, G., Frish, M. J., & Adamo, C. (2007). *Chem. Phys. Lett. 438*, 208–212
22. Jain, P. K., & Joshi, H. (2012). *J. Appl. Pharm. Sci. 6,* 236–240

23. Jensen, F. Introduction to Computational Chemistry: John Wiley & Sons: Chichester, 2007.
24. Kleinman, D. A., (1962). *Phys. Rev. 126,* 1977–1979.
25. Lacy, A., & O'Kennedy, R. (2004). *Curr. Pharm. Design, 10,* 3797–3811.
26. Lee, C., Yang, W., & Parr, R. G. (1988). *Phys. Rev. B. 37*, 785–789.
27. Lin, H. C., Tsai, S. H., Chen, C. S., Chang, Y. C., Lee, C. M., Lai, Z. Y., & Lin, Y. C. (2008). *Biochemical Pharmacology. 75*, 1416–1425.
28. Lipinski, C. A., Lombardo, F., Dominy, B., & Feeney, P. (2001). *Adv. Drug. Deliv. Rev. 46*, 3–26.
29. Manolov, I., Maichle-Moessmer, C., & Danchev, N. (2006). *Eur. J. Med. Chem. 41,* 882–890.
30. Maurya, A., Khan, F., Bawankule, D. U., Yadav, D. K., & Srivastava, S. K. Eur. (2012). *J. Pharm. Sci. 47*, 152–161.
31. Meena, A., Yadav, D. K., Srivastava, A., Khan, F., Chanda, D., & Chattopadhyay, S. K. (2012). *Chem. Biol. Drug. Des. 78*, 567–579.
32. Morais, V. M. F. Computational thermochemistry: accurate estimation and prediction of molecular thermochemical parameters. Coimbra University Press, 2011.
33. O'Boyle, N. M., Banck, M., James, C. A., Morley, C., Vandermeersch, T., & Hutchi-Son, G. R. (2011). *J. Chem. Inform. 33*, 1–14.
34. Parr, R. G., Donnelly, R. A., Levy, M., & Palke, W. E. (1978). *J. Chem. Phys. 68*, 3801–3807.
35. Parr, R. G., Sventpaly, L., & Liu, S., (1999). *J. Am. Chem. Soc. 121*, 1922–1924.
36. Parr, R. G., & Yang, W. Density Functional Theory of Atoms and Molecules; Oxford University Press: New York, 1989.
37. Parr, R. G., & Yang, W. (1984). *J. Am. Chem. Soc. 106*, 4049–4050.
38. Petit, L., Adamo, C., & Russo, N. (2005). *J. Phys. Chem. B. 109*, 12214–21.
39. Preat, J., Jacquemin, D., & Perpète, E. A. (2005). *Chem. Phys. Lett. 415*, 20–24.
40. Ramalingam, S., Karabacak, M., Periandy, S., Puviarasan, N., & Tanuja, D. (2012). *Spectrochim. Acta. A. 96*, 207–220.
41. Rauk, A. Orbital Interaction Theory of Organic Chemistry; Second Edition: John Wiley & Sons: New York, 2001.
42. Reed, A. E., & Weinhold, F. (1985). *J. Chem. Phys. 83*, 1736–1740.
43. Reed, A. E., Weinhold, R. B., & Weinhold, F. (1985). *J. Chem. Phys. 83*, 735–746.
44. Roy, R. K., Choho, K., De, P. F., & Geerlings, P. (1999). *J. Phys. Org. Chem. 12*, 503–509.
45. Roy, R. K., Proft, F. D., Geerlings, P. (1998). *J. Phys. Chem. A. 102*, 7035–7040.
46. Sadasivam, K., & Kumaresan, R. (2011). *Spectrochim. Acta A. 79*, 282–293.
47. Sajan, D., Erdogdu, Y., Reshmy, R., Dereli, Ö., Thomas, K. K., & Joe, I. H. (2011). *Spectrochim. Acta. A. 82*, 118–125.
48. Sauer, M., Hofkens, J., & Enderlein, J. Handbook of Fluorescence Spectroscopy and Imaging; Verlag GmbH & Co. KGaA, Weinheim, 2011.
49. Soon, P. S., & Gordon, W. H. (1970). *J. Phy. Chem. 74*, 4234–4240.
50. Sykes, P. A. Guidebook to Mechanism in Organic Chemistry; Sixth Edition: Longman Group Ltd: New Delhi, 1986.
51. Varkey, E. C., & Sreekumar, K. (2011). *Bull. Mater. Sci. 34*, 893–897.

52. Veber, D. F., Johnson, S. R., Cheng, H. Y., Smith, B. R., Ward, K. W., & Kopple, K. D. (2002). *J. Med. Chem. 45,* 2615–2623.
53. Vektariene, A., Vektaris, G., & Svoboda, J. (2009). *Arkivoc. 7*, 311–329.
54. Weinhold, F., & Landis, C. R. (2001). *Chem. Educ. Res. Pract. Eur. 2*, 91–104.
55. Weininger, D. J. (1988). *Chem. Inf. Comput. Sci. 28*, 31–36.
56. Yang, W., & Parr, R. G. (1985). *Proc. Natl. Acad. Sci. 82*, 6723–6726.
57. Young, D. Computational Chemistry: A Practical Guide for Applying Techniques to Real-World Problems. John Wiley & Sons Ltd: New York, 2001.
58. Zeynep, D., Kaştaş, A., & Orhan, C. B. (2015). *Spectrochim. Acta A. 139*, 539–548.
59. Zhang, R., Du, B., Sun, G., & Sun, Y. (2010). *Spectrochim. Acta A.* 75, 1115–1124.

ADDITIONAL READING

1. Cramer, C. J. Essentials of Computational Chemistry – Theories and Models; John Wiley & Sons: Chichester, 2002.
2. Guillemoles, J. F., Barone, V., Jouber, T. L., & Adamo, C. J. (2002). *Phy. Chem. A. 106*, 11354–11360.
3. http://www.molinspiration.com/cgi-bin/properties.
4. Lewars, E. Computational Chemistry – Introduction to the Theory and Applications of Molecular and Quantum Mechanics; Kluwer, Academic Publishers: Norwell, 2003.
5. Roy, R. K., Krishnamurti, S., Geerlings, P., & Pal, S. (1998). *J. Phys. Chem. A. 102*, 3746–3755.

CHAPTER 3

MOLECULAR DETERMINANTS OF TRPC6 CHANNEL RECOGNITION BY FKBP12

PENG TAO,[1] JOHN C. HACKETT,[2] JU YOUNG KIM,[3] DAVID SAFFEN,[4] CARRIGAN J. HAYES,[5] and CHRISTOPHER M. HADAD[6]

[1]*Department of Chemistry, Southern Methodist University, 3215 Daniel Avenue, Dallas, Texas 75275–0314, USA, E-mail: ptao@smu.edu*

[2]*Institute for Structural Biology and Drug Discovery, Virginia Commonwealth University, 800 East Leigh Street, Richmond, Virginia 23219, USA*

[3]*Institute for Cell Engineering, Johns Hopkins University School of Medicine, 733 N Broadway, Baltimore, Maryland 21205, USA*

[4]*Department of Cellular and Genetic Medicine, School of Basic Medical Sciences, Fudan University, 130 Dongan Rd, Shanghai 200032, P.R. China*

[5]*Department of Chemistry, Otterbein University, 1 South Grove Street, Westerville, Ohio 43081, USA*

[6]*Department of Chemistry, The Ohio State University, 100 W. 18th Ave., Columbus, Ohio 43210, USA*

CONTENTS

ABSTRACT

Transient receptor potential-canonical 6 (TRPC6) calcium channels are currently the subject of intense investigation due to their roles in modulating smooth muscle tone in blood vessels and lung airways. TRPC6 channels are also proposed to mediate physiological processes in the kidney, immune system and central nervous system. We previously reported that binding of the immunophilin FKBP12 (FK506 binding protein–12 kDa) to a TRPC6 intracellular domain is a prerequisite for the formation of a multi-protein complex involved in channel regulation. This study also demonstrated that binding of FKBP12 to TRPC6 requires prior phosphorylation of Ser768 in the putative TRPC6 binding domain. To study the elements of molecular recognition in FKBP12 for the TRPC6 intracellular domain, we performed molecular dynamics simulations in explicit solvent on model complexes containing FKBP12 and the following: (i) the unphosphorylated wild-type TRPC6 intracellular binding domain, (ii) the wild-type TRPC6 binding domain containing a phosphorylated Ser768 residue, and (iii) TPRC6 peptides in which Ser768 was replaced with Asp or Glu. Simulations using the Generalized Born/Surface Area model (MM-GB/SA) predicted favorable binding and small conformational fluctuations for the FKBP12/phosphorylation Ser768 TRPC6 peptide complex, due to the strong interactions between the phosphate group and Lys44, and Lys47 residues in the FKBP12 binding site. Decomposition of the binding free energies into each amino acid residue identified additional important structural elements necessary for this protein-protein interaction.

3.1 INTRODUCTION

Transient receptor potential-canonical (TRPC) channels are members of the mammalian TRP channel superfamily of cation channels [1, 2]. The seven known subtypes of TRPC channels (TRPC1–7) are widely expressed in cells and tissues, where they mediate the influx of extracellular Ca^{2+} and/or Na^{+} in response to the activation of cell surface receptors. These influxes regulate key cellular functions, including contraction of smooth muscle, activation of immune cells, mobility of neuronal growth cones, and cell proliferation and migration. Because many of these functions are relevant to human disease, there is currently considerable interest in developing agents that activate or inhibit TRPC channels for use as therapeutic drugs [3–6].

Native TRPC channels comprise four protein subunits, which are symmetrically organized around a central pore. Each subunit contains six transmembrane (TM) domains and a single membrane-loop domain (located between TM5 and TM6) that contributes to the channel pore. The amino- and carboxyl-termini of each subunit are located on the intracellular side of the membrane. TRPC3, TRPC6 and TRPC7 channels are structurally and functionally related and constitute a subfamily of TRPC channels [7, 8]. Each of these subtypes can form homotetrameric channels or combine with other subfamily members to form heterotetrameric channels.

Understanding the molecular mechanisms involved in the regulation of TRPC channels will be important for the identification of novel drug targets for TRPC channel-regulated processes. Moreover, these have motivated the search for post-translational modifications that alter the function of TRP channels [9] and proteins that interact with the channels [10]. As mentioned above, TRPC3/6/7 channels are regulated by PKC, which phosphorylates the channels on a conserved serine residue in the carboxyl-terminal region (Ser712 in TRPC3 [11] and Ser714/Ser768 in TRPC6A/B [12]). By contrast, phosphorylation of TRPC3 by Src [13] and TRPC6 src-family tyrosine kinases [14] is required for maximal channel activation. TRPC3/6/7 channels have also been shown to directly bind several proteins including the calcium binding protein calmodulin [15, 16], the IP3 receptor of the endoplasmic reticulum [17] and the adapter protein Homer [18, 19]. Studies by Schiling et al. [20] have shown that TRPC3/6/7 channels also contain a binding site for the immunophilin FKBP12 (FK506 binding protein–12 kDa) within the carboxyl-terminal cytoplasmic domain. Site-specific mutagenesis studies demonstrated that FKBP12 binds to the consensus sequence LPXPFYLVPSPK (X = P, V or S; Y = S or N). The serine residue within this segment is the target for PKC phosphorylation: Ser768 in the TRPC6A splice variant and Ser714 in the TRPC6B splice variant.

We previously showed that FKBP12 is a component of a TRPC6-centered protein complex that rapidly forms following activation of endogenous M_1 mAChR [12]. Data from that study suggest that the following events take place following activation of M_1 mAChR with carbachol. First, a protein complex containing M_1 mAChRs, TRPC6 channels and PKC rapidly assembles within the cell membrane. Second, PKC phosphorylates the TRPC6 channels on Ser768/Ser714. Third, phosphorylation of Ser768/Ser714 creates a binding site for FKBP12. Fourth, binding of the FKBP12 to TRPC6 results in the recruitment of the calcineurin/calmodulin to the complex. Finally, the channels are dephosphorylated by the calcineurin, releasing M_1 mAChR from the complex.

A novel aspect of the above sequence of events is the observation that TRPC6 channel phosphorylation by PKC is required for the binding of FKBP12. Evidence for this includes the observation that coimmunoprecipitation of the channels and FKBP12 is blocked when channel phosphorylation is attenuated by PKC inhibition or by substitution of Ser768/Ser714 with alanine or glycine [12]. Taken together, these studies show that phosphorylation of TRPC6 channels by PKC and the subsequent binding of FKBP12 play a central role in the regulation of TRPC6 channel trafficking and, thus, indirectly regulate TRPC6 channel activity. As described below, these studies implicate specific amino acid residues within each protein and predict that binding requires phosphorylation of Ser768/Ser714.

Molecular dynamics (MD) simulations and binding free energy calculations using implicit solvent models are powerful tools to study the interactions between biomacromolecules. It has been shown in numerous studies that simulation of protein-ligand complexes can provide detailed insight into ligand binding modes. Furthermore, their binding free energies may be accurately using a combination of molecular mechanics internal energies, solvation free energies, and vibrational entropies [21–29]. Recently, the Generalized born (GB) method was improved to produce comparable results with Poisson–Boltzmann (PB) method with much reduced computational cost [30, 31].

All of these computational advantages make calculation of binding free energy of protein complexes based on MD trajectory feasible. In particular, there is significant precedent for application of these computational methods for calculation of ligand binding free energies to FKBP12 [32–36]. In this chapter, these techniques are successfully applied to expand our understanding of the determinants of the FKBP12-TRPC6 protein-protein interaction.

3.2 COMPUTATIONAL METHODS

3.2.1 *TRPC6 PEPTIDE DOCKING*

A prerequisite for using MD to study protein-peptide interactions is a template that provides information about the location and nature of the peptide binding site on the receptor protein. To date, co-crystal structures of FKBP12 have been determined with fragments of the TGF-β receptor Type I (TGFβTRI; PDB entry: 1B6C) [37], bone morphogenetic protein receptor type–1B (PDB entry: 3MDY) [38], and the kinase domain of the type I activin receptor (PDB entry: 3H9R) [39]. In each structure, FKBP12 predominantly interacts with a leucyl-prolyl-initiated α-helix on the C-terminal side of the binding partner GS domain. The structure of FKBP12, partner α-helix, and binding mode are essentially identical in the three protein complexes. Furthermore, the amino acid sequences of the leucyl-prolyl-initiated peptides constituting these α-helices are also strongly conserved. The strong sequence and structural conservation in these protein-protein interactions are illustrated in the Figure 1. Sinkins and co-workers demonstrated that the analogous leucyl-prolyl-initiated peptide ^{759}LPVPFNLVPSP^{769} of TRPC6 mediates its interaction with FKBP12 [20]. Since this sequence has been demonstrated experimentally to mediate the FKPB12-TRPC6 interaction and the structure of FKBP12 domains of similar sequences are strongly conserved, the structure of the TRPC6 peptide was initially modeled on the α-helical structure of the TGFβTRI peptide.

To generate the initial geometries of the TRPC6 peptide, the ^{193}LPLLVQRTIAR^{203} helix was excised from the crystal structure of the FKBP12-TGF-β receptor Type I fragment complex, and the amino acids corresponding to those found in TRPC6 were introduced. In addition to the unphosphorylated and phosphorylated wild-type peptides, Ser768Asp and Ser768Glu mutants were also modeled. Each peptide was capped with methyl and acetyl groups at the N- and C-terminal ends, respectively. These peptides were then fully optimized using the AMBER ff94 force field [40]. Possible modes of peptide binding were explored using the DOCK 5.2.0 suite of programs [41], to generate initial structures for MD simulations. In doing so, the FKBP12 receptor was extracted from PDB structure 1B6C; protons were added in a manner consistent with physiological pH; and charges from the ff94 force field were applied. In docking calculations, the helical peptides were oriented into the FKBP12 binding site as a rigid body considering a maximum of 2×10^6 orientations. Torsional angles in the peptides for each binding mode were minimized to optimize the total energy score using the simplex minimizer in the DOCK suite of programs [42, 43]. Additional details of the docking methodology and energy scores for the peptides are listed in Table 3.1.

3.2.2 MD SIMULATION OF FKBP12-TRPC6 PEPTIDE COMPLEXES AND ISOLATED BINDING PARTNERS

In addition to the four FKBP12-TRPC6 peptide complexes, MD simulations of the isolated species (FKBP12 and the various TRPC6 peptides)

TABLE 3.1 DOCK Energy Scores (kcal/mol) for the Preferred Modes of TGF-β Receptor Type I and TRPC6 Peptides Binding to the FKBP12 Receptor[a]

Peptide	Van der Waals	Electrostatic	Total Energy Score
LPLLVQRTIAR[b]	–36.2	–1.4	–37.6
LPVPFNLVPSP	–36.3	–3.5	–39.8
LPVPFNLVPpSP	–28.4	–12.3	–40.7
LPVPFNLVPDP	–33.6	–5.0	–38.5
LPVPFNLVPEP	–31.1	–12.5	–43.6

[a]A Connolly solvent-accessible surface of FKBP12 was generated with a probe radius of 1.4 Å for input to the SPHGEN program, from which a set of 57 overlapping spheres defining the FKBP12 binding pocket was created. DOCK scoring grids with dimensions of 42×34×24 Å were created with the GRID program, using electrostatic potential charges from ff94 and van der Waals parameters from the ff99 force field.

[b]TGF-β receptor Type I peptide.

were also conducted. For FKBP12 alone, the initial structure was taken from the crystal structure (PDB ID: 1B6C), and the initial structures of the four peptides were prepared as noted above. All of the MD simulations were conducted with the AMBER 8 suite of programs [44]. The all-atom force field ff03 of Duan et al. [45] was used, and the simulations were conducted with explicit water solvent, represented by the TIP3P model [46]. Proteins and protein-peptide complexes were immersed in a box of water, with a minimum distance of 10 Å between the protein complex and the box surface, and included approximately 20,000 atoms. Periodic boundary conditions were applied, using the particle mesh Ewald [47–51] method for the long-range electrostatic treatment. The SHAKE bond-length constraint method [52] was applied to constrain the length of covalent bonds containing hydrogen during the simulations. A non-bonded interaction cutoff value of 8 Å was used. After initial optimization, all of the systems were equilibrated in 4000 steps, and heated from 0 to 300 K in the NVT ensemble. Then, 20 ns production runs were conducted under isothermal-isobaric ensemble (NPT) conditions, at 300 K and 1 bar. A time constant of 1.2 ps was used for heat bath coupling, and 2.0 ps was used as the relaxation time for pressure regulation [53]. The time step was 1 fs for all of the 20 ns MD simulations, and in each case, coordinates were saved every 100 steps. Details of each simulation system are listed in Table 3.2.

TABLE 3.2 MD Simulation Details of FKBP12, Probe Peptides and Their Complexes

Structure	Water Molecule Number	Box Dimension (Å)	Length of the simulation (ns)	Equilibrium Time (ps)
FKBP12	5603	58×72×58	20	4
WT	3502	55×51×50	20	4
pWT	3526	54×55×49	20	4
Ser768Asp	3423	50×57×50	20	4
Ser768Glu	3442	50×56×50	20	4
FKBP12-WT	5525	58×72×58	20	4
FKBP12-pWT	5681	58×72×60	20	4
FKBP12-Ser768Asp	5614	58×72×59	20	4
FKBP12-Ser768Glu	5643	58×72×59	20	4

3.2.3 MM/GB-SA FREE ENERGY OF BINDING CALCULATIONS

For each peptide, the free energy of binding to FKBP12 was computed using the MM-GB/SA method [54], available in the AMBER program suite. This method uses a thermodynamic cycle to calculate the free energy of binding for each ligand, in this case the TRPC6-derived peptides, to the FKBP12 receptor [55, 56]. The free energies of binding are computed using the equation:

$$\Delta G^{sol}_{binding} = \Delta G^{sol}_{complex} - \Delta G^{sol}_{receptor} - \Delta G^{sol}_{ligand} \quad (1)$$

where $\Delta G^{sol}_{binding}$ is the total free energy of binding in solution, and $\Delta G^{sol}_{complex}$, $\Delta G^{sol}_{receptor}$ and ΔG^{sol}_{ligand} are free energies in solution of the complex, receptor and ligand, respectively. The free energy in solution of each entity (ΔG^{sol}) is calculated by the following equations:

$$\Delta G^{sol} = \Delta G^{gas} + \Delta G_{solvation} \quad (2)$$

$$\Delta G^{gas} = E_{internal} + E_{vdw} + E_{electrostatic} - T\Delta S \quad (3)$$

$$\Delta G_{solvation} = \Delta G_{GB} + \Delta G_{nonpolar} \quad (4)$$

where ΔG^{gas} is the free energy in gas phase, and $\Delta G_{solvation}$ is the solvation energy. ΔG^{gas} is the sum of the internal energy ($E_{internal}$), van der Waals (E_{vdw}) and Coulombic ($E_{electrostatic}$) interaction, as grave well as entropic contributions (ΔS). The internal energy includes bond stretching, bond angle, and torsional contributions to the total molecular mechanics (MM) energies. The solvation energy $\Delta G_{solvation}$ includes polar (ΔG_{GB}) and non-polar contributions ($\Delta G_{nonpolar}$). The thermodynamic cycle for binding free energy calculation is illustrated in Figure 3.1.

For a given FKBP12 TRPC6 peptide complex, the MM-GB/SA method requires snapshots from the MD trajectories for that complex, as well as from those of FKBP12 and the peptide alone. The first 2 ns of the MD simulation were considered as an equilibration period and were discarded for the free energy of binding calculations. For each complex, 1,000 snap-shots were evenly extracted from the remaining 18 ns of MD trajectories for the free energy calculations. Water molecules were stripped from these snapshots for binding energy calculations. The contributions to the total free energy of binding include Coulombic interactions ($E_{electrostatic}$),

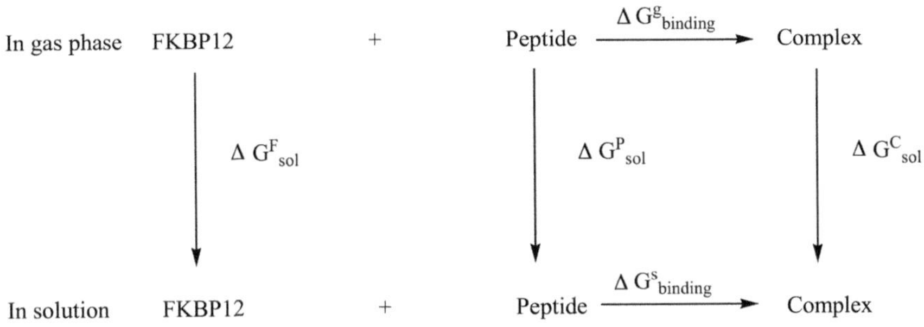

FIGURE 3.1 Thermodynamic cycle for binding free energy calculation of FKBP12 and probe peptides.

van der Waals interactions (E_{vdw}), internal energies including bond stretching, angle bending, and torsional energies ($E_{internal}$), hydrophobic effects ($\Delta G_{nonpolar}$), solvation effects (ΔG_{GB}) and entropic effects ($T\Delta S_{total}$). The entropies used in computing the composite binding free energies for the free energy calculations were calculated by the normal mode (NMODE) module available in AMBER package [57, 58]. Each snapshot was optimized in the gas-phase using conjugate gradient method with atomic pair distance-dependent dielectric model. After geometry optimization, frequencies of the vibrational modes were computed to obtain the harmonic approximation of entropy at 300K.

The contribution of each individual residue to the binding free energy was also analyzed by means of component analysis [59]. The free energy contribution of each residue G(i, j), where *i* and *j* are indices of snapshots and residues, were estimated using Eq. (5):

$$G(i, j) = E_{gas}(i, j) + G_{solvation}(i, j) - TS(i, j) \qquad (5),$$

where $G(i, j)$ is the total free energy, E_{gas} includes $E_{electrostatic}$, E_{vdw}, and $E_{internal}$, $G_{solvation}$ includes ΔG_{GB} and $\Delta G_{nonpolar}$, *i* and *j* are indices of snapshots and residues, respectively. Internal energies (bond, angle, and dihedral angle) were weighted based on the number of atoms that belong to each of the residues. Van der Waals contributions to the energy arising from atoms in a pair of residues were evenly distributed between those residues. The solvent-accessible surface area of each atom was estimated using the interaction geometry model described by Rarey et al. [60]. The electrostatic energy was decomposed based on charge distribution within the GB model [59]. The decomposition of desolvation free energies, ΔG_{GB} and $\Delta G_{nonpolar}$, was applied based on linear combination of pairwise overlaps (LCPO) method. When decomposing the binding free energy contributions into amino acid residues, the method of Fisher et al. [62] was used to calculate the translational, rotational, and vibrational entropies.

3.3 RESULTS AND DISCUSSION

3.3.1 TRPC6 PEPTIDE DOCKING

An initial study was performed to validate the docking protocol, as well as the computational definition of the FKBP12 receptor and the peptides as ligands. Docking of the ^{193}LPLLVQRTIAR^{203} peptide from TGFβTRI was performed with the computational model of the FKBP12 receptor in order to validate the docking procedure. This computational method did indeed reproduce the experimentally-derived binding mode of the TGFβTRI peptide with FKBP12, with a root-mean square deviation (RMSD) of 1.7 Å, as displayed in Figure 3.2.

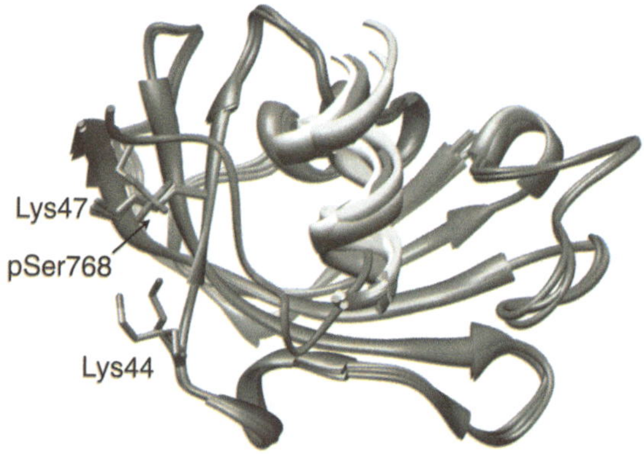

FIGURE 3.2 Superimposed crystal structures of FKBP12 with TFGβ receptor peptide (PDB code: 1BC6, LPLLVQRTIAR), bone morphogenetic protein receptor type–1B peptide (PDB code: 3MDY, LPLLVQRTIAK) and kinase domain of the type I activin receptor peptide (PDB code: 3H9R, LPFLVQRTVAR). RMSD value between docked and crystal TFGβ receptor peptide is 1.72 Å.

Given this increased confidence in our computational procedure and the experimental precedent for FKBP12 binding to α-helical domains (PDB code: 1B6C, 3H9R and 3MDY), docking of four TRPC6 peptides to FKBP12 was subsequently performed. The unphosphorylated (WT) and phosphoSer768 wild-type (pWT) TRPC6 peptides, as well as Ser768Asp and Ser768Glu mutants of TRPC6 were oriented into the FKBP12 binding pocket. From this point forward, the unphosphorylated and phosphorylated wild-type peptide will be referred as WT and pWT, respectively. The two mutants will be referred to as Ser768Asp and Ser768Glu, respectively. The top-scoring binding modes for the TGFβTRI and pWT peptides are displayed in Figure 3.2. On the basis of the DOCK energy score, the most energetically-favorable binding modes of the TRPC6 peptides reveal a different orientation relative to the experimental binding mode of the TGFβTRI peptide. The energy scores for peptides containing a negatively-charged amino acid at position 768 are dominated by a significant electrostatic contribution (Table 3.1), resulting from the binding of the anionic side chain between two surface lysine residues of FKPB12 (Lys44 and Lys47). These lysine residues do not establish interactions with the α-helical binding domain of TGFβTRI, bone morphogenetic protein receptor type–1B, or the kinase domain of the type I activin receptor (Figure 3.2). Furthermore, crystal structures indicate these residues do not contact FK506 [63].

Direct structural characterization of the delicate interactions constituting the phosphorylated TRPC6-FKBP12 binding interface pose difficulties for experiment. Thus, we employed MD simulations in explicit solvent to characterize the features of FKBP12 important for recognition of this phosphoprotein. These MD simulations were used to evaluate the stability of the protein-peptide complexes and to highlight the important residues involved in mediating these interactions. In combination with equivalent simulations of the isolated species, these simulations allowed the computation of the binding free energies of the various peptides to FKBP12.

3.3.2 RMSD OF FKBP12-PEPTIDE COMPLEXES

We computed the RMSD deviation for the FKBP12-peptide complexes over the course of each 20 ns MD simulation relative to the initial coordinates

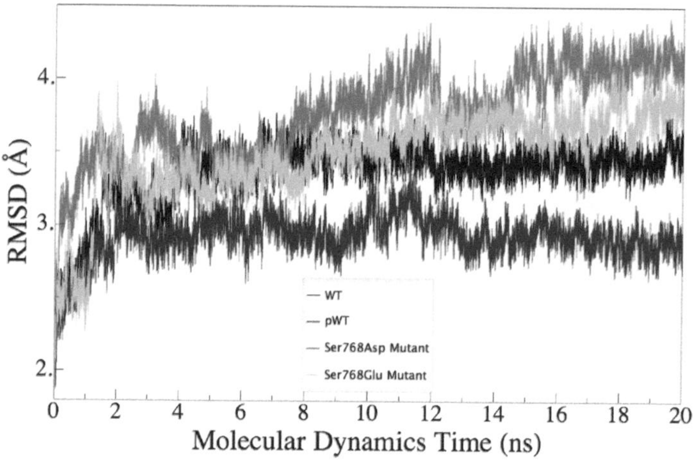

FIGURE 3.3 RMSD of the four complexes during 20 ns MD simulations.

of the production MD simulation (Figure 3.3). All of the protein atoms were included in the RMSD calculations. During the first 2 ns, the RMSD values increased to ~3 Å. Beyond this time, each protein-peptide complex remained stable for the course of the simulation, although there were notable differences in their relative flexibility as determined by decomposition of the RMSD values into each residue (*vide infra*). The average RMSD of the FKBP12-WT complex is 3.4 ± 0.1 Å. The complex containing the pWT peptide was more stable, with an average RMSD of 3.0 ± 0.1 Å from 3 to 20 ns. Apparently, phosphorylation of Ser768 restricts the conformational freedom of the protein-peptide complex, which could facilitate the formation of the multi-protein complex observed experimentally [12]. Due to the apparent importance of a negatively charged residue at position 768 of TRPC6, mutations were introduced in this position to test the hypothesis that that anionic amino acids would behave similarly to the phosphorylated peptide. The Ser768Asp and Ser768Glu peptides displayed greater RMSD (3.8 ± 0.3 Å and 3.6 ± 0.2 Å between 3 and 20 ns, respectively) in their respective MD trajectories than the WT peptides, despite their electrostatic similarity to the phosphorylated peptide. These simulations indicated that intrinsic properties of the phosphate functionality (or interactions other than electrostatic contributions of the negatively charged side chain) may contribute to TRPC6 binding.

To quantify changes in individual FKBP12 residues as a result of peptide binding, the RMSD of each residue relative to those of an isolated FKBP12 trajectory were calculated (Figure 3.3). WT- and pWT-bound FKBP12 residues demonstrate comparable values to unbound FKBP12. In contrast, regions of the Ser768Asp- and Ser768Glu-FKBP12 complexes displayed larger RMSD values than unbound FKBP12, especially in the ^{9}PGDGRTFPKRG^{19} (referred as 10 loop), ^{31}EDGKKF^{36} (30 loop) and ^{84}ATGHPGIIPPH^{94} (80–90) regions (Figure 3.4).

Gohlke and Case [64] proposed two approaches for computing binding free energies using the MM-GB/SA method. The single-trajectory approach relies upon the MD trajectories of the protein-protein complex alone, hence all of the necessary trajectory frames of each binding partner are extracted from the complex trajectories. The alternative separate-trajectory approach requires independent MD simulations of the protein-protein complex and the isolated binding partners. The computational economy of the single-trajectory approach is obvious. However, a limitation of the single-trajectory approach is that its accuracy depends on whether the binding partners undergo significant conformational changes during the binding event. When different results arise from these two approaches, the separate-trajectory approach is considered to be more reliable, since each entity is independently simulated to model its actual state before and after binding. Although both approaches were applied in the present study, the remainder

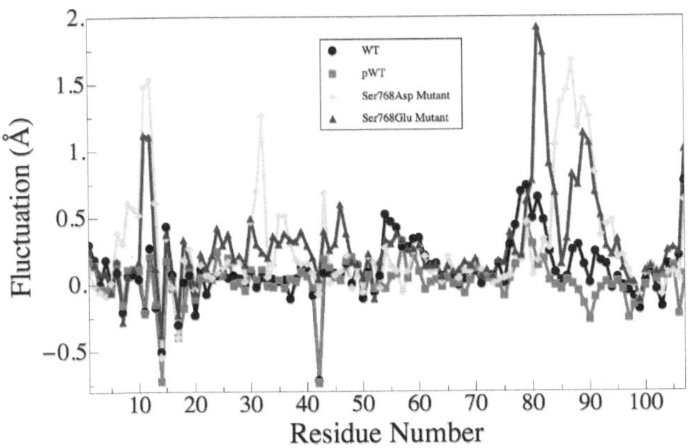

FIGURE 3.4 Average residue fluctuation of FKBP12 in MD simulations relative to isolated FKBP12.

of the discussion is primarily devoted to the results obtained considering separate trajectories of the binding partners. Unless otherwise noted, single-trajectory results are reserved for the Supporting Information.

3.3.3 FREE ENERGY OF BINDING CALCULATIONS

Calculated binding free energies of the four complexes using their respective MD trajectories are listed in Table 3.3. All of the energy terms and their corresponding standard errors calculated by the MM-GB/SA method are provided in the Supporting Information (Table 3.4). Using the separate-trajectory approach, the free energy of the pWT-FKBP12 complex was calculated to be 4.8 kcal/mol, only 0.9 kcal/mol different from result based on single-trajectory method (Table 3.3). This result indicates that the FKPB12 and pWT binding event does not involve significant overall conformational changes.

In contrast, the difference in the free energies of binding calculated by the two methods for the FKBP12-WT complex is much larger (~10 kcal/mol). The result based on separate-trajectory method is consistent with the experimental observation that FKBP12 does not bind unphosphorylated TRPC6 *in vitro* [12]. Accurate calculation of binding free energies using the single-trajectory approach requires that the isolated binding partners maintain their unbound conformations in the complex; hence it possible that phosphorylation conformationally restrains the peptide for binding to FKBP12. The free energy required to induce this conformational shift of the unphosphorylated peptide in the single-trajectory approach is absent, resulting in an erroneous prediction of the binding energy.

TABLE 3.3 Binding Free Energy (kcal/mol) of FKBP12 and Peptide Complexes[a]

TRPC6 peptide	Single-trajectory	Separate-trajectory
WT	–5.4	4.5
pWT	–5.7	–4.8
Ser768Asp	1.3	28.0
Ser768Glu	0.0	14.1

[a]Entropic contribution to the binding free energy was calculated using normal mode analysis (NMODE in AMBER).

TABLE 3.4 Binding Free Energy Components of FKBP12 and Probe Peptides Complexes

FKBP12 and Wild type TRPC6 Peptide Complex											
Contributions (kcal/mol)[a]	**FKBP12- Wild-type Complex**		**FKBP12 (bound)**		**Wild-type (bound)**		**Delta Average[c]**	**FKBP12 (unbound)**		**Wild-type (unbound)**	
	Average	**σ[b]**	**Average**	**σ[b]**	**Average**	**σ[b]**		**Average**	**σ[b]**	**Average**	**σ[b]**
$E_{electrostatic}$	–2651.3	59.6	–2502.3	60.7	–107.1	6.8	–42.0	–2521.7	63.6	–105.4	8.8
E_{vdw}	–474.5	18.4	–409.0	18.0	–8.9	4.7	–56.6	–418.1	15.5	–8.5	6.1
$E_{internal}$	2515.4	29.7	2242.1	27.9	273.3	10.0	0.0	2244.1	27.8	267.7	9.7
E_{gas}	–610.5	65.6	–669.2	66.1	157.3	11.3	–98.6	–695.7	66.7	153.8	13.5
$G_{nonpolar}$	48.5	1.0	46.9	1.0	10.1	0.2	–8.5	46.9	0.8	10.3	0.5
G_{GB}	–1328.4	56.6	–1321.8	57.2	–76.1	5.0	69.5	–1302.3	57.1	–75.7	7.3
$G_{solvation}$	–1279.9	56.0	–1274.9	56.7	–66.0	5.0	61.1	–1255.5	56.7	–65.4	7.0
$E_{gas} + G_{solvation}$	–1890.3	29.7	–1944.1	28.5	91.3	10.0	–37.5	–1951.2	28.3	88.3	10.1
TS_{total}	1348.9	10.6	1230.1	10.1	150.9	3.1	–32.1	1228.8	10.2	152.0	3.7
$\Delta G_{binding}$							**–5.4**				

Delta Average[d] (FKBP12 and Wild type TRPC6 Peptide Complex): $E_{electrostatic}$ –24.2; E_{vdw} –47.9; $E_{internal}$ 3.6; E_{gas} –68.5; $G_{nonpolar}$ –8.7; G_{GB} 49.7; $G_{solvation}$ 41.1; $E_{gas} + G_{solvation}$ –27.5; TS_{total} –31.9; $\Delta G_{binding}$ **4.5**

FKBP12 and Phosphorylated Wild type TRPC6 Peptide Complex												
Contributions (kcal/mol)[a]	**FKBP12- Phosphorylated Wild-type**		**FKBP12 (bound)**		**Phosphorylated Wild-type (bound)**		**Delta Average[c]**	**FKBP12 (unbound)**		**Phosphorylated Wild-type (unbound)**		**Delta Average[d]**
	Average	**σ[b]**	**Average**	**σ[b]**	**Average**	**σ[b]**		**Average**	**σ[b]**	**Average**	**σ[b]**	
$E_{electrostatic}$	–2901.6	73.3	–2508.1	61.1	–160.9	38.0	–232.6	–2521.7	63.6	–184.8	22.1	–195.1
E_{vdw}	–466.4	17.2	–410.3	16.1	–6.6	4.8	–49.4	–418.1	15.5	–6.1	5.3	–42.2
$E_{internal}$	2528.4	29.9	2242.2	28.2	286.2	10.6	0.0	2244.1	27.8	282.7	9.7	1.7

TABLE 3.4 (Continued)

E_{gas}	–839.5	73.5	–676.2	64.0	118.7	34.7	–282.0	–695.7	66.7	91.8	22.2	–235.6
G_{GB}	–1375.8	65.0	–1321.7	55.3	–301.3	36.0	247.2	–1302.3	57.1	–277.1	20.4	203.7
$G_{solvation}$	–1326.6	64.7	–1275.0	55.0	–290.6	35.6	239.0	–1255.5	56.7	–266.3	20.3	195.2
$E_{gas} + G_{solvation}$	–2166.2	29.7	–1951.2	28.1	–171.9	10.1	–43.1	–1951.2	28.3	–174.5	9.5	–40.4
TS	1350.7	9.9	1229.7	10.0	158.5	4.7	–37.4	1228.8	10.2	157.8	4.0	–35.6
$\Delta G_{binding}$							**–5.7**					**–4.8**

FKBP12 and TRPC6 Peptide with Mutation Ser768Asp Complex

Contributions (kcal/mol)[a]	FKBP12-Mutant (Ser768Asp)		FKBP12 (bound)		Mutant (Ser768Asp) (bound)		Delta Average[c]	FKBP12 (unbound)		Mutant (Ser768Asp) (unbound)		Delta Average[d]
	Average	σ[b]	Average	σ[b]	Average	σ[b]		Average	σ[b]	Average	σ[b]	
$E_{electrostatic}$	–2681.0	82.0	–2461.0	91.2	–81.1	6.7	–138.9	–2521.7	63.6	–79.9	8.6	–79.4
E_{vdw}	–461.5	19.8	–406.2	17.4	–10.7	4.7	–44.5	–418.1	15.5	–8.5	6.1	–35.0
$E_{internal}$	2520.3	29.2	2244.7	27.9	275.6	9.0	–0.0	2244.1	27.8	266.6	8.9	9.7
E_{gas}	–622.1	84.9	–622.5	91.9	183.8	10.8	–183.4	–695.7	66.7	178.3	13.6	–104.7
$G_{nonpolar}$	49.9	1.1	47.7	1.0	9.8	0.2	–7.6	46.9	0.8	10.8	0.7	–7.8
G_{GB}	–1349.1	70.5	–1354.2	79.2	–153.4	6.1	158.5	–1302.3	57.1	–150.5	7.3	103.8
$G_{solvation}$	–1299.1	70.2	–1306.4	79.0	–143.6	6.0	150.9	–1255.5	56.7	–139.7	7.0	96.1
$E_{gas} + G_{solvation}$	–1921.3	31.9	–1928.9	29.8	40.1	9.0	–32.5	–1951.2	28.3	38.6	10.1	–8.6
TS	1349.1	10.8	1231.4	9.8	151.6	2.6	–33.9	1228.8	10.2	156.9	4.1	–36.7
$\Delta G_{binding}$							**1.3**					**28.0**

TABLE 3.4 (Continued)

	FKBP12 and TRPC6 Peptide with Mutation Ser768Glu Complex											
Contributions (kcal/mol) [a]	**FKBP12-Mutant (Ser768Glu)**		**FKBP12 (bound)**		**Mutant (Ser768Glu) (bound)**		**Delta Average**[c]	**FKBP12 (unbound)**		**Mutant (Ser768Glu) (unbound)**		**Delta Average**[d]
	Average	**σ**[b]	**Average**	**σ**[b]	**Average**	**σ**[b]		**Average**	**σ**[b]	**Average**	**σ**[b]	
$E_{electrostatic}$	–2661.6	70.5	–2474.5	67.8	–105.0	8.7	–82.1	–2521.7	63.6	–94.8	9.7	–45.1
E_{vdw}	–459.1	16.7	–395.1	15.5	–11.9	4.5	–52.2	–418.1	15.5	–8.0	5.4	–33.0
$E_{internal}$	2515.0	29.3	2239.8	27.6	275.2	9.5	–0.0	2244.1	27.8	271.1	9.5	–0.1
E_{gas}	–605.7	76.6	–629.7	73.8	158.3	12.8	–134.3	–695.7	66.7	168.3	12.7	–78.3
$G_{nonpolar}$	50.2	1.0	48.8	0.9	2.6	9.4	–8.9	46.9	0.8	10.8	0.5	–7.5
G_{GB}	–1406.2	66.6	–1350.0	63.9	10.3	0.2	105.4	–1302.3	57.1	–168.2	8.3	64.4
$G_{solvation}$	–1356.0	66.0	–1301.2	63.3	–161.6	8.3	96.5	–1255.5	56.7	–157.4	8.2	56.9
$E_{gas} + G_{solvation}$	–1961.7	30.7	–1931.0	29.4	7.1	9.4	–37.8	–1951.2	28.3	10.9	10.1	–21.3
TS	1352.0	10.7	1234.4	10.6	155.3	2.9	–37.7	1228.8	10.2	158.7	4.0	–35.5
$\Delta G_{binding}$							**–0.0**					**14.1**

[a] $E_{electrostatic}$: Coulombic energy; E_{vdw}: van der Waals energy; $E_{internal}$: internal energy; $E_{gas} = E_{electrostatic} + E_{vdw} + E_{internal}$; $G_{nonpolar}$: nonpolar solvation free energy; G_{GB}: polar solvation free energy; $G_{solvation} = G_{nonpolar} + G_{GB}$; TS_{total}: total entropy contribution by normal mode analysis; $\Delta G_{binding} = E_{gas} + G_{solvation} - TS_{total}$.

[b] Standard error of average values.

[c] Calculation based on trajectory of complex only.

[d] Calculation based on separated trajectories of complex, FKBP12 and TRPC6 peptides.

Surprisingly, the mutants Ser768Asp and Ser768Glu are not predicted to have thermodynamically-favorable binding energies with FKBP12. This observation suggests these mutants cannot constitutively mimic the phosphoserine necessary for FKBP12 binding.

3.3.4 SEPARATE-TRAJECTORY FREE ENERGY OF BINDING DECOMPOSITION ANALYSIS

The free energies of binding based on the separate-trajectory approach were decomposed into each residue with the entropic contribution computed with the method of Fisher et al. (Figures 3.5–3.7) In the pWT complex, several peptide residues (Leu759, Pro760, Val761, and Asn764) contribute significantly to the binding free energies. Notably, the N-terminal LP residues are conserved in of the FKBP12 recognition sequences. pSer768 also contributes a positive contribution to the binding free energy, although it is approximately one-third (+2.1 kcal/mol) of that

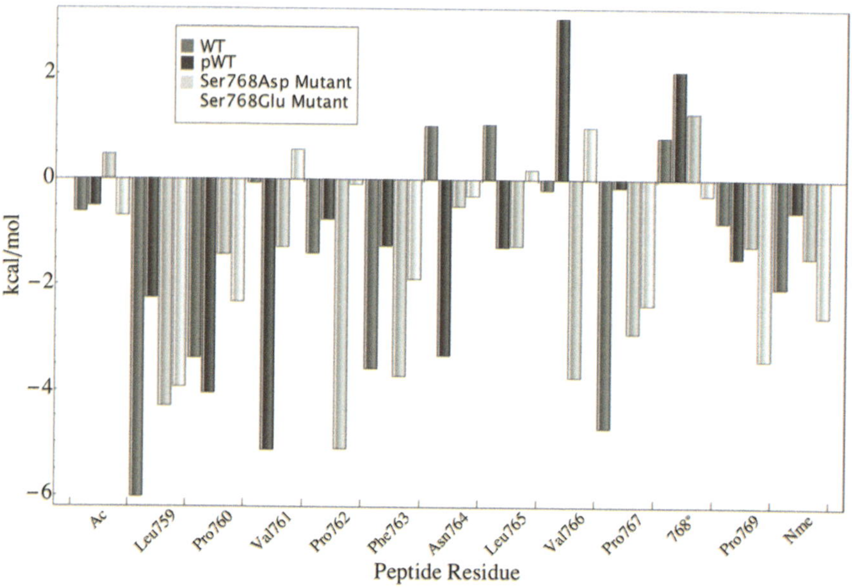

FIGURE 3.5 Decomposition of binding free energies into single residues of peptide based on separated MD trajectories.

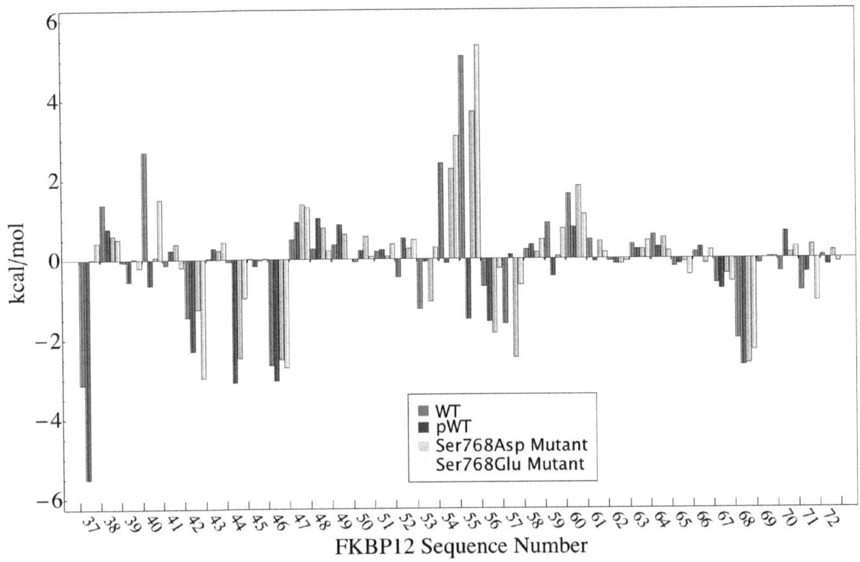

FIGURE 3.6 Decomposition of binding free energies into single residues of FKBP12 (residues 37 to 72) based on separated MD trajectories.

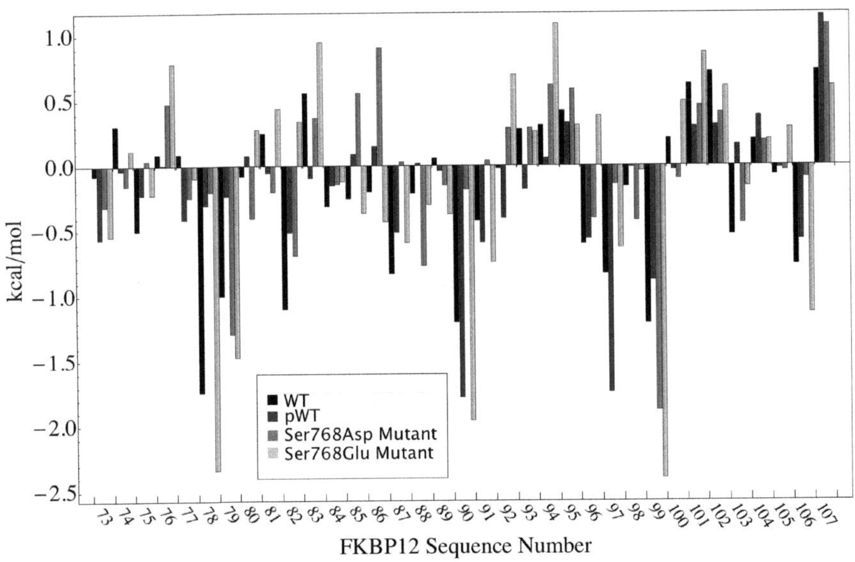

FIGURE 3.7 Decomposition of the binding free energies (from separate trajectory approach) into FKBP12 residues 73–107.

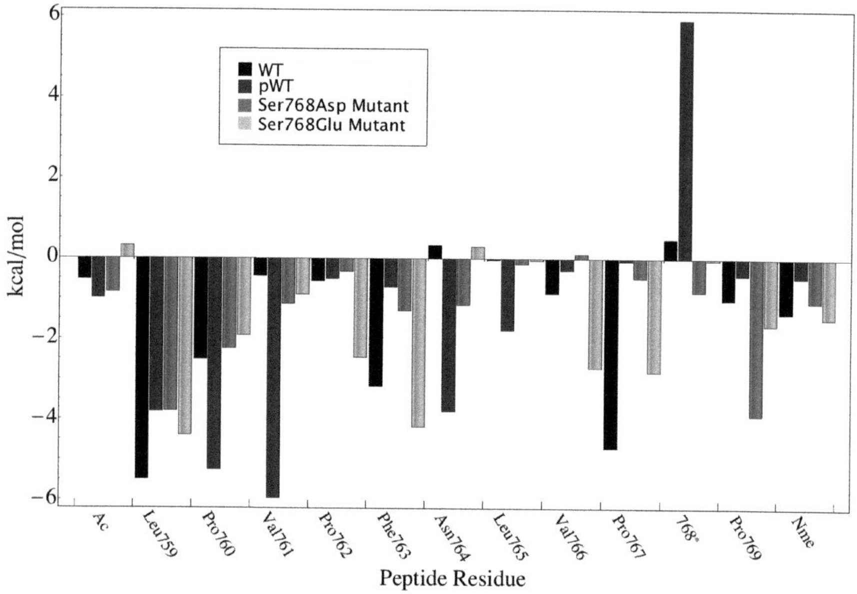

FIGURE 3.8 Decomposition of the binding free energies into single residues in the TRPC6 peptides. Data are based on single MD trajectory.

calculated in the single-trajectory approach (Figure 3.8). The unfavorable contribution (+0.9 kcal/mol) of FKBP12 Lys47 (Figure 3.6) was surprising, and may be due to the limited frames used for the binding free energy calculation (*vide infra*). Nevertheless, the negative contribution of FKBP12 Lys44 (−3.1 kcal/mol) offsets the small positive contributions of peptide pSer768 and FKBP12 Lys47. The results of the energetic decomposition analysis are rather different for the other three complexes. It is interesting to note that the decomposed contributions of some residues from WT and Ser768Asp peptides to binding free energies are somewhat more favorable in separate-trajectory results than in single-trajectory results (Figures 3.5 and 3.8). However, contributions from FKBP12 to binding free energies in these two complexes are more unfavorable in separate-trajectory results than in single-trajectory results.

In the pWT complex, both hydrophobic and hydrophilic residues from FKBP12 make significant contributions to binding affinity. Hydrophobic residues are evenly arranged at the bottom of the FKBP12 binding site (Figure 3.9), forming a hydrophobic pocket to host the peptide ligand.

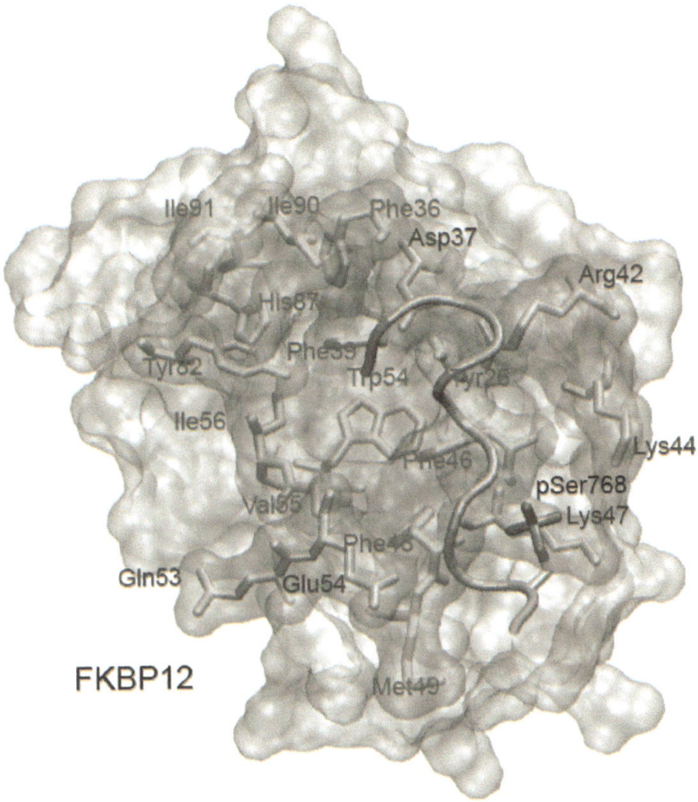

FIGURE 3.9 Residues with significant contribution to thetotal binding free energy at the binding site of FKBP12. Hydrophobic residues: Tyr26, Phe36, Phe39, Phe46, Phe48, Met49, Trp54, Val55, Ile56, Tyr82, His87, Ile90, and Ile91. Hydrophilic residues: Asp37, Arg42, Lys44, Lys47, Glu54, and Gln53.

Phe36, Val55, Ile56, Trp59, and Phe99 are located at the bottom of the binding pocket. His87, Ile90, and Ile91 (from the 80–90 loop) form a hydrophobic wall on one side, while Phe46, Phe48 and Met49 form another hydrophobic wall on the opposite side. Overall, these residues form a large, "U-shaped" hydrophobic pocket to host the N-terminal end of the peptide. A number of hydrophilic residues with hydrogen-bonding and ion pair interaction capabilities are arranged above the rim of this hydrophobic cavity (Figure 3.9). Asp37, Arg42, Lys44 and Lys47 are located on one side, while Gln53 and Glu54 are on the other side of the hydrophobic cavity.

3.3.5 INTERACTIONS AMONG FKBP12 LYS44, LYS47 AND PEPTIDE RESIDUE 768

Despite the apparently unfavorable energetic contribution of Lys47 predicted by decomposition of the separate-trajectory data, single-trajectory data and pair-wise distance analyzes support both Lys residues being important for interaction with pSer768 (Figure 3.10). In the FKBP12-WT complex (Figure 3.10A), Ser768 does not have a stable interaction either Lys residues. By contrast, the pSer768 maintains stable interactions with the Lys residues throughout the MD simulation (Figure 3.10B). In the early stages of the MD simulation, the peptide phosphoserine residue is strongly coupled to FKBP12 Lys47. The pSer768-Oγ to Lys47-Nε distance remains ~ 5 Å, while the pSer768-Oγ to Lys44-Nε and Lys44-Nε to Lys47-Nε distances are much longer (>10 Å). The two FKBP12 lysines residues approach one another and at 7.1 ns achieve a minimum distance of 3.8 Å. The negative charge of the phosphate group apparently screens

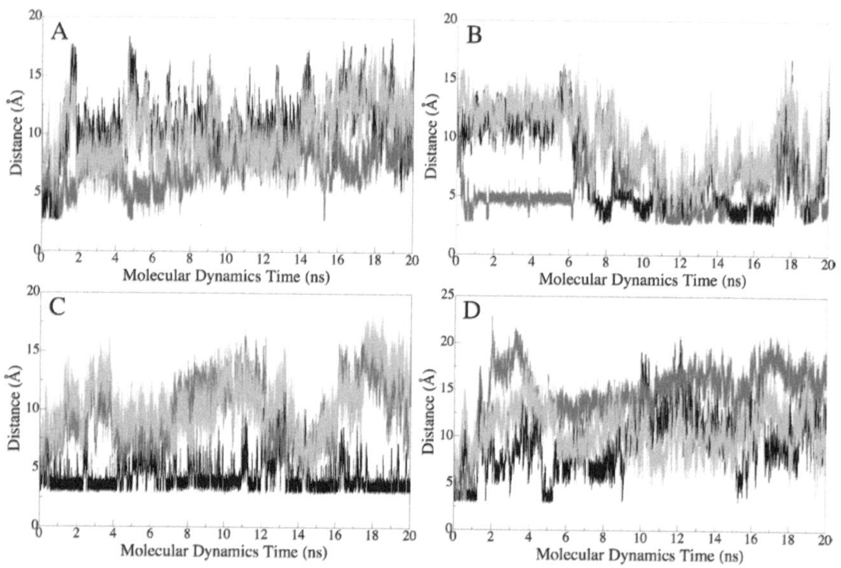

FIGURE 3.10 Atomic pair distance analysis for Lys44, Lys47 and Residue 768 from probe peptides. A. WT complex (Lys44-Nε:Ser768-Oγ, black; Lys47-Nε:Ser768-Oγ, dark gray; Lys44-Nε:Lys47-Nε, light gray); B. pWT complex (Lys44-Nε:pSer768-Oγ, black; Lys47-Nε:pSer768-Oγ, dark gray; Lys44-Nε:Lys47-Nε, light gray); C. Ser768Asp complex, (Lys44-Nε:Asp768-Cγ, black; Lys47-Nε: Asp768-Cγ, dark gray; Lys44-Nε:Lys47-Nε, light gray); D. Ser768Glu complex, (Lys44-Nε:Glu768-Cδ, black; Lys47-Nε:Glu768-Cδ, dark gray; Lys44-Nε:Lys47-Nε, light gray).

the repulsion between the lysines, allowing them to approach each other and transfer the salt bridge. The close approach of the two lysine amino groups allow transfer of the ion pair interaction with the peptide phosphoserine to FKBP12 Lys44. This switching occurs four times throughout the 20 ns MD simulation. This process is illustrated in Figure 3.11. After each switch, either FKBP12 Lys44 or Lys47 remains in close contact with the peptide phosphoserine residue, indicating that the two lysine residues are equally important when interacting with the phosphate group.

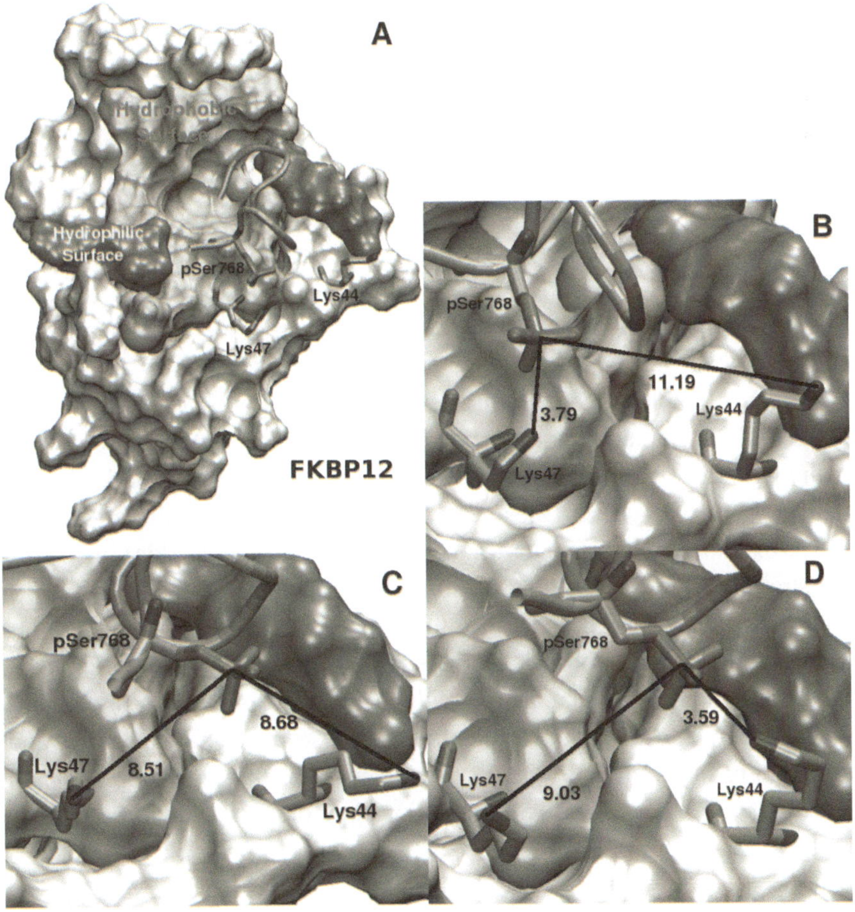

FIGURE 3.11 Transient ionic interaction between the pWT phosphoSer768 and FKBP12 Lysines 44 and 47: (a) surface of FKBP12, (b) phosphate group forming salt bridge with Lys47, (c) transition state during switch, (d) phosphate group forming salt bridge with Lys44. Distances are in Å.

In the Ser768Asp mutant complex, the peptide Asp768 residue has a close contact with the Lys44 of FKBP12. This observation supports our hypothesis that the carboxylate side chain is capable of mimicking the anionic character of the phosphoserine by forming an apparent ion pair. However, the carboxylate side chain cannot recruit both lysines for interaction when bound to FKBP12. (Figure 3.10C) After 5 ns of the MD simulation involving the Ser768Asp peptide, the FKBP12 Lys47 amino group approaches γ-carbon of peptide Asp768 (Asp768-Cγ), with a concomitant increase in the Lys44-Nε-Asp768-Cγ distance. Between 5 to 7 ns, these side chains remain within 5 to 8 Å. Despite the approach of FKBP12 Lys47 several times during the simulation, the apparent salt bridge involving the carboxylate unit is not transferred to Lys44. Beyond 7 ns, FKBP12 Lys47 returns to the previous configuration. An analogous trajectory consistent with another 'switch' of the cationic partner occurs between 14 and 15 ns, but is also unsuccessful. Indeed, the anionic side chain in the Ser768Asp peptide forms an apparent salt bridge with one of the Lys residues, however; the carboxylate unit is not sufficient to recruit both FKBP12 Lys44 and 47 as cationic partners, as is observed in the pWT peptide. The dual recruitment of these lysine residues appears to be a prerequisite for favorable binding affinity.

The Ser768Glu mutant (Figure 3.10D) behaves quite differently from the Ser768Asp mutant. First, peptide Glu768 cannot maintain close contact with FKBP12 Lys44. Although the Lys44-Nε and δ carbon of peptide Glu768 (Glu768Cδ) can approach one another within 3 Å to form salt bridge briefly at 5 ns of simulation, the dynamics of FKBP12 Lys44 and peptide Glu768 are not strongly coupled through the course of our simulation. Instead, this distance fluctuates between 5 to 15 Å afterward. The distance from Glu768-Cδ to Lys47-Nε is much longer than to Lys44-Nε. Peptide Glu768 appears to be ineffective in involving interaction with both lysine residues. From comparing the behavior of the two Ser768 mutants, we see that while glutamate and aspartate residues differ by only one side-chain methylene unit, this small structural difference causes large deviations in protein-protein interactions.

3.4 CONCLUSIONS

In this study, we have applied MD simulations to study interactions between FKBP12 and TRPC6 channel peptides. Experiments suggest phosphorylation of Ser768 in the TRPC6 intracellular domain is a prerequisite for

FKBP12 binding and subsequent formation of a multiprotein complex. The FKBP12-pWT complex is the most stable and the least flexible among the studied peptide complexes. Both Ser768Asp and Ser768Glu complexes have larger RMSD than those in the pWT complex. The calculated free energies of binding showed that the pWT peptide but not WT peptide has a thermodynamically favorable binding affinity with FKBP12, consistent with experimental data. Neither the Ser768Asp nor Ser768Glu mutants demonstrate a thermodynamically favorable binding affinity with FKBP12. These calculations indicate these mutants cannot constitutively mimic the phosphoserine residue, which is necessary for FKBP12 binding.

The decomposition of binding free energies into individual residues of the FKBP12 and TRPC6 peptide revealed a specific binding pocket composed of both hydrophobic and hydrophilic residues for the recognition and interaction of FKBP12 with the TRPC6 intracellular domain. About twenty residues, including both hydrophobic and hydrophilic residues contributed significantly to the FKBP12-TRPC6 binding energy. Hydrophobic residues formed a "U" shaped binding pocket to recognize hydrophobic residues in the TRPC6 binding domain. Lys44 and Lys47, surrounding the rim of the hydrophobic cavity, apparently contribute the key elements for recognition of the phosphopeptide.

ACKNOWLEDGMENTS

The authors are indebted to the Ohio Supercomputer Center for computational resources.

KEYWORDS

- **FKBP12**
- **free energy**
- **MM-GB/SA**
- **molecular dynamics**
- **molecular recognition**
- **TRPC6**

REFERENCES

1. Ramsey, I. S., Delling, M; & Clapham, D. E. (2006). *Annu. Rev. Physiol. 68*, 619.
2. Wu, L.J; Sweet T. B., & Clapham, D. E. (2010). *Pharmacol. Rev. 62*, 381,
3. Li, S., Westwick, J., & Poll, C. T. (2002). *Trends Pharmacol. Sci. 23*, 63.
4. Ong, H. L., & Barritt, G. J. (2004). *Respirology 9*, 448.
5. Harteneck C., & Gollasch, M. (2011). *Curr. Pharm. Biotechnol. 12*, 35
6. Morelli M. B., Amantini, C., Liberati, S; Santoni, M., & Nabissi, M. (2013). *CNS Neurol. Disord. Drug Targets 12*, 274
7. Trebak, M., Vazquez, G., Bird, G. S., & Putney, J. W. Jr. (2003). *Cell Calcium 33*, 451.
8. Dietrich, A., Kalwa, H., Rost, B. R., & Gudermann, T. (2005). *Pflugers Arch. 451*, 72.
9. Yao, X., Kwan, H. Y., & Huang, Y. (2005). *Neurosignals, 14*, 273.
10. Kiselyov, K., Kim, J. Y., Zeng, W., & Muallem, S. (2005). *Pflugers Arch. 451*, 116.
11. Trebak, M., Hempel, N., Wedel, B. J., Smyth, J. T., Bird, G. S., & Putney, J. W. (2005). *Mol. Pharmacol. 67*, 558.
12. Kim, J. Y., & Saffen, D. (2005). *J. Biol. Chem. 280*, 32035.
13. Vazquez, G., Wedel, B. J., Kawasaki, B. T., Bird, G. S., & Putney, J. W. Jr., (2004). *J. Biol. Chem. 279*, 40521.
14. Hisatsune, C., Kuroda, Y., Nakamura, K., Inoue, T., Nakamura, T., Michikawa, T., Mizutani, A., & Mikoshiba, K. (2004). *J. Biol. Chem. 279*, 18887.
15. Zhang, Z., Tang, J., Tikunova, S., Johnson, J. D., Chen, Z., Qin, N., Dietrich, A., Stefani, E., Birnbaumer, L., & Zhu, M. X. (2001). *Proc. Natl. Acad. Sci. USA, 98*, 3168.
16. Tang, J., Lin, Y., Zhang, Z., Tikunova, S., Birnbaumer, L., & Zhu, M. X. (2001). *J. Biol. Chem. 276*, 21303.
17. Kiselyov, K., Xu, X., Mozhayeva, G., Kuo, T., Pessah, I., Mignery, G., Zhu, X., Birnbaumer, L., & Muallem, S. (1998). *Nature 396*, 478.
18. Yuan, J. P., Kiselyov, K., Shin, D. M., Chen, J., Shcheynikov, N., Kang, S. H., Dehoff, M. H., Schwarz, M. K., Seeburg, P. H., Muallem, S., & Worley, P. F. (2003). *Cell 114*, 777.
19. Kim, J. Y., Zeng, W., Kiselyov, K., Yuan, J. P., Dehoff, M. H., Mikoshiba, K., Worely, P. F., & Muallem, S. (2006). *J. Biol. Chem. 281*, 32540.
20. Sinkins, W. G., Goel, M., Estacion, M., & Schiling, W. P. (2004). *J. Biol. Chem. 279*, 34521.
21. Wang, W., Lim, W. A., Jakalian, A., Wang, J., Luo, R., Bayly, C. I., & Kollman, P. A., (2001). *J. Am. Chem. Soc. 123*, 3986.
22. Wang, J., Morin, P., Wang, W., & Kollman, P. A. (2001). *J. Am. Chem. Soc. 123*, 5221.
23. Swanson, J. M., Henchman, R. H., & McCammon, J. A. (2004). *Biophys. J. 86*, 67.
24. Srinivasan, J., Miller, J., Kollman, P. A., & Case D. A. *J. Biomol. Struct. Dyn.* (1998). *16*, 671.
25. Reyes, C. M; & Kollman, P. A., *J. Mol. Biol.* (2000). *297*, 1145.
26. Luo, C., Xu, L., Zheng, S., Luo, X., Shen, J., Jiang, H., Liu, X., & Zhou, M. (2005). *Proteins 59*, 742.
27. Huo, S., Massova, I., & Kollman, P. A. (2002). *J. Comput. Chem. 23*, 15.
28. Gouda, H., Kuntz, I. D., Case, D. A., & Kollman, P. A. (2003). *Biopolymers 68*, 16.
29. Fogolari, F., Brigo, A., & Molinari, H. (2003). *Biophys. J. 85*, 159.
30. Kollman, P. A., Massova, I., Reyes, C., Kuhn, B., Huo, S., Chong, L., Lee, M., Lee, T., Duan, Y., Wang, W., Donini, O., Cieplak, P., Srinivasan, J., Case, D. A., & Cheatham, T. E. III (2000). *Acc. Chem. Res. 33*, 889.

31. Onufriev, A., Bashford, D., & Case, D. A. (2004). *Proteins 55*, 383.
32. Fujitani, H; Tanida, Y; Ito, M; Jayachandran, G; Snow, C. D., Shirts, M. R., Sorin, E. J., & Pande, V. S. (2005). *J. Chem. Phys. 123*, 084108/1–084108/5.
33. Wang, J; Deng, Y; & Roux, B. (2006). *Biophys. J. 91*, 2798.
34. Deng, Y; & Roux, B. (2009). *J. Phys. Chem. B 113*, 2234.
35. Xu, Y., & Wang, R. (2006). *Proteins*, *64*, 1058.
36. Lamb, M. L., Tirado-Rives, J., & Jorgensen, W. L. (1999). *Bioorg. Med. Chem. 7*, 851.
37. Huse, M., Chen, Y. G., Massague, J., & Kuriyan, J. (1999). *Cell 96*, 425.
38. Crystal structure of the cytoplasmic domain of the bone morphogenetic protein receptor type-1B (BMPR1B) in complex with FKBP12 and LDN–19(3189). DOI:10.2210/pdb3mdy/pdb.
39. Crystal structure of the kinase domain of type I activin receptor (ACVR1) in complex with FKBP12 and dorsomorphin, DOI:10.2210/pdb3h9r/pdb.
40. Cornell, W. D., Cieplak, P., Bayly, C. I., Gould, I. R., Merz, K. M., Ferguson, D. M., Spellmeyer, D. C., Fox, T., Caldwell, J. W., & Kollman, P. A. (1995). *J. Am. Chem. Soc. 117*, 5179.
41. DOCK 5.20. Molecular Design Institute, University of California San Francisco. For more information see: http://dock.compbio.ucsf.edu.
42. Meng, E. C., Shoichet, B. K., & Kuntz, I. D. (1992). *J. Comput. Chem. 13*, 505.
43. Ewing, T. J. A., Makino, S., Skillman, A. G., & Kuntz, I. D. (2001). *J. Comput-Aided Mol. Design 15*, 411.
44. Case, D. A., Darden, T. A., Cheatham, III, T. E., Simmerling, C. L., Wang, J., Duke, R. E., Luo, R., Merz, K. M., Wang, B., Pearlman, D. A., Crowley, M., Brozell, S., Tsui, V., Gohlke, H., Mongan, J., Hornak, V., Cui, G., Beroza, P., Schafmeister, C., Caldwell, J. W., Ross, W. S., & Kollman, P. A. (2004). AMBER 8, University of California, San Francisco.
45. Duan, Y., Wu, C., Chowdhury, S., Lee, M. C., Xiong, G., Zhang, W., Yang, R., Cieplak, P., Luo, R., Lee, T., Caldwell, J., Wang, J., & Kollman, P. (2003). *J. Comput. Chem. 24*, 1999.
46. Jorgensen, W. L., Chandrasekhar, J., Madura J. D., Impey, R. W., & Klein, M. L. (1983). *J. Chem. Phys. 79*, 926.
47. Darden, T., York, D., & Pedersen, L. (1993). *J. Chem. Phys. 98*, 10089.
48. Essmann, U., Perera, L., Berkowitz, M. L., Darden, T., Lee H., & Pedersen, L. G. *J. Chem. Phys.* (1995). *103*, 8577.
49. Crowley, M. F., Darden, T. A., Cheatham, T. E. III; & Deerfield, D. W. II. (1997). *J. Supercomput. 11*, 255.
50. Sagui, C., & Darden, T. A. (1999). *AIP Conf. Proc.*, *492*, 104.
51. Toukmaji, A., Sagui, C., Board, J., & Darden, T. (2000). *J. Chem. Phys. 113*, 10913.
52. Ryckaert, J. -P., Ciccotti, G., & Berendsen, H. J. C. (1977). *J. Comput. Phys. 23*, 327.
53. Berendsen, H. J. C., Postma, J. P. M., Gunsteren, W. F. van; DiNola, A., & Haak, J. R. (1984). *J. Chem. Phys. 81*, 3684.
54. Srinivasan, J., Cheatham, III, T. E., Cieplak, P., Kollman P., & Case, D. A. (1998). *J. Am. Chem. Soc. 120*, 9401.
55. Wang, W., & Kollman, P. A. (2000). *J. Mol. Biol. 303*, 567.
56. Zhong, H., & Carlson, H. A. (2005). *Proteins 58*, 222.
57. Case, D. A. (1994). *Curr. Opin. Struct. Biol. 4*, 285.
58. Kottalam, J., & Case, D. A. (1990). *Biopolymers 29*, 1409.

59. Gohlke, H., Kiel, C., & Case, D. A. (2003). *J. Mol. Biol. 330*, 891.
60. Rarey, M., Kramer, B., Lengauer, T., & Klebe, G. (1996). *J. Mol. Biol. 261*, 470.
61. Weiser, J., Shenkin, P. S., & Still, W. C., (1999). *J. Comput. Chem. 20*, 217.
62. Fischer, S., Smith, J. C., & Verma, S. C. (2001). *J. Phys. Chem. B 105*, 8050.
63. Griffith, J. P., Kim, J. L., Kim, E. E., Sintchak, M. D., Thomson, J. A., Fitzgibbon, M. J., Fleming, M. A., Caron, P. R., Hsiao, K., & Navia, M. A. (1995). *Cell 82*, 507.
64. Gohlke, H., & Case, D. A. (2004). *J. Comput. Chem. 25*, 238.

CHAPTER 4

IN SILICO DESIGN OF PDHK INHIBITORS: FROM SMALL MOLECULES TO LARGE FLUORINATED COMPOUNDS

RITA KAKKAR

Computational Chemistry Group, Department of Chemistry, University of Delhi, Delhi–110007, India, Tel.: +91-11-27666313; E-mail: rkakkar@chemistry.du.ac.in

CONTENTS

ABSTRACT

Pyruvate dehydrogenase complex (PDC) is one of the largest enzyme complexes in mammals. It regulates the decarboxylation of pyruvate, a key step in metabolism. It depletes the carbohydrate reserves in the body.

To regulate this enzyme, there is another enzyme, pyruvate dehydrogenase kinase (PDHK), which phosphorylates PDC, rendering it inactive. In turn, too much activity of this enzyme reduces metabolism, leading to diseases like diabetes. Therefore, there needs to be a delicate balance between the two enzymes and one needs to inhibit PDHK under some conditions in order to activate PDC. PDHK is actually an enzyme of four types named PDHK1, PDHK2, PDHK3 and PDHK4, all of which have similar amino acid sequences, yet differ in their activities and tissue distribution. Of these, PDHK2 is the most widely distributed in mammalian tissues, and we have focused on this in our work. This chapter describes our work on finding inhibitors of this enzyme. Pyruvate is the most obvious inhibitor of this enzyme, since it is the substrate of PDC. Another similar small molecule is dichloroacetate (DCA), which has proven effective for controlling cancer, but is toxic in itself. Starting from the coordinates of the DCA-PDHK2 complex available in the protein data bank, we explored the interactions within this complex. Using drug databases, we looked for similar non-toxic molecules that could dock similarly into the DCA binding site, but could not find any suitable molecule. Since pyruvate also binds into the same site, we also performed virtual screening to find molecules similar to pyruvate, and found some candidates, which were experimentally found to be more potent than DCA without the toxicity. We continued looking for more potent inhibitors based on different strategies, which are described in this chapter.

4.1 INTRODUCTION

The power of computational chemistry in solving everyday problems, be it in the fields of life sciences or materials sciences, is growing rapidly. The growth of this field of chemistry can be ascribed not only to the design of better algorithms to solve numerical problems, but also to the increasing computational power. The contributions of numerous computer scientists, chemists, physicists, biologists and mathematicians to this rapidly developing field cannot be ignored. Who would have thought even two decades ago that today it would be possible to carry out *ab initio* computations on enzymes and nanostructures on the humble PC? And that too with accuracy that sometimes surpasses experiment. More and more properties are being added to the list of results available from computational chemistry software. However, a word of caution here. It is not enough to churn out numbers – it is also important to understand how to interpret the numbers and design further computer experiments. For this, a basic knowledge of chemistry is essential – understanding of inter- and intramolecular interactions, hydrogen bonding, etc. My intention is to say that, instead of using these software as black boxes, we should try to understand what the numbers are trying to tell us.

In this chapter, I take you through my still ongoing journey to the quest for potent PDHK inhibitors. Here, I confine myself to a part of the research work, i.e., that based on the natural physiological inhibitors of kinase activity. This journey started about ten years ago, when we started work on pyruvic acid, a very important molecule in biology, since it is an essential component of the metabolism process. In our quest to understand the structure and reactivity of this molecule, we carried out several studies [13–15] on the conformational stability, properties and decarboxylation reaction of the molecule, as this is the reaction involved in the metabolism of carbohydrates. Pyruvate dehydrogenase, an enzyme responsible for metabolism, catalyzes the conversion of pyruvate to acetic acid, complexed with coenzyme-A, the entire complex being named acetyl-CoA. The pyruvate dehydrogenase enzyme is actually a complex of several different activities, called the pyruvate dehydrogenase complex (PDC) and is one of the largest multienzymes found in living cells. The decarboxylation reaction is coupled (Scheme 1) to the reduction of nicotinamide adenine dinucleotide (NAD^+) to NADH [23].

This reaction depletes the carbohydrate reserves in mammals, and hence, to conserve these resources under conditions of fasting or pathological conditions associated with insulin resistance, such as obesity and diabetes, the PDC activity is reduced [6, 16, 20, 28]. This happens by the action of another enzyme, the pyruvate dehydrogenase kinase (PDHK), which phosphorylates PDC, rendering it inactive. In humans and other mammals, there are at least four PDHK enzymes (PDHK1, PDHK2, PDHK3 and PDHK4), all of which are similar in their amino acid sequence, but very different in their activities, tissue distribution and regulation [3, 8, 24]. Of these, PDHK2 is the most widely distributed amongst tissues [3, 32] and we focus on it in this work.

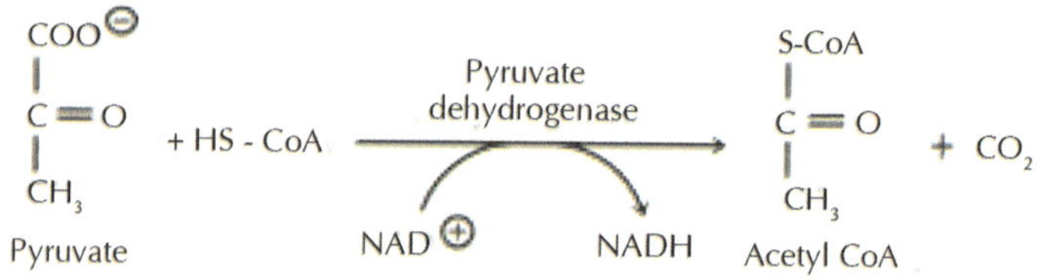

SCHEME 1 Reactants and products amounts of the PDC reaction.

In turn, the PDHKs are regulated by the amounts of reactants and products of the PDC reaction shown in Scheme 1. Thus, the products of this reaction, i.e., acetyl-CoA and NADH activate PDHK, promoting phosphorylation and thus inactivation of PDC [21]. Conversely, the reactants (pyruvate and CoA) inhibit PDHK, hence promoting PDC activity. In certain conditions, this is desirable. For example, cancer cells have inactivated PDC; therefore, inhibition of PDHK is suggested as a possible line of treatment for cancer. Other diseases that can be targeted with this kind of treatment are heart ischemia and insulin resistant diabetes. Therefore, the pharmaceutical industry is keen to identify potent inhibitors of the PDHKs.

From the foregoing discussion, it is clear that pyruvate and CoA can be taken as models to design PDHK inhibitors. Let us first discuss on our search for inhibitors based on pyruvate.

4.2 INHIBITORS BASED ON PYRUVATE

As stated above, pyruvate is an important physiological inhibitor of kinase activity. Another similar molecule, dichloroacetate (DCA), which is a well-known activator of PDC [2, 30], has also been used in cancer therapy, but clinical trials on this molecule had to be dropped prematurely because of its toxicity [25, 26]. However, because of its effectiveness against cancer cells, it continued to be used clandestinely. Therefore, the need of the hour is to find more potent inhibitors without the accompanying toxicity.

Towards this end, we undertook a study to identify small molecule inhibitors similar to pyruvate and DCA [10], to compute their absorption, distribution, metabolism, and excretion (ADME) properties, and short-list those expected to be non-toxic to humans and more potent than DCA for further biological evaluation. The crystal structure of pyruvate complexed with PDHK2 could not be determined, but Knoechel et al. [18] determined that of DCA complexed with PDHK2. It was found that DCA binds at the same site as pyruvate. We first characterized this site, and then docked similar molecules at this site and calculated their docking scores using a variety of procedures. The Schrödinger suite was used for these studies.

4.2.1 THE DCA/PYRUVATE BINDING SITE

We first carried out a protein refinement. The starting coordinates of the DCA-PDHK2 complex were taken from the crystal structure [18] deposited in the Research Collaboratory for Structural Bioinformatics (RCSB) Protein Data Bank (http://www.rcsb.org) (PDB ID: *2bu8*). Most crystal structures taken from X-ray diffraction are incomplete since they lack hydrogen atoms (XRD is not able to provide hydrogen positions with accuracy because of the low atomic number and hence small peaks due to hydrogen) and hence the protonation and tautomeric states of the residues. Hydrogen atoms were added to the structure at the most likely positions of hydroxyl and thiol hydrogen atoms, protonation states and tautomers of His residues, and Chi 'flip' assignments for Asn, Gln and His residues were selected. Co-crystallized water molecules beyond 5 Å of the active site were removed. The protein structure was first optimized using a molecular mechanics calculation using the OPLS–2005 force-field [9]. Minimizations were performed until the average root mean square deviation of the non-hydrogen atoms reached 0.3 Å.

DCA is predicted to be in the ionized state (–1) at the physiological pH and the overall charge on the complex is thus +6. Indeed, since the pK_a value of DCA is 1.48, it is very corrosive, and only its salts have been used in therapy. In agreement with Knoechel et al. [18], we found that the residues lining the DCA binding pocket of human PDHK2 are Leu53, Tyr80, Ser83, Ile111, Tyr112, Arg114, His115, Asp117, Arg154, Ile157, Arg158, and Ile161. We ran into a difficulty here. During the protein refinement, the IMPACT minimization made both the CO bonds of the carboxylate group of DCA equal in length (1.251 Å). However, in the original *2bu8* structure, the two CO bonds of the carboxylate group have different bond lengths; one has a typical carbonyl bond length of 1.182 Å and the other is close to the single C-O bond length in C-OH (1.324 Å). Because of its low pK_a, it is unlikely that DCA exists in anything but the ionized form in the human body. Therefore, the asymmetry must arise from differences in the environment of the two C-O groups.

To resolve this issue and to gain better insight into the mechanism of DCA recognition by PDHK2, we investigated charge polarization effects. Since pyruvate and DCA exist as negatively charged ions at the physiological pH, their binding to PDHK is governed by charge polarization effects, which cannot be adequately treated by classical molecular mechanics (MM) methods. *Ab initio* quantum mechanics (QM) is required to study interactions involving charged species in a protein environment. However, even with today's computer technology, full QM calculations of entire proteins are still not within reach. Mixed QM/MM calculations provide the ideal solution by separating out the reactive core, which can be accurately described with QM, while treating the remainder of the complex more efficiently with MM. We used QSite, the QM/MM component of Schrödinger, to treat the DCA molecule using density functional theory (B3LYP/6–31G(d)) [1] and the rest of the protein by the OPLS–2005 force-field [9]. As expected, the QM/MM minimization again restored the difference in the CO bond lengths. It was found that, not only is there a hydrogen bond with Arg154, another one with Arg158 is present for the longer C-O bond (1.256 Å). The other oxygen is involved in hydrogen bonding with Tyr80 and His115. This C-O bond is shorter (1.244 Å), indicating that the Arg form stronger hydrogen bonds. Klyuyeva et al. [17] who examined the model of Knoechel et al. [18] using alanine scanning mutagenesis, also showed the importance of His115, but they failed to establish the role of the arginines in DCA recognition. According to our results, they do play an important role. Figure 4.1 shows the optimized structure of DCA in the protein binding site.

4.2.2 PYRUVATE DOCKING

In order to validate the docking procedure, we first removed the ligand (DCA) from the protein and re-docked it into the protein, using the extra precision (XP) mode of Glide [5] as our docking engine. A grid having coordinates 56.5, 44.6 and 80.9 Å was created with its centroid at the ligand and a grid of

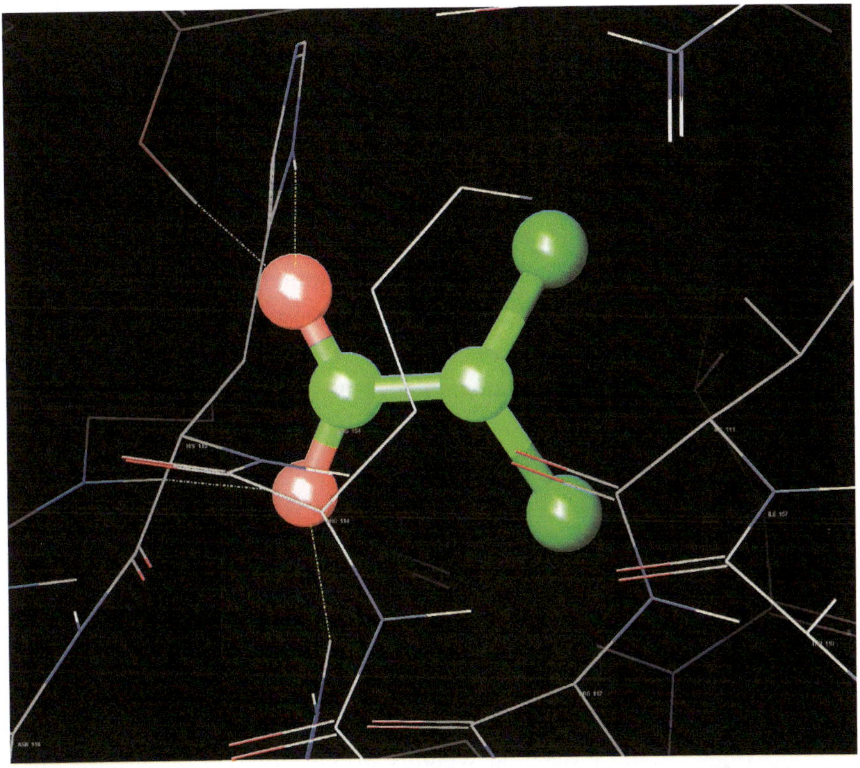

FIGURE 4.1 QSite optimized DCA-bound PDHK2 showing extensive hydrogen bonding (yellow lines) of the oxygens.

size 48×48×48 Å^3 was created. No constraint was placed on the docking site. On superposition with the original structure, the root mean square deviation (RMSD) of the atomic positions was found to be only 0.270 Å. The carboxylate oxygens were found to form hydrogen bonds with Arg154 and Arg158. Pyruvate is also expected to dock at the same position and it was found that, in this case, there are no hydrogen bonding interactions with the protein. The Glide XP scores for the two ligands are –5.67 and –4.69, respectively, showing that DCA docks better and can help replace the pyruvate moiety, which does not appear to form hydrogen bonds with the protein.

After validating our docking procedure, it was the turn to find more potent inhibitors than DCA and pyruvate. For this, we performed a virtual screening of small drug-like molecules, since both DCA and pyruvate are small and fit into a small binding site.

4.2.3 SMALL DRUG-LIKE MOLECULES

For this purpose, we searched for small drug-like molecules. Since pyruvate and DCA have molecular weights of 87 and 128 g mol^{-1}, respectively, we limited our search to drug-like molecules having a molecular weight upto 250 g mol^{-1}. Smiles representations of 798 such molecules were downloaded from *www.zincdocking.org* and they were subjected to the LigPrep protocol, which searches for tautomers and rotamers of the ligands and predicts the most likely ones and their ionization states in the pH range 7±2. The ligand library consisted of molecules exhibiting a neutral state, positively charged state and negatively charged ones.

The 798 molecules were then individually docked into the DCA docking site without imposing any constraints. Barring two, all the rest found suitable docking poses and their Glide XP scores ranged from –5.76 to 5.57. Only the 80 top-ranked ligands (10%) were selected for the next step, the last ligand in this selection having a Glide XP score of –4.68, close to that of pyruvate. The original set consisted of ligands in all possible ionization states – neutral, positively and negatively charged at physiological pH, but only the neutral ones (except one negatively charged ligand) found their way to the top 10% list. Most of the ligands that were top rankers were the ones having molecular weights in the range 72–88 g mol^{-1}, the average being 84 g mol^{-1}. For the full set, the range was 57–90, and the average was 80. Hence the lower molecular weight compounds (< 72 g mol^{-1}) did not figure in the top 10% list. Similarly, for the full set, the molecular volume ranged between 257 and 456 $Å^3$ and the average volume was 368 $Å^3$, while for the selected molecules, the range was 329 to 446 $Å^3$, with an average of 392 $Å^3$. Again, the optimum volume for docking was in the higher volume range. What was surprising was that instead of negatively charged ions, the top rankers were small alcohols and esters.

The next step was to look for the best scoring function to describe the affinity of a ligand to the binding site. There is no hard and fast rule to decide which of the scoring functions performs best. One has to decide the best scoring function for a given dataset on the basis of how well it correlates with the ligand properties. We subjected the complexes to various minimization procedures, and their other properties were generated. Specifically, eMBrAcE, Prime/MM-GBSA scoring, and Liaison scoring were performed. Basically all of these compute the binding affinity as the difference in energies of the energy minimized complex and the free ligand and protein.

$$\Delta E = E_{complex} - E_{ligand} - E_{protein}$$

The difference is in the force-field and treatment of solvation. Only the best scoring pose for each ligand was taken into consideration.

In eMBrAcE (Multi-Ligand Bimolecular Association with Energetics), the OPLS_2005 force-field with a constant dielectric electrostatic treatment of 1.0 [29, 32] is used. Traditional MM methods are used to calculate the ligand–receptor interaction energies (G_{ele}, G_{vdW}, G_{solv}) by a GB/SA method [22] for the electrostatic part of solvation energy and solvent-accessible surface for the non-polar part of solvation energy. A conjugate gradient minimization protocol with default values is used in all minimizations. The full effects of solvation and relaxation are taken into account.

The prime MM-GBSA method combines OPLS molecular mechanics (MM) energies (E_{MM}), surface generalized Born (SGB) solvation model for polar solvation (G_{SGB}), and a nonpolar solvation term (G_{NP}). The G_{NP} term comprises the nonpolar solvent accessible surface area and van der Waals interactions.

The Liaison program uses a linear interaction approximation (LIA) model. The LIA model is an empirical method fitted to a set of known binding free energies. Liaison runs MM simulations for the ligand-receptor complex, and for the free ligand and free receptor using the SGB continuum solvation model. The simulation data and empirical binding affinities are analyzed to generate the Liaison parameters, which are subsequently used to predict binding energies for other ligands with the same receptor. The empirical function used by Liaison for the prediction of binding affinities is as follows:

$$\Delta G = \alpha\left(\left\langle U^{b}_{vdW}\right\rangle - \left\langle U^{f}_{vdW}\right\rangle\right) + \beta\left(\left\langle U^{b}_{elec}\right\rangle - \left\langle U^{f}_{elec}\right\rangle\right) + \gamma\left(\left\langle U^{b}_{\mathrm{cov}}\right\rangle - \left\langle U^{f}_{\mathrm{cov}}\right\rangle\right)$$

In this equation, ⟨ ⟩, *b* and *f* represent the ensemble average, the bound form, and the free form of the ligand, respectively. Parameters α, β and γ are the coefficients. U_{vdW}, U_{elec} and U_{cav} are the van der Waals, electrostatic and cavity energy terms in the SGB model, respectively.

Table 4.1 gives the Pearson correlation matrix for the various scoring functions, and it can be seen that there is no significant correlation between any pair of functions, and we must therefore look for the scoring function amongst these that best describes the biological activity.

In the absence of experimental data on the biological activities of the ligands, we tried to correlate the scoring functions with various ligand-based descriptors. Since none of these scoring functions correlate with each other, each one should be independently tested for dependence on the ligand parameters. The biological activity is expected to correlate with the ligand-based descriptors; hence, the score that displays the best correlation with the ligand structure parameters is also expected to give the best description of the biological activity.

The various ligand-based descriptors used in the present study include the structural parameters, as well as the properties calculated in QikProp. It may be pointed out that many of the structure based parameters are highly correlated with each other. For example, the number of rings is correlated with the number of aliphatic rings and number of aromatic rings, so some of these descriptors were removed from the regression analysis.

Taking the various scoring functions like Glide Score, Glide Energy, eMBRaCe energies, Liaison Score and Prime-MM/GBSA scores in turn, we found that the best correlation is with the Liaison score ($R^2 = 0.967$). The respective R^2 values for Glide Score (0.917), Glide Energy (0.843), eMBrAcE (0.890) and Prime MM-GBSA (0.911) are only slightly smaller. Since the regression coefficient is largest for the LiaScore, we concentrated our attention on this score. Figure 4.2 gives the residual plots for LiaScore. It can be seen that the distribution is fairly normal and the peak is at zero residual. The normal probability plot of the residuals

TABLE 4.1 Pearson Correlation Matrix for the Various Scoring Functions in Schrödinger

	Glide Score	Glide Energy	eMBrAcE	LiaScore
Glide Energy	–0.065			
eMBrAcE	–0.035	0.085		
LiaScore	0.226	–0.092	0.034	
Prime MM-GBSA	0.050	0.509	0.208	–0.328

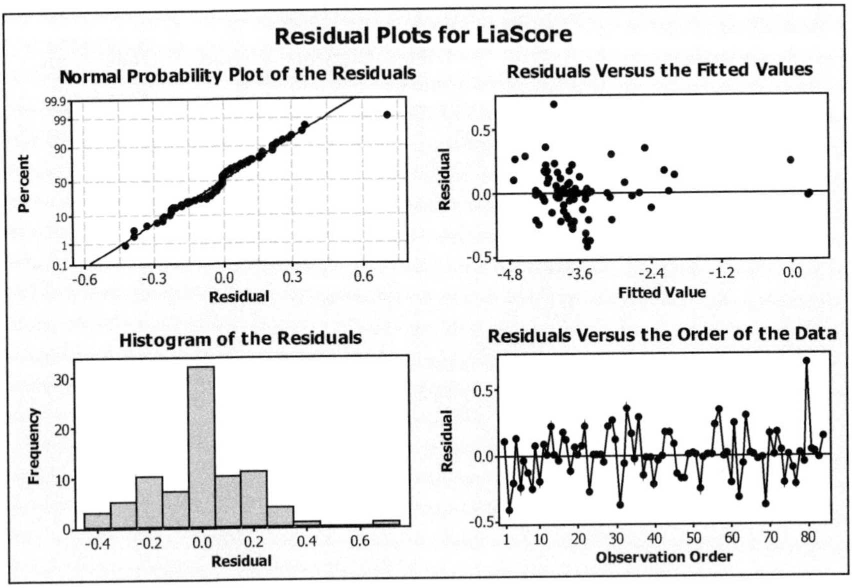

FIGURE 4.2 Residual plots for LiaScore.

is linear. No pattern in the plot of the residuals versus the fitted values is detected, nor do the residuals depend upon the observation order.

In order to decrease the number of predictors for better interpretation, we carried out a step-wise reduction. The LiaScore can be expressed as:

$$\text{LiaScore} = 3.767 - 1.26\ \text{Num chiral centers} - 0.603\ \text{H} - 0.42\ \text{accptHB} - 0.60\ \text{logKp} - 0.56\ \text{IP(eV)} + 0.0084\ \text{FOSA} + 1.16\ \text{C} - 0.50\ \text{polrz} + 1.69\ \text{aliphatic ketone} - 0.66\ \text{Neutral carbonyls} - 0.43\ \text{Ethers} \quad (1)$$

with R^2 is 0.631. Increase in the number of chiral centers, hydrogen atoms, hydrogen bond acceptors, neutral carbonyls and ether groups increase binding. Similarly, increase in log*Kp* (skin permeability), ionization potential, and polarizability also favors binding. However, increase in carbon atoms and aliphatic ketones decreases binding, as does increase in hydrophobic component of surface area (FOSA).

We then rescored the ligands according to their LiaScores. We also estimated the pIC_{50} values from the LiaScores. The top eight scorers (Table 4.2) were then taken for further analysis. It is found that two of the

TABLE 4.2 The Eight Ligands with the Highest Liaison Scores

S. No.	Structure	Glide Score	LiaScore	Prime ΔG_{bind}	Glide energy	pK_d
1	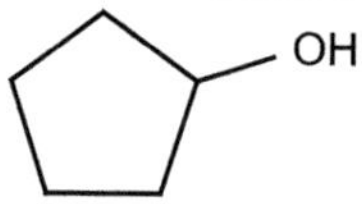 cyclopentanol	–5.76	–4.62	–13.10	–12.98	–3.39
2	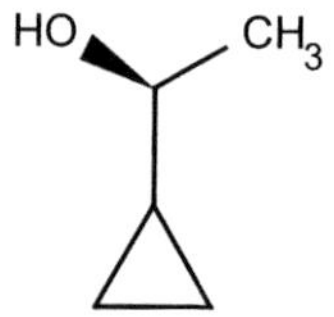 (1R)-1-cyclopropylethan-1-ol	–5.54	–4.56	–9.35	–13.74	–3.34
3	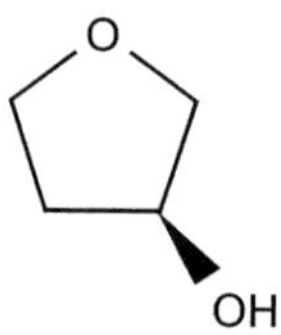 (3R)-oxolan-3-ol	–5.06	–4.44	–9.98	–14.33	–3.25
4	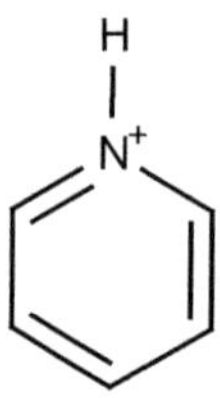 pyridin-1-ium	–4.82	–4.35	2.36	–8.38	–3.19
5	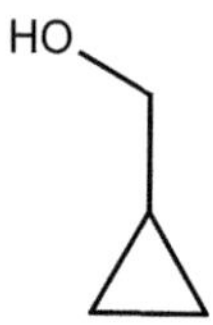 Cyclopropylmethanol	–4.94	–4.32	–12.55	–14.62	–3.17
6	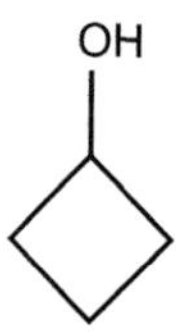 cyclobutanol	–5.01	–4.28	–12.51	–13.97	–3.14

TABLE 4.2 (Continued)

S. No.	Structure	Glide Score	LiaScore	Prime ΔG_{bind}	Glide energy	pK_d
7	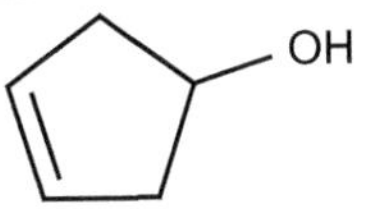 cyclopent-3-en-1-ol	–5.40	–4.26	–7.54	–12.16	–3.12
8	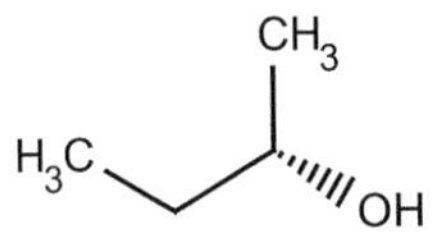 (2S)-butan-2-ol	–4.98	–4.23	–14.19	–14.57	–3.10

ligands are the same but with different conformations. It is heartening to note that the top four ligands according to the LiaScore also figure among the top eight for Glide Score and eMBrAcE and Prime.

Interestingly, the eight molecules are all small alcohols, except the pyridinium ion at the fourth position. The higher scoring ligands are small ring alcohols. They all score better than the pyruvate ion, which has a Glide Score of –4.69. However, although the pyridinium ion has good Glide Score and LiaScore, meaning that it docks well, it has a positive value for the Prime free energy of binding, and was thus rejected.

For relating the calculated free energies of binding with the biological activity, the latter must be somehow quantified in a form that is also related to free energy changes. Biological activity is often reported as ED_{50} (dose/concentration required to achieve 50% of maximal response), IC_{50} (concentration required to achieve 50% inhibition), or LD_{50} (dose/concentration resulting in death of 50% of a population). Through simple receptor theory [4], the values of ED_{50}, IC_{50}, and LD_{50} can be shown to be equal to the dissociation equilibrium constant of the drug and its receptor, K_d, when non-competitive kinetics apply. The logarithm of K_d, or equivalently a value such as IC_{50}, is therefore proportional to the free energy change of drug binding.

$$K_d = \frac{[\text{receptor}][\text{drug}]}{[\text{receptor} - \text{drug}]}$$

$$\Delta G_{binding} = -RT \ln K_d \qquad (2)$$

The pK_d values estimated from the Liaison free energies of binding are also given in Table 4.2. Except cyclopentanol, all the molecules exhibit a smaller Glide Score than DCA (–5.67). We, therefore, looked for other drug-like molecules which could exhibit better potency than DCA. The obvious choice is pyruvate-like molecules that have higher potency than pyruvate, so that they can compete successfully with it for binding to PDHK2.

4.2.4 PYRUVATE-LIKE MOLECULES

We next investigated drug-like molecules that resemble pyruvate. About 83 such molecules that found docking poses were identified. Most of these are negatively charged like the pyruvate ion and have the pyruvate $-COCOO^-$ grouping. Their Glide docking scores range from –7.74 to 5.01. Thirty-six of these have docking scores above –4.69, the value obtained for pyruvate, and are therefore good candidates for further evaluation.

As expected, these molecules exhibit better ligand binding to PDHK2. Taking all the structure-related parameters into account, we obtained regression equations. The calculated R^2 values are 0.955 (Glide Score), 0.961 (Glide Energy), 0.953 (eMBrAcE total energy without constraints), 0.979 (LiaScore), and 0.990 for Prime MM-GBSA $\Delta G_{binding}$. The graphs related to the Prime MM-GBSA free energy of binding are displayed in Figure 4.3. It can be seen that the residuals are fairly normally distributed around the 0.0 value and all statistical "goodness of fit" tests are satisfied.

The partial least squares with 10 components give a straight line graph passing through the origin, signifying the good correlation (Figure 4.4).

To understand the dependence of the Prime MM-GBSA free energies of binding on the various structure-related parameters, we performed stepwise regression, and obtained the following regression equation:

Prime-MM/GBSA = –137.27 + 15.73 Num charged acceptor groups + 132 Charged amines + 0.0179 QPPCaco + 1.35 FISA + 0.219 MW + 7.30 Aliphatic carbons – 6.2 Num chiral centers + 3.82 stars + 46.7 QPlogKp + 0.84 dipole – 7.7 EA + 40 QPlogBB + 0.151 SASA (3)

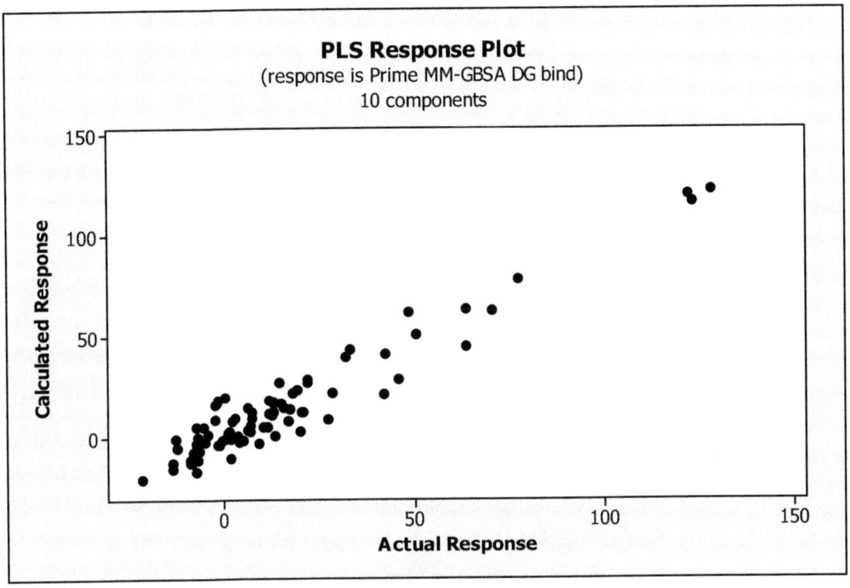

FIGURE 4.3 Residual plots for Prime MM-GBSA ΔG_{bind}

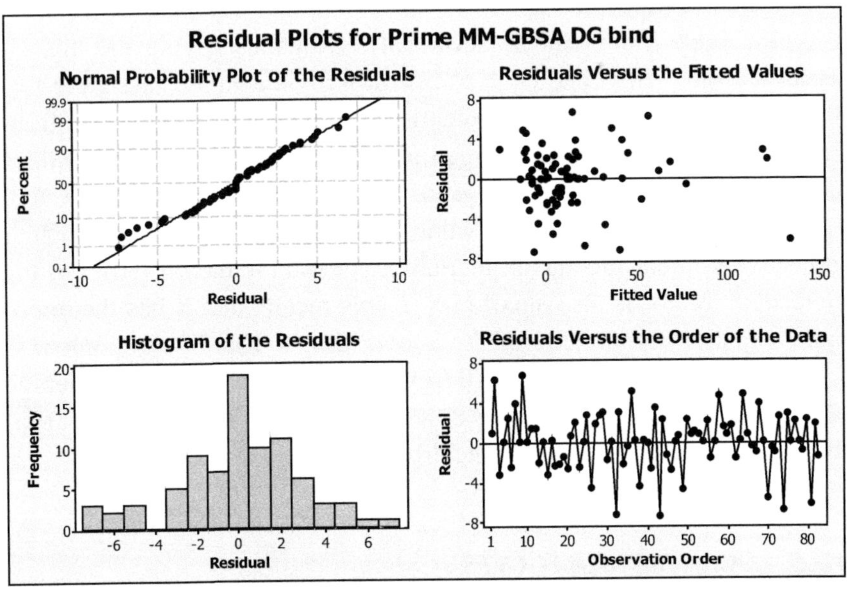

FIGURE 4.4 Partial least squares response plot for Prime MM-GBSA ΔG_{bind}

The regression coefficient is 0.948. Unlike the equation for the previous set of molecules, which showed dependence only on shape and size parameters, this equation shows an interesting dependence on molecular and ADME properties too. First and foremost, increase in the number of charged acceptor groups and charged amines in the ligand is not conducive to stronger binding. However, increase in the apparent CACO–2 permeability (in mm s^{-1}) decreases the magnitude of the (negative) free energy of binding. Increase in hydrophilic component of the solvent-accessible surface area (FISA), molecular weight and number of aliphatic carbons all decrease the free energy of binding. On the other hand, chirality helps in stronger binding. Next in importance is logKp, the skin permeability in cm h^{-1}, which should be high. The ligand should have a small dipole moment, but high electron affinity favors stronger binding. These quantum mechanical quantities were calculated by the PM3 method [27]. Both QPlogBB (brain/blood barrier) and solvent accessible surface area (SASA) decrease the magnitude of the free energy of binding. Hence, the ligand should be nonpolar with a more negative energy of the lowest unoccupied molecular orbital (LUMO), i.e. it should be a good electron acceptor.

From Table 4.3 it can be seen that the ligands having the highest scores are the methyl and ethyl esters of pyruvic acid. Of these, the former stands out as one having exceptional binding affinity. As seen from Eq. (3), ligands having no charged acceptor groups or charged amines are preferred, and these two ligands satisfy this condition. The Caco, FISA, logKp, logBB and SASA values are 531, 134, –3.9, –0.46 and 269, respectively. It has the maximum number of stars in the set (10), and the molecular weight is also small (102). This means that it has the maximum violations from 95% of the known drugs. The dipole moment is 5.56 D and the electron affinity (0.61 eV) is also smaller than the average (0.71 eV). However, most of the molecules showing higher binding affinity violate the range of 95% of the drugs.

4.2.5 DCA-LIKE MOLECULES

There are 1001 molecules in this category in the drug bank, but, except for one molecule, no ligand could find a good pose. The reason is that most of

TABLE 4.3 The Ligands Having the Highest Free Energies of Binding

No.	Structure	Glide Score	LiaScore	Prime MM-GBSA Δ*Gbind*	Glide energy	p*Kd*
1	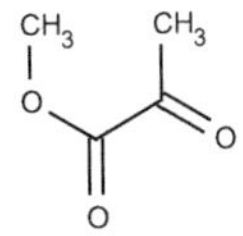 methyl 2-oxopropanoate	–4.42	–4.49	–21.83	–21.95	–3.83
2	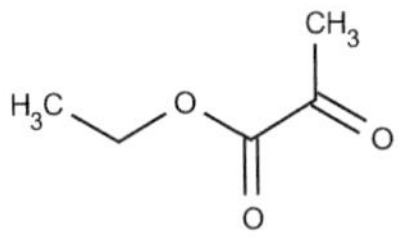 ethyl 2-oxopropanoate	–4.17	–3.10	–13.73	–22.21	–2.41
3	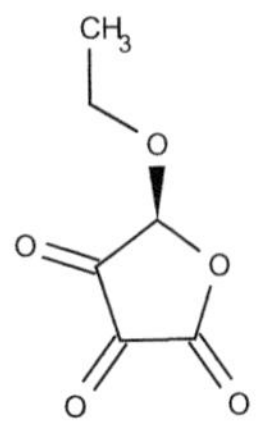 (5S)-5-ethoxyoxolane-2,3,4-trione	–5.45	–5.51	–13.56	–22.97	–2.38
4	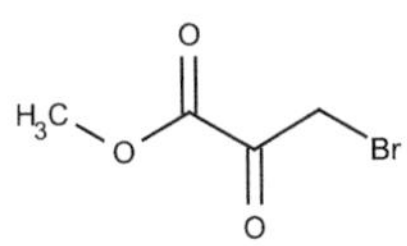 methyl-3-bromo-2-oxopropanoate	–4.42	–3.75	–12.67	–21.72	–2.22
5	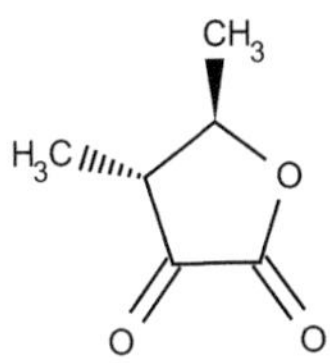 (4R,5R)-4,5-dimethyloxolane-2,3-dione	–5.39	–5.33	–12.38	–7.66	–2.17
6	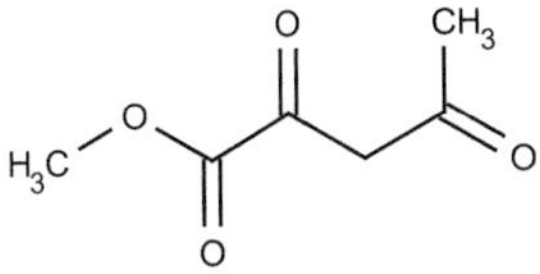 methyl 2,4-dioxopentanoate	–4.60	–4.28	–8.88	–20.97	–1.56

the molecules in this category are too large to fit into this docking site. The only molecule that finds a docking pose is 6,7-dihydro–3H-purin–6-one. However, a number of its tautomers, rotamers and ionized forms also find suitable docking poses, as shown in Table 4.4.

From Table 4.4, it is obvious that the free energy of binding for all tautomers is positive. Therefore, none of these ligands needs to be considered any further.

4.2.6 DISCUSSION

A virtual library consisting of small drug-like molecules and pyruvate-like molecules was created and analyzed for its docking score and binding affinity. The five molecules with the highest binding affinities are found to be methyl 2-oxopropanoate (–21.83), (2S)-butan–2-ol (–14.19), ethyl 2-oxopropanoate (–13.73), (5S)–5-ethoxyoxalane–2,3,4-trione (–13.56), and cyclopentanol (–13.10). Two of these are esters of pyruvic acid and two are small aliphatic alcohols. None of the molecules violates Lipinski's Rule of Five.

The most active molecule is expected to be the methyl ester of pyruvic acid. From our QSAR studies, we had seen that the active molecule should be nonpolar, with a small number of aliphatic carbons, and hence the methyl ester seems to be the best choice. A potential possibility of hydrolysis to pyruvic acid is ruled out, as the ester form is predicted to be the major microspecies at the physiological pH of 7.4. However, it has no primary metabolites and metabolism and excretion could be major issues. Moreover, this molecule is small compared to 95% of the known drug molecules, as is evident from its small molecular weight, 102 (average range 130–725), SASA, 292 (300–1000), molecular volume, 421 Å^3 (500–2000), and polarizability, 10 Å^3 (13–70). Its ionization potential, 10.6 eV (7.9–10.5), logP (hexadecane/gas), 2.9 (4.0–18.0), and logP (ocatanol/gas), 4.8 (8.0–35.0) also do not fall in the normal range. The second molecule, (2S)-butan-2-ol, and the third, the ethyl ester, also suffer on the same account. The fourth molecule, (5S)–5-ethoxyoxolane–2,3,4-trione, has a slightly higher ionization potential (10.6 eV) than the average drug molecule, and also has some aqueous solubility issues, as

TABLE 4.4 Tautomers of 6,7-dihydro–3H-purin–6-one, Along with Their Glide Scores and Prime MM-GBSA ΔG_{bind}

No.	Structure	Glide GScore	Prime MM-GBSA ΔG_{bind}
1	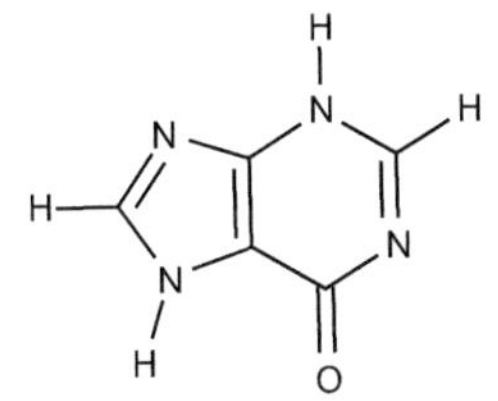 6,7-dihydro-3H-purin-6-one	–4.83	1.57
2	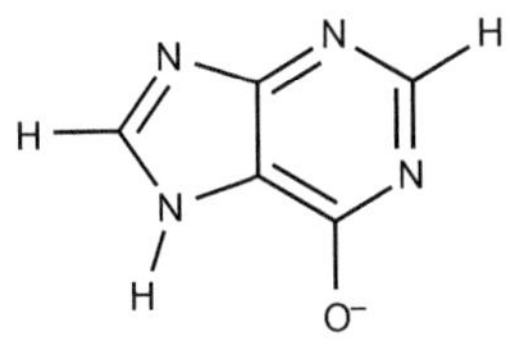 7H-purin-6-ol anion	–4.75	28.74
3	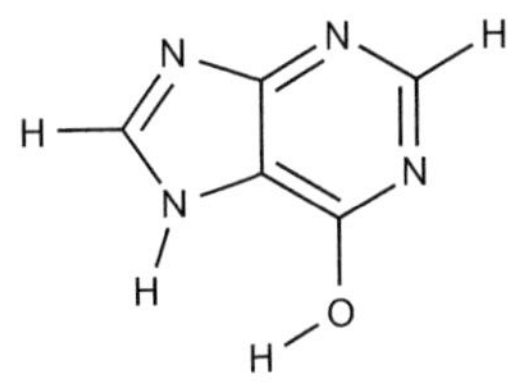 7H-purin-6-ol	–4.66	8.26
4	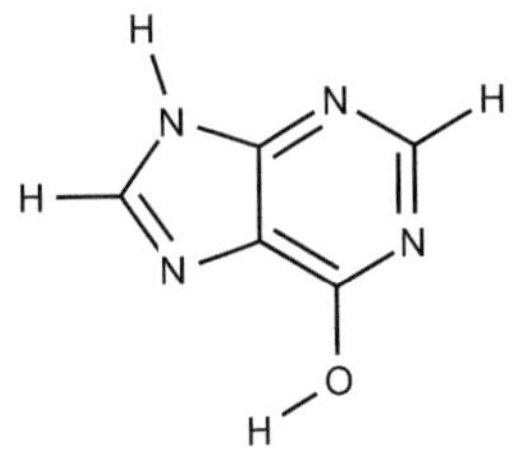 9H-purin-6-ol	–4.51	5.27
5	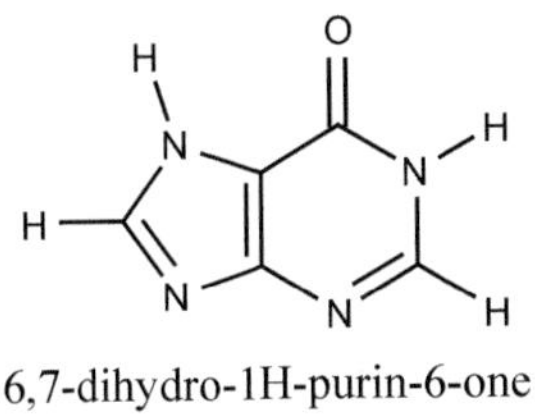 6,7-dihydro-1H-purin-6-one	–4.09	17.61

TABLE 4.4 (Continued)

No.	Structure	Glide GScore	Prime MM-GBSA ΔG_{bind}
6	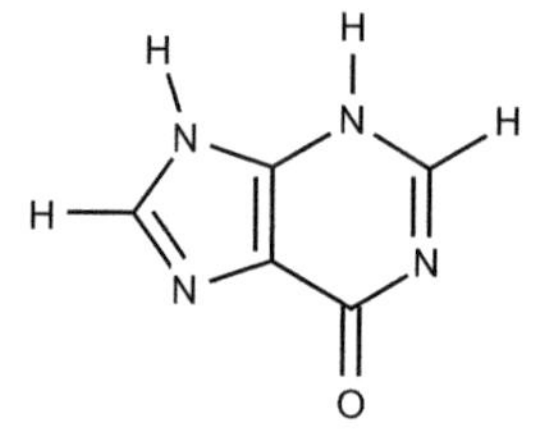 6,9-dihydro-3H-purin-6-one	–4.05	3.23
7	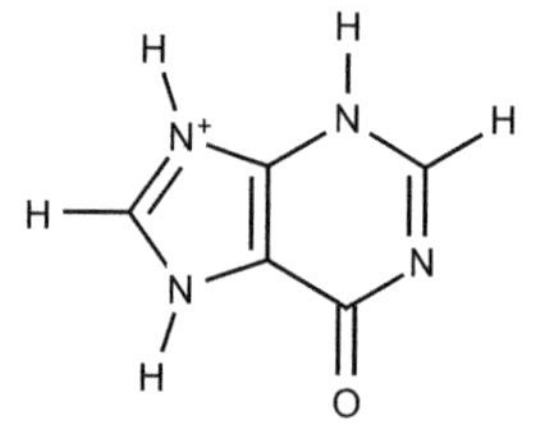 6,7-dihydro-3H-1,3,7,9λ^4-purin-6-one	2.34	19.59

its logS for aqueous solubility is predicted as 2.0 (average –6.5 to 0.5), and its log*P* for octanol/water is –2.8, while the average drug molecule has values between –2.0 and 6.5. Its log*K* for hsa protein binding (–2.8) is also lower than the average drug molecule –1.5 to +1.5, and there are no primary metabolites. However, this drug has been screened by the NCI Yeast Anticancer Drug Screen for its ability to inhibit the growth of selected yeast strains altered in DNA damage repair or cell cycle control and found inactive. Cyclopentanol exhibits eleven violations of the normal drug range.

Another promising compound is methyl–3-bromo–2-oxopropanoate, which shows a small number of violations and has a relatively high binding free energy of –12.67 kcal mol^{-1}. The only violations are related to the molecular size and polarizability, which are slightly smaller than the usual drug molecule. Its likely metabolism pathway is α-hydroxylation of the carbonyl group. This compound, methyl bromopyruvate, has also been under investigation by the National Cancer Institute.

(4R,5R)-4,5-dimethyloxolane–2,3-dione also has no metabolism pathway and needs to be discarded. The next molecule, methyl 2,4-dioxopentanoate, can also metabolize via α-hydroxylation of a carbonyl group, and has very slight deviations from the normal drug range. Its ionization potential (10.8 eV), log for aqueous solubility (0.7) and logK hsa (–1.6) are only slightly different from the range observed for 95% of drug molecules. The methylene protons are slightly acidic, with a predicted pK_a value of 8.36. As seen from Figure 4.5, at the physiological pH, about 10% of the anion is present.

Therefore, the most promising molecules are (2S)-butan–2-ol (**1**), ethyl 2-oxopropanoate (**2**) and, particularly, methyl–3-bromo–2-oxopropanoate (3), which show the least number of violations of the 95% range. These compounds were purchased from the Sigma-Aldrich Co. and their biological evaluation performed.

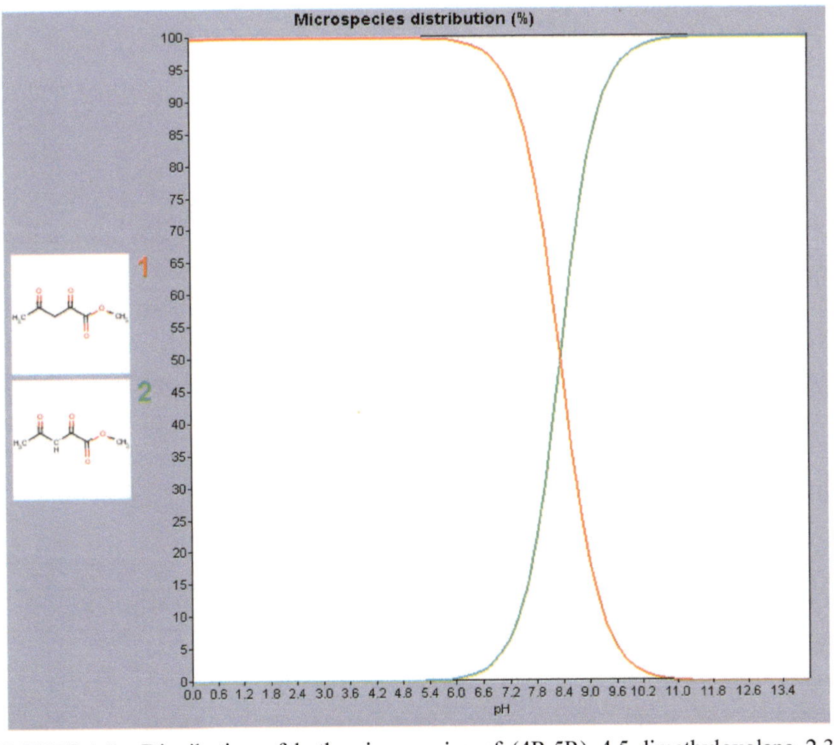

FIGURE 4.5 Distribution of both microspecies of (4R,5R)–4,5-dimethyloxolane–2,3-dione. Chart generated by ChemAxon's Marvin Sketch.

4.2.7 EXPERIMENTAL

PDC containing intrinsic kinase activity was incubated for 45 min at 37°C in a buffer (40 mM Mops (pH 7.10/0.5 mM EDTA/30 mM KCl/1.0 mM $CaCl_2$/0.25 mM acetyl-CoA/0.05 mM NADH/3 mM dithiothreitol/10 mM NaF) at a concentration of approximately 75 μg mL^{-1}. Subsequently, the PDHK reaction was performed in a total volume of 100 μL at 37°C and was initiated by dilution of the PDC mixture four-fold in 1.5x buffer containing 50 μM ADP and 100 μM ATP. After 5 min the reaction was turned off by the addition of stopping buffer (50 mM ADP/50 mM pyruvate).

The PDC activity was assayed in a total volume of 200 μL at 37°C by addition of 100 mM Tris (pH 7.5)/0.60 mM EDTA/0.75 mM $CaCl_2$/2.0 mM thiamine production of NADH at 340 nm as the NADH-related change in absorbance at 340 nm.

PDHK inhibition was calculated as follows:

$$Inhibition(\%) = 100\frac{A_{ATPadded} - A_{ATP}}{A_{noATP} - A_{ATP}}$$

Table 4.5 compares the measured activities of the three compounds with that of DCA under the same conditions. The values are the mean ± SD of independent experiments.

Though these results were successful in discovering more potent non-toxic inhibitors of PDHK, e.g., compounds **2** and **3**, there is room for further improvement. Also, the docking scores are not too good. Towards this end, we tried to adopt a different strategy using the other substrate of the pyruvate reaction, i.e. CoA. A synthetic alternative is the allosteric inhibitor, Pfz3 (*N*(2-aminoethyl)–2-{3-chloro–4-[(4-isopropylbenzyl) oxy] phenyl} acetamide).

TABLE 4.5 Percentage PDHK2 Inhibition at 1 μM Concentration

Compound	IC_{50} (μM)	IC_{50} (μM)	Mean K_i	% PDHK inhibition
DCA	>1000	126±6.02	NS	64.03%
[1]	4.3±0.12	11.8±0.86	2.9±0.98	51.91%
[2]	9.6±1.37	46.2±5.24	5.1±0.02	77.68%
[3]	6.1±0.94	16.0±1.05	8.6±1.14	81.34%

4.3 Pfz3

This inhibitor was found in screening inhibition of PDHK associated with porcine PDHK2. It has been suggested that this inhibitor may be acting as an analog of CoA (structural comparison, Scheme 2).

Pfz3 binds in an extended site at one end of the R domain. Since the binding site of Pfz3 is distant to the ATP-binding site and the proposed catalytic machinery, Knoechel et al. [18] postulated that it inhibits PDHK2 by an allosteric mechanism. The binding pocket is located at one end of the 4-helix bundle. It is very lipophilic and extends approximately 15 Å into the core of the domain [18].

4.3.1 DOCKING MODE OF Pfz3

To study the molecular basis of interaction and affinity of binding of Pfz3, the docking site of Pfz3 was analyzed in detail [7]. We found that the –NH_2 group should exist in the protonated form at pH values 7 ± 2, which includes the physiological pH. The protonated state is lower in energy than the neutral one by 3.4 kcal mol^{-1}. The refined structure (Figure 4.5) of the ligand in complex with PDHK2, shows the amino end protruding out, and that the rest of the structure embedded in the receptor is almost planar. Figure 4.6 also shows the hydrogen bonding interactions. Specifically,

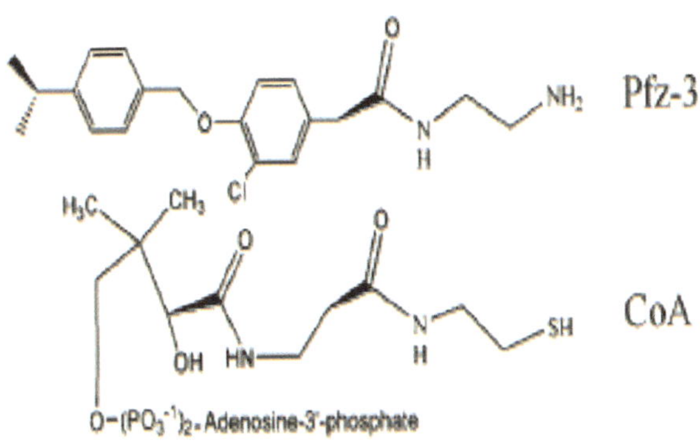

SCHEME 2 Comparison of CoA and Pfz3 structures.

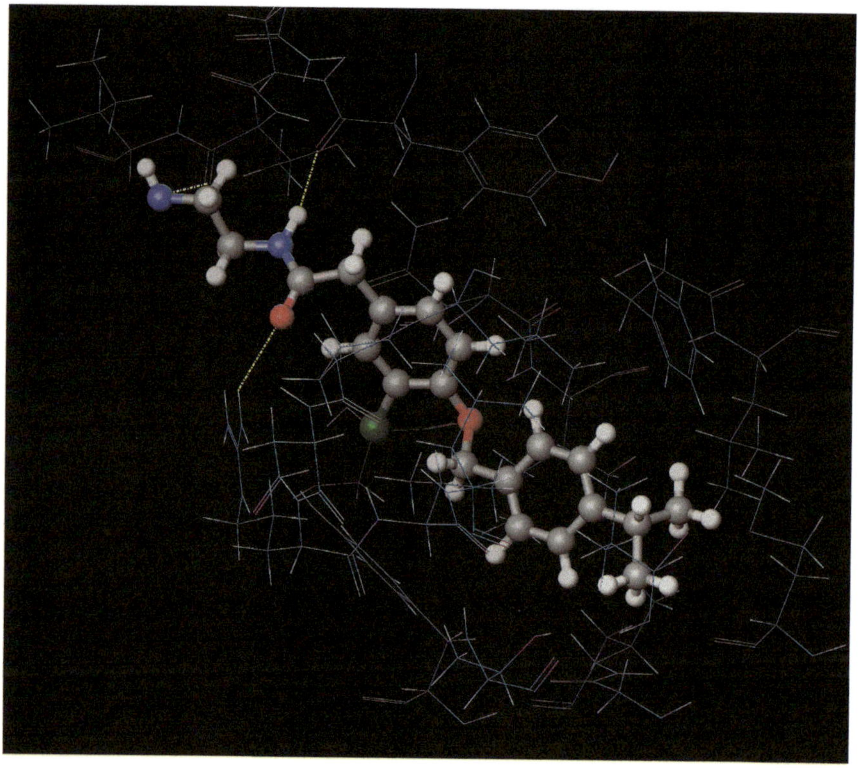

FIGURE 4.6 Hydrogen bonding between the ligand and receptor sites.

there is a C=O---H-N interaction with Arg66, whereas the peptide N-H and that of the end amino group form hydrogen bonds with the carbonyls of Tyr374 and Thr376, respectively.

It can be seen that only the amino end of the ligand is involved in hydrogen bonding with the amino acid residues of PDHK2. The receptor grid for docking was first generated. This is shown in Figure 43.7, where the pink squares represent hydrophobic cells. It can be seen that the isopropyl end of the molecule is embedded in the hydrophobic region.

To validate the procedure, we first removed the ligand Pfz3 and redocked it into the receptor. It was found that this structure, when superimposed on the crystal structure, has an RMSD of only 0.26 Å compared with the original structure. The largest deviation is for the NH_2 group of Gly319, which undergoes a flip in the docked structure.

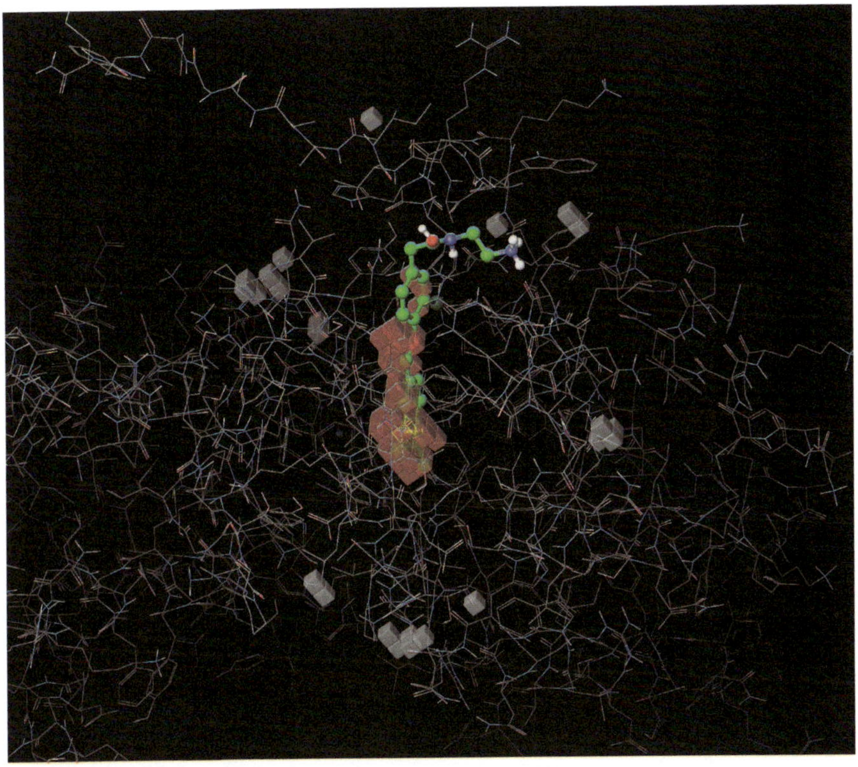

FIGURE 4.7 Receptor grid used for docking of Pfz3 analogs.

We now used this knowledge about the interactions involved to design better inhibitors based on Pfz3. To prepare the compound libraries, we considered the following substitutions:

1. Replacement of the amino group by $-NHCH_3$ or $-N(CH_3)_2$. Since the NH_2 protons are involved in hydrogen bonding with the C=O of Thr376, this substitution is likely to decrease the binding energy and was so rejected.
2. Replacement of the carbonyl group by the thiocarbonyl group may also reduce the binding energy since there is a hydrogen bond of the carbonyl group with Arg66, but, on the other hand, sulfur compounds exhibit biological activities.
3. Replacement of Cl^- by another halogen was considered. It was found that all other halogens improve the binding energy, but Br^- gives the best binding energy results.

4. Finally, the isopropyl group fits loosely in the lipophilic region. Halogens, particularly fluorine, increase the hydrophobicity of a group, and hence all the hydrogens of the isopropyl group were replaced by fluorines to improve lipophilic binding.
5. A methyl substituent on the aromatic ring close to the isopropyl end may further improve lipophilic binding. Calculations showed that substitution at the ortho position results in greater increase in the binding activity as compared to substitution at the meta position.

Accordingly, taking Scheme 3 as a template, we prepared a compound library consisting of the compounds given in Table 4.6.

The docking results of these ligands are given in Table 4.7. The ranking of ligands is based on their Glide scores. All the 16 ligands accepted poses with the receptor (PDHK2). The difference in Glide scores among all the 16 ligands is very small (–15.39±1.18), suggesting that they dock at the same site.

Comparing P1 and P2, we find that replacement of a hydrogen on the phenyl ring by a methyl group increases the Glide Score. Similarly, replacement of the $–CH(CH_3)_2$ group by $–CF(CF_3)_3$, as in P3, improves the score considerably. In fact, of the top nine ligands, eight have this grouping. A marginal improvement is observed on replacement of –Cl by –Br, as in P5, but substitution of the carbonyl oxygen by sulfur decreases the score.

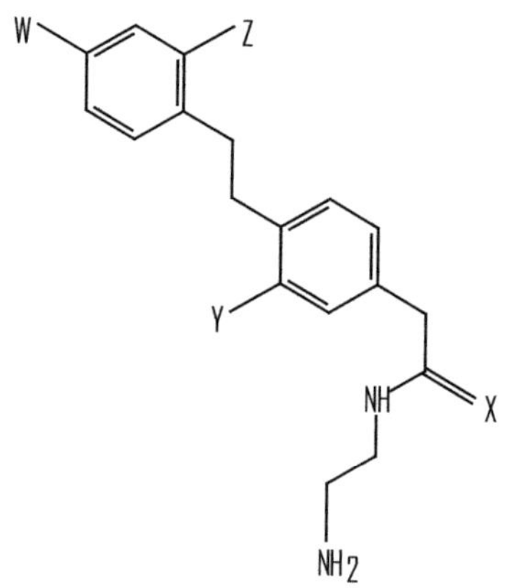

SCHEME 3 Template for compound library.

TABLE 4.6 Structures of the Compounds Used in This Work

Compound	X	Y	Z	W
P1 (Pfz3)	O	Cl	$CH(CH_3)_3$	H
P2	O	Cl	$CH(CH_3)_3$	CH_3
P3	O	Cl	$CF(CF_3)_3$	H
P4	O	Cl	$CF(CF_3)_3$	CH_3
P5	O	Br	$CH(CH_3)_3$	H
P6	O	Br	$CH(CH_3)_3$	CH_3
P7	O	Br	$CF(CF_3)_3$	H
P8	O	Br	$CF(CF_3)_3$	CH_3
P9	S	Cl	$CH(CH_3)_3$	H
P10	S	Cl	$CH(CH_3)_3$	CH_3
P11	S	Cl	$CF(CF_3)_3$	H
P12	S	Cl	$CF(CF_3)_3$	CH_3
P13	S	Br	$CH(CH_3)_3$	H
P14	S	Br	$CH(CH_3)_3$	CH_3
P15	S	Br	$CF(CF_3)_3$	H
P16	S	Br	$CF(CF_3)_3$	CH_3

Analysis of the components of the Glide score clearly reveals that the binding site is very lipophilic and lipophilic interactions dominate (Table 4.8). For example, for the highest ranked ligand, P4, the largest contribution is the van der Waals energy term (–41.53). The lipophilic contact term is also large (–11.02), and the Coulomb energy term is –8.02. However, the contributions of the other terms – hydrogen bonding (–1.90), penalty for buried polar groups (0.0), penalty for freezing rotatable bonds (0.47), and polar interactions at the active site (0.24) are negligible completely. On an average, the lipophilic contact term contributes 70% to the Glide Score, showing that this is the dominant interaction controlling the ranking of a ligand. The second contact term, the vdW interaction energy, contributes –39.77 kcal mol^{-1} on an average. The average of the hydrogen bonding terms is –1.98 kcal mol^{-1}, corresponding to roughly three hydrogen bonds. An analysis of the terms in Table 4.8 reveals that the higher ranked ligands score better in the van der Waals terms. It is thus obvious why the large groups $–CF(CF_3)_3$ and $–CH_3$ (in place of –H) improve the

TABLE 4.7 The Docking Results (kcal mol^{-1}) of Pfz3 and Its Analogs in the Original Crystal Structure of PDHK2 (*2bu7*) Using Glide-XP

Rank	Ligand	Glide score	E_{model}	$E_{internal}$
1	P4	–16.57	–78.24	5.19
2	P15	–16.51	–81.30	23.51
3	P16	–16.35	–58.91	2.64
4	P8	–16.22	–73.82	14.80
5	P14	–15.96	–73.02	9.53
6	P11	–15.84	–81.30	9.73
7	P3	–15.68	–77.27	2.28
8	P12	–15.60	–64.47	36.05
9	P7	–15.49	–75.42	5.82
10	P6	–15.47	–76.62	8.36
11	P2	–15.43	–76.74	4.59
12	P10	–15.38	–72.15	7.63
13	P5	–15.22	–78.48	9.33
14	P13	–15.18	–77.02	10.81
15	P1 (Pfz3)	–15.08	–71.70	11.13
16	P9	–14.21	–69.40	7.99

score by improving the contact at the site. They simultaneously increase the lipophilic contact term. Sulfur, on the other hand, decreases the hydrogen bond contribution.

All the 16 Pfz3 analogs are found to be good binders of PDHK2 (docking score –15.39±1.18), with Pfz3 itself having the second lowest score (–15.08). This proves that its analogs could be potential drugs for second-generation drug development. However, these are unknown molecules and their ADME properties ought to be determined to ascertain their toxicity.

4.3.2 *PREDICTED ADME PROPERTIES*

These properties consist of principal descriptors and physiochemical properties with a detailed analysis of the predicted polarizability (Polrz), predicted octanol/gas partition coefficient (logPoct), water/gas partition coefficient (logPw), octanol/water partition coefficient (logPo/w),

TABLE 4.8 Decomposition of the Glide Score into the Various Energy Terms

Ligand	*lipo*	*hbond*	E_{burp}	E_{vdW}	E_{coul}	E_{rotb}	E_{site}
P1	–10.78	–2.30	–0.59	–34.65	–8.57	0.74	0.16
P2	–10.95	–2.40	–0.50	–36.83	–9.31	0.69	0.13
P3	–10.31	–1.75	0.00	–41.54	–7.41	0.49	0.17
P4	–11.02	–1.90	0.00	–41.53	–8.02	0.47	0.24
P5	–10.49	–2.53	–0.30	–41.40	–8.85	0.60	0.15
P6	–11.01	–2.49	–0.20	–38.71	–8.72	0.56	0.16
P7	–10.35	–1.50	0.00	–40.75	–6.69	0.42	0.05
P8	–11.32	–1.73	0.00	–38.80	–7.54	0.40	0.30
P9	–10.51	–1.49	–0.49	–40.86	–5.05	0.68	0.32
P10	–10.76	–2.42	–0.39	–42.10	–8.35	0.64	0.18
P11	–10.58	–1.90	0.00	–43.03	–7.26	0.46	0.10
P12	–11.09	–1.67	0.00	–33.42	–6.96	0.44	0.32
P13	–10.64	–2.40	–0.19	–42.28	–8.16	0.56	0.16
P14	–12.17	–1.51	–0.10	–37.58	–8.17	0.53	0.13
P15	–10.72	–1.90	0.00	–41.87	–7.08	0.40	0.19
P16	–11.12	–1.72	0.00	–40.89	–6.61	0.32	0.29

IC_{50} value for blockage of HERG K^+ channels (logHERG), apparent Caco–2 cell permeability in mm/s (PCaco), brain/blood partition coefficient (logBB), apparent MDCK cell permeability (PMDCK), skin permeability (logKp), and %human absorption in the intestines (QP%). Caco–2 cells are a model for the gut-blood barrier, whereas MDCK cells are considered to be good mimics for the blood-brain barrier. The predicted solubilities, molecular weights, ionization potentials and electron affinities were also evaluated (Tables 4.9a–c).

The acceptability of the proposed drug molecules was also analyzed in terms of Lipinski's rule of five [19]. This is a rule of thumb to evaluate drug likeness, or determine if a chemical compound with a certain pharmacological or biological activity has properties that would make it a likely orally active drug in humans. Poor absorption or permeation are more likely when a ligand violates Lipinski's rule of five, i.e. has more than five hydrogen donors, the molecular weight is over 500, the logP is over 5, and the sum of N's and O's is over 10.

TABLE 4.9A Descriptors Calculated for Pfz3 and Its Analogs by QikProp Simulation

Ligand	Polrz/Å^3	logPC16	logPoct	logPw	logPo/w	logS	CIlogS
P1	37.55	12.49	19.17	12.71	3.05	–2.93	–3.60
P2	39.10	12.71	19.63	12.53	3.28	–3.34	–3.93
P3	41.49	12.16	21.93	12.16	4.51	–5.17	–6.58
P4	43.17	12.57	22.29	12.26	4.71	–5.39	–6.92
P5	39.67	13.18	19.89	12.92	3.38	–3.58	–4.64
P6	44.09	14.47	21.34	13.27	4.01	–4.75	–4.98
P7	41.49	12.08	22.35	12.13	4.50	–5.22	–7.65
P8	42.82	12.42	22.11	12.82	4.58	–4.75	–7.99
P9	39.77	13.39	19.93	9.56	4.62	–5.33	–4.81
P10	41.61	13.58	20.38	9.28	5.01	–5.49	–5.14
P11	42.59	12.57	21.68	9.52	5.93	–6.52	–7.81
P12	44.78	13.03	22.57	9.39	6.27	–7.01	–8.15
P13	40.58	13.52	20.08	9.45	4.89	–5.35	–5.86
P14	39.82	12.95	20.04	9.17	4.56	–5.04	–6.19
P15	42.94	12.67	22.56	9.55	5.97	–6.93	–8.88
P16	44.16	12.89	22.72	9.38	6.11	–7.02	–9.22

The rule describes molecular properties important for a drug's pharmacokinetics in the human body, including its ADME. However, the rule does not predict if a compound is pharmacologically active [19]. In this study, out of 16 ligands, 12 structures showed allowed values for the properties analyzed and exhibited drug-like characteristics based on Lipinski's rule of 5, as it allows up to 1 violation. Only the analogs P11, P12, P15 and P16 do not show drug-like characteristics owing to their high molecular weights.

4.3.3 DISCUSSION

Natural and prepared compounds (Table 4.6) were evaluated using docking and their binding energy with PDHK2. The docking results showed that structurally homologous inhibitors bind in a very similar pattern at the active site of PDHK2, which suggests that homologous inhibitors have similar binding patterns in and modes of interaction with PDHK2, and have

TABLE 4.9B Predicted ADME Properties of the Compounds

Ligand	logHERG	PCaco	logBB	PMDCK/ mm s^{-1}	logKp	IP/eV	EA/eV
P1	–4.83	99.32	–0.63	181.83	–4.21	9.42	0.35
P2	–4.76	92.60	–0.96	167.67	–4.40	9.33	0.34
P3	–5.22	141.94	–0.10	3121.41	–4.05	9.14	0.73
P4	–5.16	122.59	–0.19	2417.28	–4.22	9.16	0.72
P5	–5.25	97.33	–0.69	191.51	–4.16	9.50	0.36
P6	–5.82	89.07	–0.84	166.61	–4.16	9.39	0.40
P7	–5.28	142.28	–0.12	2996.07	–4.05	9.68	0.77
P8	–4.81	85.28	–0.24	1813.66	–4.42	9.69	0.85
P9	–6.76	362.53	–0.26	767.92	–3.63	8.28	1.00
P10	–6.38	453.17	–0.11	1017.94	–3.56	8.64	0.90
P11	–6.47	428.08	0.37	>10000	–3.54	8.77	1.21
P12	–6.48	468.24	0.37	>10000	–3.55	8.73	1.00
P13	–6.58	489.55	–0.06	1214.41	–3.38	8.62	0.87
P14	–6.22	352.56	–0.29	560.98	–3.84	8.04	1.15
P15	–6.79	420.03	0.31	>10000	–3.57	8.26	1.21
P16	–6.59	430.73	0.29	>10000	–3.64	8.60	1.15

a similar inhibitory mechanism. The most potent inhibitor should have the best interaction with PDHK2. It can be seen that substitution of functional groups at positions X, Y, Z and W leads to an increase in the binding affinity of modified analogs which is even more intense than that of Pfz3. For example, substitution of O, Cl, $CF(CF_3)_3$ and CH_3 in P4; S, Br, $CF(CF_3)_3$, and H in case of P15; S, Br, $CF(CF_3)_3$, and CH_3 in the P16; and O, Br, $CF(CF_3)_3$, and CH_3 in the P8 analog at X, Y, Z and W positions in the scaffold structure of Pfz3 obtained the best results with docking scores = –16.57, –16.51, –16.35 and –16.22; FEB = –22.09, –23.82, –15.78 and –19.74 kcal mol^{-1}. Further, ADME screening provided a peer analysis for the final selection of potential candidates from the compound library generated. Based on the overall analysis we can conclude that the analog P4 (with Glide score: –16.57, E_{model} = –78.24 kcal mol^{-1}, Table 4.7) is the most potent analog. It exhibits effective binding in the active site of PDHK2, and could be used for second-generation drug development. The next two analogs, P15 and P16, also exhibit high Glide Scores, but do not qualify Lipinki's rule of 5, as

TABLE 4.9C Principal Descriptors Calculated for Pfz3 and Its Analogs by QikProp Simulation

Ligand	QP%	Mol. Wt.	PSA	N & O	Donor HB	Rule of 5
P1	81.78	360.883	69.852	4	3	0
P2	78.64	374.909	70.011	4	3	0
P3	83.87	486.816	68.508	4	3	0
P4	81.38	500.843	70.063	4	3	1
P5	82.79	405.334	69.693	4	3	0
P6	83.26	419.36	70.178	4	3	0
P7	84.01	531.267	70.958	4	3	1
P8	78.56	545.294	70.576	4	3	1
P9	89.73	376.943	54.895	3	3	0
P10	90.05	390.97	50.586	3	3	1
P11	90.66	502.877	50.872	3	3	2
P12	90.03	516.904	50.643	3	3	2
P13	92.92	421.394	50.437	3	3	0
P14	86.05	435.421	56.382	3	3	0
P15	90.17	547.328	53.72	3	3	2
P16	88.75	561.355	52.576	3	3	2

both show two violations. Another candidate that can be considered is P8, which has a Glide Score of –16.22 and its E_{model} = –73.88 kcal mol^{-1}. Both P4 and P8 have similar substitutions, except that in the latter Cl is replaced by Br. The results indicate that substitution of the isopropyl group with its fluoro analog has the greatest effect on the GlideScore.

The combined approach of docking–ADME screening used in this work has the power to express the binding affinity of a large set of ligands in the receptor as well as to validate them as potential candidates for second generation drug discovery.

4.4 CONCLUSIONS

Overall, we may say that this step-wise approach to drug discovery has yielded good results in the form of analogs of Pfz3, which owes its activity to its structural similarity to CoA. Most of the proposed modifications in the original ligand have yielded more potent analogs. One particular modification suggested that fluorine substitution greatly enhances lipophilic binding and should be considered for further modifications. Such synthetic modifications, such as in AZ12, have been carried out, leading to enhanced biological activity. We searched for even more potent inhibitors based on AZ12 and found a few likely candidates. The results have been published elsewhere [11, 12].

ACKNOWLEDGMENTS

Part of the work on Pfz3 was also performed by Dr. Pragya Gahlot in the Department of Chemistry, University of Delhi. The author thanks University of Delhi's "Scheme to Strengthen Doctoral Research by Providing Funds to Faculty" for financial support.

KEYWORDS

- **dichloroacetate**
- **PDHK enzymes**
- **pyruvate dehydrogenase complex**
- **pyruvate dehydrogenase kinase**

REFERENCES

1. Becke, A. D. (1988). *Phys. Rev. A*; *38*, 3098–3100.
2. Bersin, R. M., & Stacpoole, P. W. (1997). *Am. Heart J. 134*, 841–855.
3. Bowker-Kinley, M. M., Davis, W. I., Wu, P., Harris, R. A., & Popov, K. M. (1998). *Biochem. J. 329*, 191–196.
4. Cheng, Y., & Prusoff, W. H. (1973). *Biochem. Pharmacol. 22*, 3099–3108.
5. Friesner, R. A., Banks, J. L., Murphy, R. B., Halgren, T. A., Klicic, J. J., Mainz, D. T., Repasky, M. P., Knoll, E. H., Shelley, M., Perry, J. K., Shaw, D. E., Francis, P., & Shenkin, P. S. (2004). *J. Med. Chem. 47*, 1739–1749.
6. Fuller, S. J., & Randle, P. J. (1984). *Biochemistry 219*, 635–646.
7. Gahlot, P., & Kakkar, R. (2011). *Intl. R. J. Pharmaceuticals 1*(1), 33–41.

8. Gudi, R., Bowker-Kinley, M. M., Kedishvili, N. Y., Zhao, Y., & Popov, K. M. (1995). *J. Biol. Chem. 270*, 28989–28994.
9. Jorgensen, W. L., Maxwell, D. S., & Tirado-Rives, J. (1996). *J. Am. Chem. Soc. 118*, 11225–11236.
10. Kakkar, R. (2011). *Intl. R. J. of Pharmaceuticals 1*(2), 50–58.
11. Kakkar, R., Arora, R., Gahlot, P., & Gupta, D. (2014). *J. Comput. Sci. 5*(4), 558–567.
12. Kakkar, R., Azad, N., & Gahlot, P. (2012). *International Review of Biophysical Chemistry (IREBIC) 3*(5), 163–167.
13. Kakkar, R., Chadha, P., & Verma, D. (2006a). *Internet Electron. J. Mol. Des. 5*, 27–48.
14. Kakkar, R., Pathak, M., & Gahlot, P. (2008). *J. Phys. Org. Chem. 21*, 23–29.
15. Kakkar, R., Pathak, M., & Radhika, N. P. (2006b). *Org. Biomol. Chem. 4*, 886–895.
16. Kelley, D. E., Mokan, M., & Mandarino, L. J. (1992). *Diabetes 41*, 698–706.
17. Klyuyeva, A., Tuganova, A., & Popov, K. M. (2007). *FEBS Lett. 581*(16), 2988–2992.
18. Knoechel, T. R., Tucker, A. D., Robinson, C. M., Phillips, C., Taylor, W., Bungay, P. J., Kasten, S. A., Roche, T. E., & Brown, D. G. (2006). *Biochemistry 45*, 402–415.
19. Lipinski, C. A., Lombardo, F., Dominy, B. W., & Feeney, P. J. (2001). *Adv. Drug Del. Rev. 46*, 3–26.
20. Orfali, K. A., Fryer, L. G., Holness, M. J., & Sugden, M. C. (1993). *FEBS Lett. 336*, 501–505.
21. Randle, P. J. (1995). *Proc. Nutr. Soc. 54*, 317–327.
22. Reynolds, C. H. (1995). *J. Chem. Inf. Comput. Sci. 35*, 738–742.
23. Roche, T. E., & Hiromasa, Y. (2007). *Cell. Mol. Life Sci. 64*, 830–849.
24. Rowles, J., Scherer, S. W., Xi, T., Majer, M., Nickle, D. C., Rommens, J. M., Popov, K. M., Harris, R. A., Riebow, N. L., Xia, J., Tsui, L.-C., Bogardus, C., & Prochazka, M. (1996). *J. Biol. Chem. 271*, 22376–22382.
25. Stacpoole, P. W., Henderson, G. N., Yan, Z., & James, M. O. (1998a). *Environ. Health Perspect. 106* (Suppl 4), 989–994.
26. Stacpoole, P. W., Henderson, G. N., Yan, Z., Cornett, R., & James, M. O. (1998b). *Drug. Metab. Rev. 30*, 499–539.
27. Stewart, J. J. P. (1989). *J. Comput. Chem. 10*(2), 209–220.
28. Sugden, M. C., Orfali, K. A., & Holness, M. J. (1995). *J. Nutr. 125*, 1746S–1752S.
29. Todorov, N. P., Mancara, R. L., & Monthoux, P. H. (2003). *Chem Phys. Lett. 369*, 257–263.
30. Whitehouse, S., Cooper, R. H., & Randle, P. J. (1974). *Biochem. J. 141*, 761–774.
31. Wu, P., Blair, P. V., Sato, J., Jaskiewicz, J., Popov, K. M., & Harris, R. A. (2000). *Arch. Biochem. Biophys. 381*, 1–7.
32. Wu, X., Milne, J. L. S., Borgnia, M. J., Rostapshov, A. V., Subramaniam, S., & Brooks, B. N. R. (2003). *J. Struct. Biol. 141*, 63–76.

PART II

COMPUTATIONAL CHEMISTRY METHODOLOGY IN MATERIALS SCIENCE

CHAPTER 5

THE SMART CYBERINFRASTRUCTURE: SPACE-TIME MULTISCALE APPROACHES FOR RESEARCH AND TECHNOLOGY

DANIELE LICARI, GIORDANO MANCINI, ANDREA BROGNI, ANDREA SALVADORI, and VINCENZO BARONE

Piazza dei Cavalieri 7, 56126 Pisa, Italy, E-mail: vincenzo.barone@sns.it

CONTENTS

5.1 INTRODUCTION: WHY WE NEED A CYBERINFRASTRUCTURE FOR MOLECULAR SCIENCES

5.1.1 EVOLUTION OF MOLECULAR MODELING

The development of quantum chemistry and molecular mechanics methods in the middle of the 20th century and the birth of computers and of computer based simulations in the same years lead swiftly to the application of these methodologies to address computational chemistry problems: the first computer simulation of a liquid was performed in 1953 [1] while the *ab initio* Hartree–Fock calculations on diatomic molecules were carried out in 1956 at MIT [2], using a basis set of Slater orbitals. The idea of a *in silico* chemical lab not restricted to theoreticians and developers of new methods, which was visionary only a few years before came into being by the end of the '60, when the first generation of "black box" computer programs developed to address quantum chemistry problems, the first being POLYATOM [3]. It was followed rapidly by others such as Gaussian70, IBMOL [4] and others; one of these, Gaussian09 [7] is today the most worldwide adopted quantum chemistry program. It is noteworthy that the first steps of computer molecular graphics were also made at MIT and ORNL (such as ORTEP-III [6, 7]) in the same period: molecular graphics was recognized since the beginning both as an independent field of research in chemistry and as fundamental tool for the understanding of data coming from simulations or experiments. A decade later terms such as "computational chemistry" or "molecular modeling" were well established: the *Journal of Computational Chemistry* was first published in 1980. The field of computational chemistry was celebrated by Nobel Prizes in 1998 (Walter Kohn, "for his development of the density-functional theory" and John Pople, "for his development of computational methods in quantum chemistry" (The Nobel Prize in Chemistry 1998 http://www.nobelprize.org/nobel_prizes/chemistry/laureates/1998/index.html)] and in 2013 [Martin Karplus, Michael Levitt and Arieh Warshel received the 2013 Nobel Prize in Chemistry for "the development of multiscale models for complex chemical systems" (Nobel Prize in Chemistry 2013 http://www.nobelprize.org/nobel_prizes/chemistry/laureates/2013/)]. Given the existence of Moore's Law [11] and the continuous increase of computer power (see Figure 5.1) it was possible for computational chemistry to

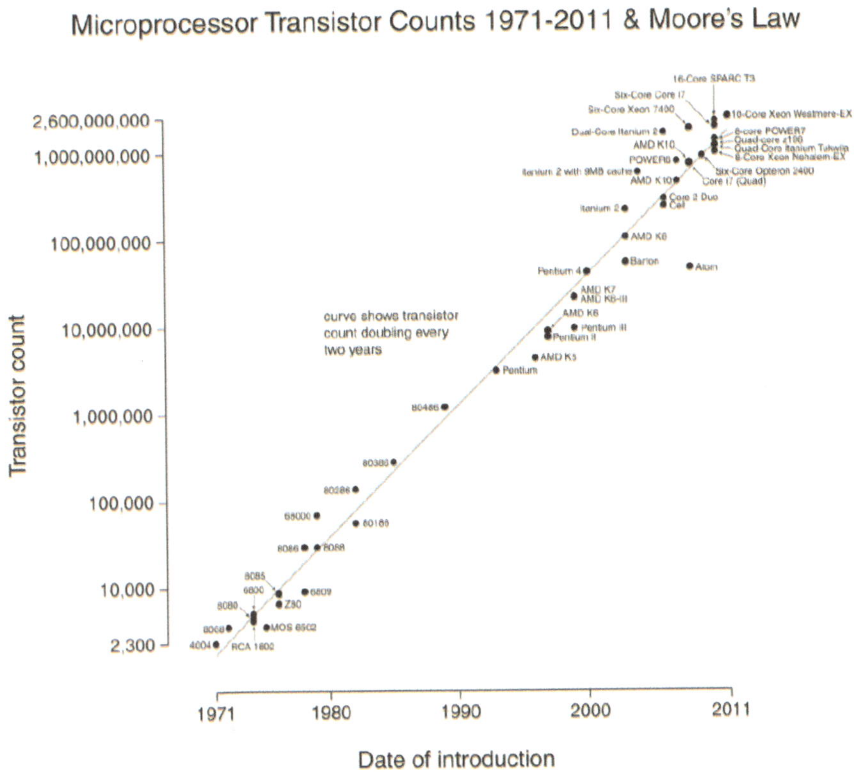

FIGURE 5.1 Moore's Law: the Number of Total Transistor (and, Consequently of Flops) Present in each Chip Doubles Almost Yearly.

undergo an equally fast paced evolution in both the theoretical finesse of the models with which the systems under investigation were studied and their size and complexity as Table 5.1 shows in the case molecular mechanics and biochemical systems [12].

Another point deserving mention in Table 5.1, and which holds true also for quantum chemistry applications, is the extension in the time domain of the phenomena studied as well as the sheer physical size increase of the systems simulated (in terms of number of atoms and/or basis functions or any other relevant criterion). Quantum mechanical or classical simulation can now be performed on thousands or hundreds of thousands of atoms and multi-scale methods allow us to describe in a unified framework phenomena spanning several orders of magnitude in time and space.

TABLE 5.1 Evolution of Size and Complexity of Biological Phenomena Studied by Computational Chemistry Methods

Year	System	Simulated time
1957	Bidimensional rigid disks	10 ps
1964	Monoatomic Liquids	5 ps
1971	Molecular Liquids	10 ps
1971	Melted Salts	10 ps
1975	Simple polymers	20 ps
1977	Small protein in vacuo	20 ps
1982	Simple membrane model	20 ps
1983	Protein crystal	2 ps
1986	DNA in water	100 ps
1989	Complex DNA-Protein	100 ps
1993	Protein/DNA in solution	100 ps
1996	Protein/DNA in solution	10 ns
2000	Protein/DNA in solution	100 ns
2004	Protein/DNA in solution	1 μs
2010	Protein/DNA in solution	1 ms

In addition, computer software for molecular modeling, in particular "black box" programs composed by hundreds of thousands of lines of code, are nowadays used primarily by non-computational specialists. Such a trend may be observed, for example by the number of references to 'DFT' or 'TD-DFT' or 'Gaussian' found on Scopus (Figure 5.2):

This has been possible for three fundamental reasons which hold true for both commercial (e.g Gaussian [7] and academic software (e.g., MolPro and GAMESS):

1. Some applications which with the previous hardware generation were restricted to run on supercomputers are now feasible also for personal workstations or even laptops;
2. User (graphical) interfaces have been developed in order to run/understand/manage/export the data feeded in and obtained by the use of computational chemistry software; however it must be observed that the integration and power of these interfaces usually

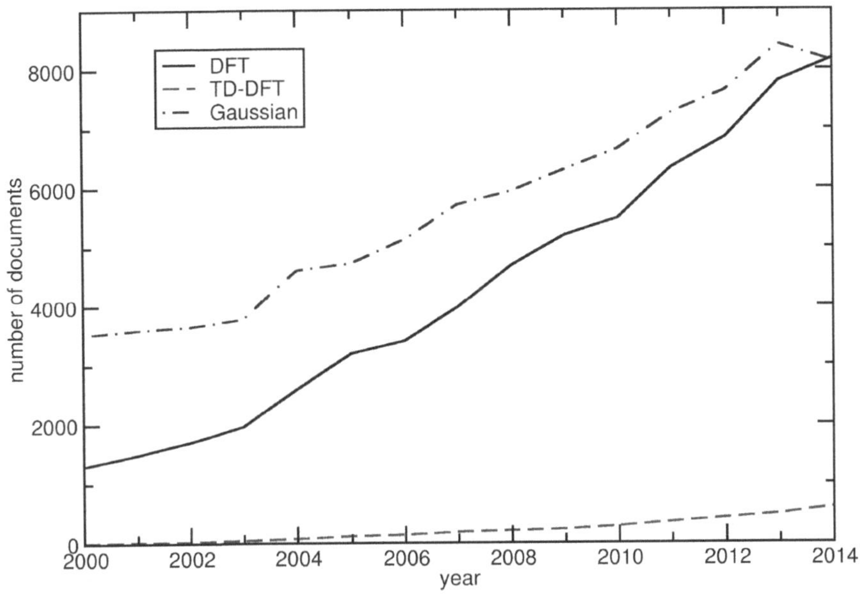

FIGURE 5.2 References to 'DFT' or 'TD-DFT' or 'Gaussian' in chemistry journals found in Scopus between 2000 and 2014.

remains behind that of the "computational engine". Very often users work with text parsers, spreadsheets and script in some high level interpreted language (e.g., Perl, Python) to perform all the tasks needed;

3. There are several supercomputing centers worldwide that manage these software on supercomputers with millions of processors and allow their use to academic or non-academic researchers as a service.

The situation sketched above corresponds to an increasing number of researchers working world wide over an increasing number of problems of increasing complexity, and accumulating an increasing amount of scientific information, a situation by far not unique to molecular modeling or chemistry in general. The repetition of the word "increasing" in the sentence could create the perception of a Babel's tower situation of irreversible disorder if some appropriate change in the way of managing research is not made in order to facilitate data integration, enabling collaboration among the various actors involved and converting data accumulation into scientific breakthroughs. This fact, which could be regarded "just" as important may become critical in a time of grant-driven and project-driven research which often groups into a common effort specialists from several fields, making necessary not only to possess a way to integrate knowledge but also to "transcend specific disciplinary perspectives" [15]. The irreducible multidisciplinary character of modern research has been effectively summed up in a famous image. It showed how complex and intertwined were the connections between scientific disciplines using a map of keywords from 800,000 papers from 776 different areas.

From all the data shown in the precedent paragraphs it is clear that an incremental progress in the capabilities of the "traditional" molecular modeling tool boxes may not be sufficient to ensure an acceptable rate of scientific productivity and to convert this gargantuan amount of information into real knowledge and comprehension. One tool to achieve this conversion, is the integration of existing and new research tools and protocols in what is defined as a *cyberinfrastructure* to carry out this type of *data intensive* research. This is by no means a simple technological upgrade: just like at some point computer became powerful enough to help the spawning of computational sciences and thus changed the way in which overall research is conduct, the widespread development of advanced cyberinfrastructures is likely to have a "cultural" effect on research in several disciplines as already proposed by Hey [19] and synthesized in the book *The Fourth Paradigm* (Hey et al., 2009). In particular, this is true in molecular sciences: the typical bottom up approach (see Section 'Space and Time Domain of Molecular Modeling: Method Integration') used for investigating nanoscopic phenomena involves the integration of knowledge and skills from different areas which need to be integrated in some way.

5.1.2 WHAT IS A CYBERINFRASTUCTURE?

In addition to the "information crisis" cited above, there is yet another problem that we have to consider: even if Moore's Law will still hold for a few years more, this will occur in a way which has changed Information

Technology. The number of transistors on a chip is still growing but they are now distributed in a number of cores yielding limited advancement in single processor peak performance. This is primarily caused by heat dissipation problems on the same core: the power/volume ratio of modern processors (W/m^3) would otherwise exceed that of a nuclear power plant. This change in CPU evolution has a profound effect on computational chemistry since we cannot expect a scale up in system size (time and space) that depends on a serial application unless the problem and the software may be turned in a (massively) parallel application which has always a considerable cost and may not be feasible for theoretical or practical reasons (e.g., a big existing code-base). However, there is also a positive side in the "many core" era: since technological stacks such as web oriented applications, databases and data mining techniques, grid based applications and high performance networks may now be deployed even on a relatively small scale with acceptable effort and cost. These technologies constitute the backbone of massive cyberinfrastructures (CI) in several disciplines such as high energy physics (e.g., The Worldwide LHC Computing Grid | WLCG, http://wlcg.web.cern.ch), genomics (e.g., the Ensembl Genome Browser, http://www.ensembl.org), protein crystallography (e.g., the RCS Protein Databank, http://www.rcsb.org/pdb/home/home.do). Another cyberinfrastructure related to chemistry is the CombeChem project (http://www.combechem.org) related to combinatorial chemistry and drug discovery and which tries to adopt some of the features of eScience paradigm described by Hey [19]. One of the existing cyberinfrastructures with the highest specific emphasis on molecular modeling is the Computational Chemistry Grid (GridChem, https://www.gridchem.org) project which tries to integrate hardware resources distributed across the US to deliver computational resources to a wide community and integrating "traditional" computational chemistry tools with data management technologies and workflow facilities.

Indiana University has developed one of the most widely cited definitions of cyberinfrastructure: "a set of computing systems, data storage systems visualization environments, and people all linked together by software and high performance networks, to improve research productivity and enable breakthroughs not otherwise possible". The term 'cyberinfrastructure' was coined in the late 1990s, it was used in a press briefing on

Presidential Decision Directives 63 by the President of the United States on May 22, 1998. But only in 2003 was popularized by The National Science Foundation (NSF) with the publication of "Revolutionizing Science and Engineering Through Cyberinfrastructure: Report of the National Science Foundation Blue-Ribbon Advisory Panel on Cyberinfrastructure" by Atkins et al. [25] described the technology substrate of Cyberinfrastructure, involving the following components:

- **High-end general-purpose computing centers** that provide supercomputing capabilities to the community at large.
- **Data repositories** that are well curated and that store and make available to all researchers large volumes and many types of data, both in raw form and as associated derived products.
- **Digital libraries** that contain the intellectual legacy of researchers and provide mechanisms for sharing, annotating, reviewing, and disseminating knowledge in a collaborative context.
- **High-speed networks** that connect computing resources, data repositories, and digital libraries.
- **Software and services** for specific community (e.g., chemical data management and analysis).

The National Science Foundation (NSF) has made cyberinfrastructures a central theme in its plans for developing and delivering tools to enhance scientific discovery, and has set out an aggressive set of plans for development of cyberinfrastructure as a national discovery environment [26].

In 2004, the NSF funded a project for the development of a CI for computational chemistry community. GridChem, cited above, provides a collection of grid-based resources to routinely run chemical physics applications and integrates a desktop environment into an infrastructure for a specific community of users. The GridChem client, is a Java desktop application that provides an interface to integrate the hardware, software and middleware resources necessary to solve quantum chemistry problems using grid technologies. Currently the GridChem Science gateway integrates pre- and post- processing components with the high performance applications and provides data management infrastructures. The authentication system is based on proxy certificates that are used for various services where authentication is required.

The national cyberinfrastructure in the USA is not a single entity. Extreme Science and Engineering Discovery Environment (XSEDE, https://www.xsede.org). The Open Science Grid (OSG, http://www.opensciencegrid.org) projects are the middleware of the infrastructure, providing access to CI computers and clusters. Service organizations use the resources of XSEDE and OSG to create unified user experiences for different research communities. Other NSF Cyberinfrastructure Projects includes:

- EarthCube (http://earthcube.org) is a new research initiative of the U. S. National Science Foundation, with the mission to develop community-guided cyberinfrastructure integrating data, models and other resources across geoscience disciplines.
- ICME, i.e., the Integrated Computational Materials Engineering (National Research Council, 2008) cyberinfrastructure provides storage, access, and computational capabilities for study a variety of materials used in multiscale engineering and design.
- nanoHUB.org (https://nanohub.org) is a science and engineering gateway comprising community-contributed resources and geared toward educational applications, professional networking, and interactive simulation tools for nanotechnology.
- Vhub.org (https://vhub.org) is a CI for collaborative volcano research and risk mitigation.
- Ocean Observatories Initiative (http://oceanobservatories.org) manages and integrates data from all OOI sensors, linking marine infrastructure to scientists and users.
- GriPhyN (Grid Physics Network in ATLAS http://www.usatlas.bnl.gov/computing/grid/griphyn/) for sharing high-energy physics data from single, large data sources.

5.1.3 CHAPTER SUMMARY

Obviously, pure "number crunching" software (e.g., GROMACS [35] or Orca [36]) is a key component of a cyberinfrastructure; however its integration in it depends on the implementation of the data fruition and management aspects and for this reason this chapter will deal mainly with the other parts which constitute a cyberinfrastructure.

In general such a cyberinfrastructure may be considered as the integration of computational chemistry techniques and tools with the technologies listed in Section 5.1.2. For this reason the chapter is organized as follows: in Section 5.2 a very brief resume of what is molecular modeling and what are its physical basis is given, trying to explain the reasons to create a cyberinfrastructure. Section 5.3 deals with some of the technologies which underpin the development of a cyberinfrastructure for molecular modeling focusing on data fruition and molecular graphics. Section 5.4 shows a possible implementation of these technologies illustrating a number of case studies.

5.2 MOLECULAR MODELING: SIMULATING THE INTERACTIONS BETWEEN MOLECULES, CHEMICAL SUBSTANCES, AND LIGHT

The purpose of this section is to list in the first part (very briefly) the various types of molecular modeling approach and to sketch in the second one how they are complementary used and why their integration in a dedicated CI may be of great interest. We will not give a detailed presentation of the mathematical foundations of the various methods listed since it is not the focus of this chapter and because there is plenty of excellent textbooks on the various topics [36–41].

The term Molecular Modeling includes a set of algorithms and techniques based on the principles of quantum mechanics and statistical thermodynamics applied to investigate the properties of chemical substances and reactions. From a methodological point of view it is a branch of computational sciences and it is composed by software, the algorithms it includes and the mathematical models on which the algorithms are based. The approach of a computational chemist as compared to an experimental one may be summarized in the flowchart shown in Figure 5.3.

Molecular modeling is applied to make predictions about a wide number of properties, such as:

- predict molecular geometries and conformations, i.e., the (different) possible shapes of molecules and their relative energies;
- predict chemical reactivity by individuation of electrophilic or nucleophilic sites in different conditions;

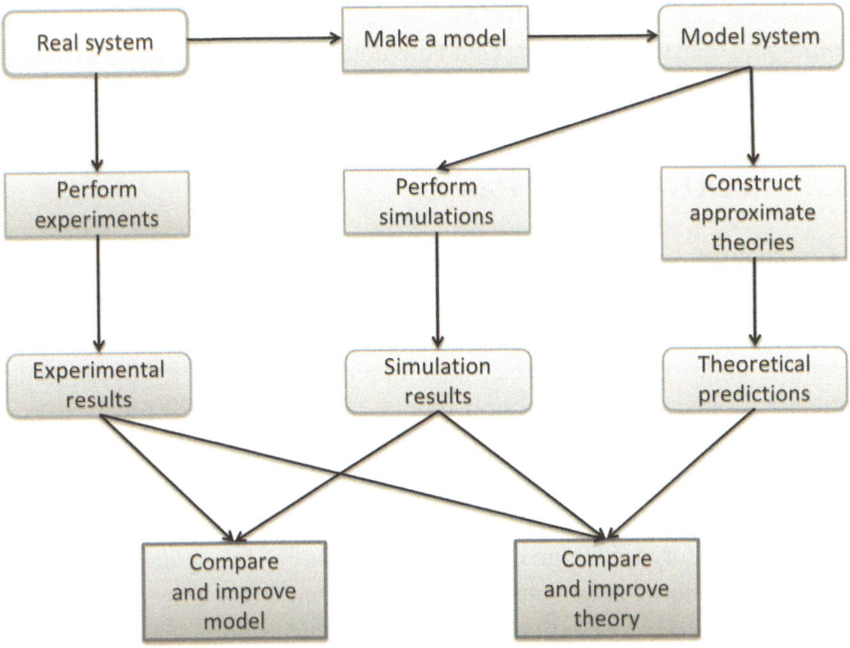

FIGURE 5.3 Computational workflow as compared to an experimental one in molecular sciences.

- predict the bulk properties of a substance, e.g., the density vs. temperature in a given phase or the dielectric constant;
- predict how the environment may affect the properties of a molecule, e.g., by changing the dipole moment;
- prediction of IR, microware, UV or NMR spectra;
- prediction of the mechanical properties of large molecular assemblies over a given span of time, e.g., the folding or unfolding of proteins or the interactions between drugs and their targets.

Molecular modeling is useful in several types of applications, including drug discovery, structural biology, organic synthesis, nanotechnology and material science and it is often used to predict properties or to make mechanistic hypotheses used for the interpretation of experimental data.

It is often (not always) cheaper than experimental methods and in many cases it does not involve the risks and hazards connected with the former. Based on the type of approximations introduced to solve Schrödinger's equation molecular modeling methods may be divided in three very broad categories: classical (Molecular Mechanics, MM), wave function (e.g., *ab initio*) and density functional (DFT) methods which may be applied over a wide and intertwined range of space and time domains.

5.2.1 MOLECULAR MECHANICS

In molecular mechanics methods, electrons are included in completely implicit way and Schrödinger's equation is replaced by a classical description of the motion of atoms. These atoms are actually point charges endowed with masses and other parameters to approximate nuclear repulsion and dispersion forces; bonds are described by means of harmonic or Morse springs; bond angles and torsions are included by means of additional potential terms. The forces with which the atoms interact depend only on their positions and on a reduced set of parameters called a force field; no chemical reactions can take place and purely quantum phenomena (e.g., tunneling) are completely disregarded. The computational cost of typical MM calculation usually scales as N^2 or better depending on the type of approximations used to deal with long range effects.

MM methods are often applied in the framework of Molecular Dynamics (MD) or MonteCarlo (MC) simulations in which the low computational cost is coupled with the ability of these techniques to explore the phase space of a system and thus to achieve a good statistical description of complex systems with many degrees of freedom (see Table 5.2). Molecular docking is also usually based on MM type approximations plus additional techniques to improve the conformational search.

TABLE 5.2 Computational Cost Scaling with the Respect to the Number of Atoms for Hartree Fock and Post Hartree Fock Computations

Method	Scaling Behavior
HF	N^4
MP2	N^5
MP3, CISD, CCSD	N^5
MP4, CCSD(T)	N^7
MP5, CISDT	N^8

5.2.2 WAVE FUNCTION METHODS

Electronic structure methods are based on the law of quantum mechanics rather than classical physics and they instead of depends on a number of empirical parameters (the force field). They are based, in principle, on limited number of fundamental physical constants such as the electronic and nuclear masses and charges and Planck's constant. Quantum mechanics states that all the properties of a molecule may be obtained by solving Schrödinger equation and obtaining the system's energy and wave function. However, the Schrödinger equation can be solved exactly in a very limited number of scenarios. Thus approximations are used; the less serious these are, the "higher" the level of the *ab initio* calculation is said to be. There are two major classes of electronic structure methods: semiempirical methods, which solve an approximate form of the Schrödinger equation that depends on having appropriate parameters available for the type of chemical system under investigation (from which the definition of semiempirical) and *ab initio* methods, are based solely on the laws of quantum mechanics. Electronic structure methods, in particular, highly correlated methods, are able to deliver a whole lot of information about chemical reactivity, spectroscopic properties and other effects that are completely beyond the scope of MM. However this comes at high computational cost, as shown in Table 5.2: the basic electronic structure approach, the Hartree Fock method (which is seldom an appropriate choice to describe e.g., organic molecules or reactions), scales as N^4 and the cost of post Hartree Fock methods grows steadily.

This implies that the use of ES methods is restricted the applications involving a relatively small number of atoms (and electrons!) over (for time dependent phenomena) a limited time span.

5.2.3 DENSITY FUNCTIONAL METHODS

The basis for Density Functional Theory (DFT) is the proof by Hohenberg and Kohn that the ground state electronic energy is determined completely by the electron density ρ, i.e., there exists a one-to-one correspondence between the electron density of a system and the energy. This functional may be divided into four parts: kinetic energy, Coulombic attraction between the electron and between the electrons and nuclei and attraction between the nuclei and an exchange correlation functional. The first three functionals are known and easily described, but the Exchange-correlation functional is not. Thus, while the H-K theorem proves that a functional must exist, it's exact composition is non-trivial and presently unknown. In place of an exact solution, approximations are made to the Exc term which give rise to different levels of theory. DFT methods use the same basis functions as HF methods and apply a similar convergence (SCF) optimization protocol. DFT methods often scale as N^4 or better, the same cost of HF while giving usually better results and thus yielding a good balance between the accuracy of the results and the computational cost.

5.2.4 SPACE AND TIME DOMAIN OF MOLECULAR MODELING: MULTISCALE APPROACHES

The main point of the precedent paragraphs is to underline the different degrees of accuracy of physical information, amount of statistical sampling and complexity of the studied systems that can be attained with different molecular modeling approaches. In other word the same complex molecular system can be studied from a number of different perspectives to address different physical properties: in short, with molecular modeling is possible to describe a set of phenomena spanning several orders of magnitude over time and space as shown. Figure 5.4 shows a "bottom-up" envisioning of a complete molecular modeling framework in which the higher level, first principle approaches are used to build a solid and physically consistent basis that can constitute the reference for the application on less detailed models which are used to study more complex phenomena.

If the proper corrections are made to switch continuously between the different layers that compose the complete physical model it is possible to integrate multiple computational techniques into a unified computational protocol, i.e., applying a *multiple scale* approach over time or space or both. This unification can take place either with a unique suite of computer codes (e.g., ONIOM [42] calculations in Gaussian) or wrapping up different specialized software at a higher level "meta package" taking advantage of widespread used codes and concentrating efforts in the communication and integration part (a strategy used for example by ChemShell,

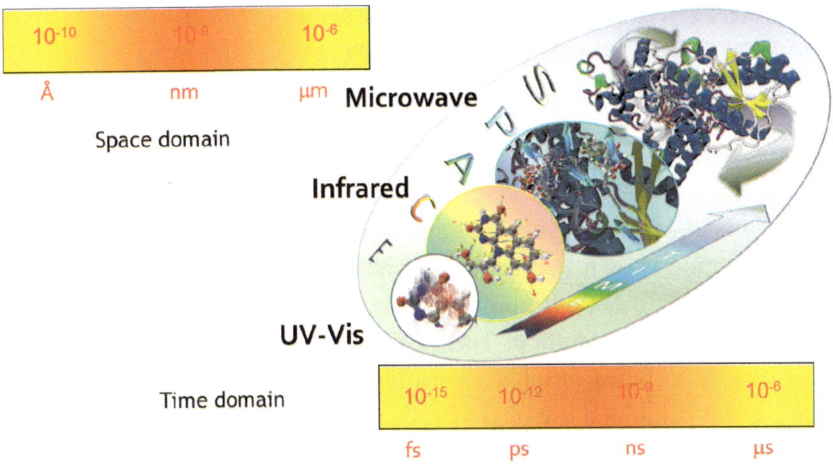

FIGURE 5.4 Pictorial representation of time and space domains of phenomena that can be studied with molecular modeling: a bottom up perspective. On the fastest and smallest end there are interactions between electrons and light (e.g., UV/visible spectroscopic transitions) which are usually investigated by means of *ab-initio* or DFT methods; nuclear motion within molecules, i.e., molecular vibrations is connected with IR spectroscopy and is still studied by quantum chemistry approaches by may also be described by classical methods if very accurate force fields are available; the internal degrees of freedom of large macromolecules over timescales from nanoseconds to hundreds of nanoseconds constitute the target of MD or MC studies with classical atomistic force fields, while larger systems (e.g., protein assemblies) or longer phenomena (phase transitions, protein unfolding) may involve further approximations such as the use of coarse graining techniques.

http://www.stfc.ac.uk/SCD/research/app/ccg/40495.aspx); both approaches have their own advantages and drawback. There is another aspect of method integration which concern the different or integrated tools used to perform preliminary (pre-processing) or final (post-processing) analysis over computation data since even this may involve some kind of "scaling up" e.g., if calculating a bulk microscopic property or using velocities autocorrelation functions to estimate infrared spectra. The main point of this discussion is to show that a research project involving a significative amount of molecular modeling will likely involve several experts in different fields (including experimentalists and computational experts) either just applying existing tools of exploring new methodologies (i.e., conducting some software development work in computational chemistry) a situation that could greatly benefit from the development of a dedicated cyber infrastructure.

5.3 THE ELEMENTS OF A CYBERINFRASTRUCTURE FOR MOLECULAR MODELING AND COMPUTATIONAL SPECTROSCOPY

5.3.1 DATA FRUITION

The graphical display of information has been one of the oldest forms of communication known to man, having its origins in cave drawings dated as early as 30,000 B.C. It allows to communicate clearly and quickly a large amount of information. Tufte [45] gave the definition of the data visualization as "data graphics visually display measured quantities by means of the combined use of points, lines, a coordinate system, numbers, symbols, words, shading, and color". It's much easier for our brain to recognize patterns and comprehend an image rather than numbers and words. The increasing accessibility and quantity of data requires effective ways to analyze and communicate the information that datasets contain in simple, easy-to-understand formats. Visualization serves two major purposes, data presentation and data analysis and thus, in the context of a *data intensive* research environment, data sharing. Since most of the research activity is communicated in form of short "manuscripts" data presentation is one of the most important components of research since visualizing data is the most effective way to communicate it to others. Using visualization for data analysis allows us to discover hidden information from datasets. Raw data becomes useful only when we apply methods of deriving insights from it. Large data sets present specific scaling problems due to dimensionality and heterogeneity of sources and items, and deriving knowledge from these data set is impossible without visual tools.

The human ability to process visual information can augment analysis, especially when analytic results are presented in iterative and interactive ways. Visual analytics, the science of analytical reasoning enabled by interactive visual interfaces, can be used to synthesize the information content and derive insight from massive, dynamic, ambiguous, and even conflicting data. Suitable fully interactive visualizations enhance one's ability to interpret and analyze data. Researchers can thus detect the expected and discover the unexpected, uncovering hidden associations and deriving knowledge from information. As an added benefit, their insights are more easily and effectively communicated to others. Kelleher and Wagener [48] drew up the 10 guidelines to follow for an effective data presentation (see Figure 5.5):

1. create the simplest graph that conveys the information you want to convey.
2. consider the type of encoding object and attribute used to create a plot.
3. focus on visualizing patterns or on visualizing details, depending on the purpose of the plot.

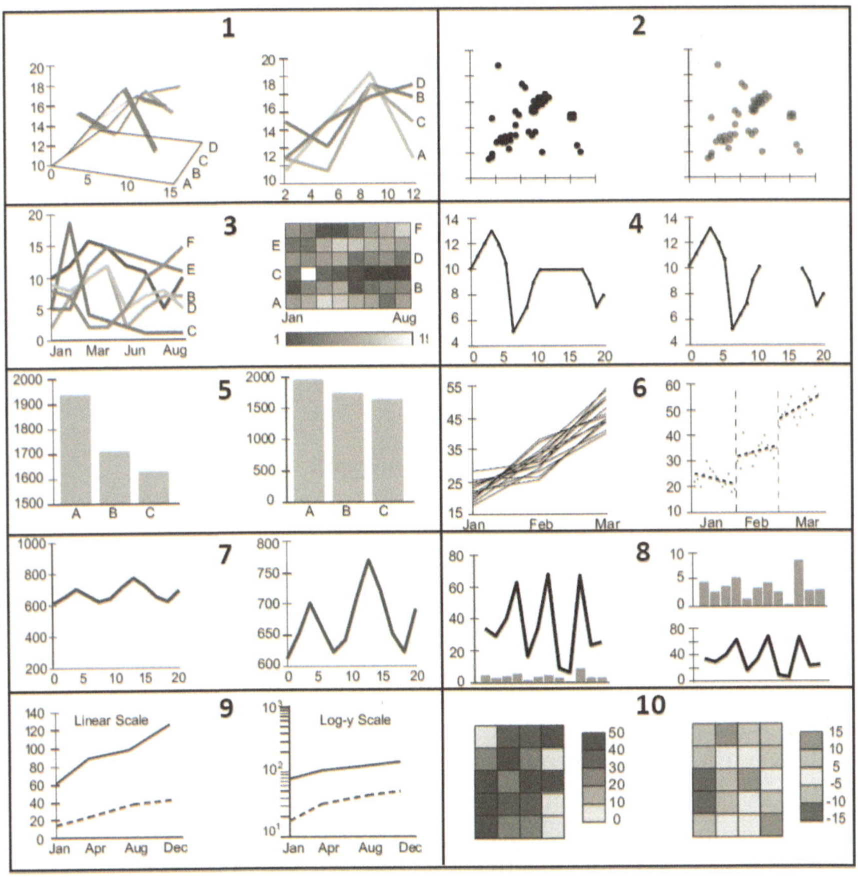

FIGURE 5.5 Visual examples of the Kelleher's guidelines. (Adapted from Kelleher, C., & Wagener, T. (2011). Environ. Model. Softw., 26, 822–827. © With permission from Elsevier.)

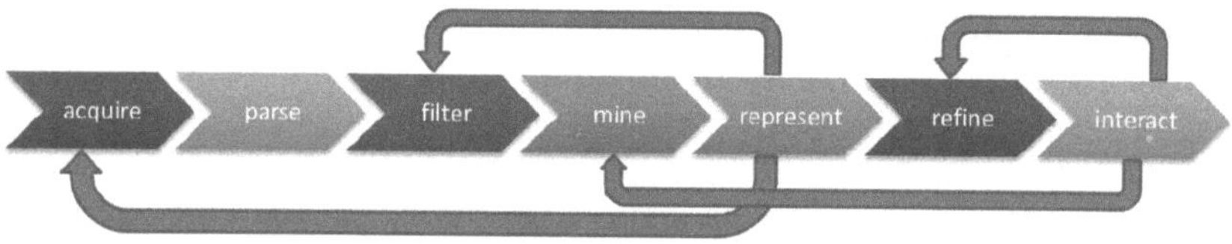

FIGURE 5.6 The seven stages of visualization [48].

4. select meaningful axis ranges.
5. data transformations and carefully chosen graph aspect ratios can be used to emphasize rates of change for time series data.
6. plot overlapping points in a way that density differences become apparent in scatter plots.
7. use lines when connecting sequential data in time-series plots.
8. aggregate larger datasets in meaningful ways.
9. keep axis ranges as similar as possible to compare.
10. select an appropriate color scheme based on the type of data.

Ben Fry [49] describes the seven stages typically involved in understanding data:

- Acquire: obtain the data, whether from a file on a disk or a source over a network.
- Parse: provide some structure for the data's meaning, and order it into categories.
- Filter: remove all but the data of interest.
- Mine: apply methods from statistics or data mining as a way to discern patterns or place the data in mathematical context.
- Represent: choose a basic visual model, such as a bar graph, list, or tree.
- Refine: improve the basic representation to make it clearer and more visually engaging.
- Interact: add methods for manipulating the data or controlling what features are visible.

It was illustrated the process for developing a useful visualization, from acquiring data to interacting with it. An equally important step for an effective is the right choosing of the most adequate visualization design pattern. It will immediately tell users about how the data is organized and what you are trying to communicate about the data. The visualization pattern depends also on the type of data you want to see. Figure 5.7 shows

Data Type	Data patterns
Independent quantities	Bar charts
Continuous quantities	Line graphs Stacked area charts
Proportions	Pie charts
Correlations	Scatterplots Bubble charts
Hierarchies	Tree diagrams
Networks	Diagram maps
Cartographic	Maps
Flows	Sankey diagrams
Heterogeneous	Combination

FIGURE 5.7 Data visualization patterns.

the list of data visualization patterns which can be combined in several patterns to create a single message in which each pattern can then be used to represent specific variables in the data.

The design of the individual elements play an important role in communicating information to others. There are certain visual features in design that will work pre-attentively (Kelleher e Wagener, 2011).These attributes are processed in parallel by the low-level visual system and caught immediately our eye when we look at visualization. Then, the brain filters and processes what is important. Information that has the highest salience (a stimulus that stands out the most) or relevance to what a person is thinking about is selected for further and more complete analysis by conscious (attentive) processing. These attributes come into play implicitly when we analyze any visualization. In Figure 5.8 a list of the pre-attentive attributes.

5.3.2 *MOLECULAR GRAPHICS*

As cited in the Introduction together with "computational engines" molecular graphics were developed since the birth of molecular modeling, evolving in a sort of extension of the "classical" chemical language

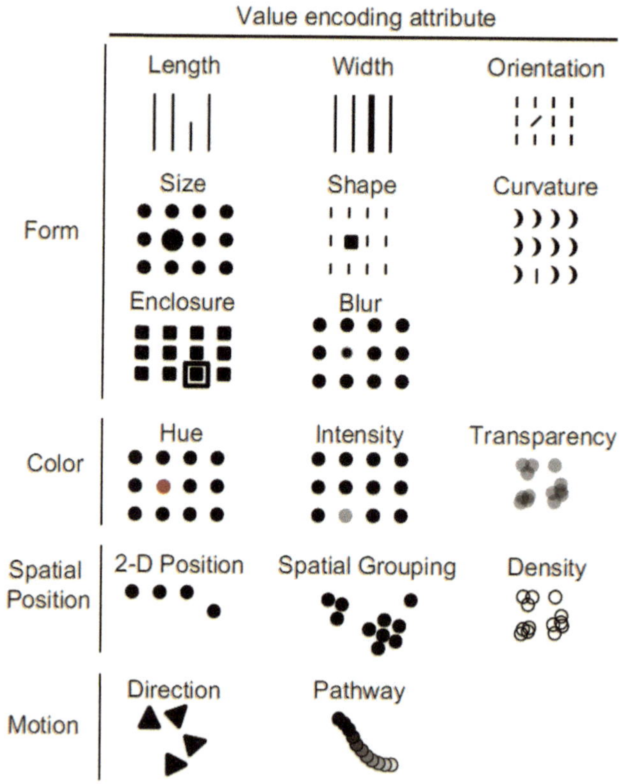

FIGURE 5.8 Different encoding objects can be used in data visualization. (Reprinted from Kelleher, C., & Wagener, T. (2011). Environ. Model. Softw., 26, 822–827. © With permission from Elsevier.)

based on formulas [50]. Nowadays, many good software packages are available for interactive 3-D molecular graphics with focus one several areas of molecular modeling, including biochemistry and structural biology (e.g., VMD [51] or UCSF Chimera [52]), quantum chemistry (Molden, GaussView and ECCE [55]), and crystallography and material science (XCrysDen [56] and PLATON [57]). Many of them provide the ability to extend their functionality through scripts or plug-ins and in some cases may act as front-end for a computational software (e.g., GaussView for Gaussian or VMD [51] for NAMD [58]). The following list includes some of the most commons conventions used in modern molecular graphics to represent molecules and molecular properties, each one coupled with an example from Figure 5.9, obtained with the Caffeine molecular viewer [59].

- All-atoms models: atoms and bonds are represented by combinations of cylinders and spheres with different coloring, shading and dimensions according to their nature and/or to highlight some particular subset of atoms.
- "Ribbon" models: for large macromolecules, instead of drawing all the atoms it may be convenient to highlight a supramolecular

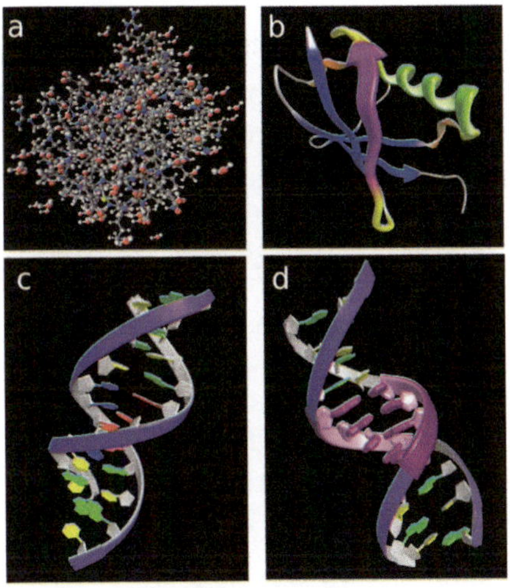

FIGURE 5.9 Different representation conventions for biopolymers: (a) Ball-and-stick representation of the Ubiquitine protein from a crystallographic structure at 1.8 Å (PDB ID 1UBQ). Atom colors, radii and bond thickness are based on the atomic number. (b) Ribbon representation of the same protein; the secondary structure motifs are easily observable and ribbon thickness and color are customized to highlight different secondary structures (thin: coil; thick: α helices, in green or 3_{10} helices, in orange). β-sheets (blue or violet) are represented as arrow pairs with each arrow pointing from the N-terminal to the C-terminal residue in the sheet; color and thickness may also be exploited to distinguish between several regions with the same secondary structure (e.g., the first violet β-strand as compared to the others). Transition between different secondary structure types is highlighted by color and thickness gradients. (c) Ribbon representation of B-DNA duplex of 13 base pairs; the phosphate backbone is drawn with a purple/gray ribbon, narrower or wider at the end to show the different composition typical of terminal nucleosides; bases are drawn to highlight purines or pyrimidines and each base has a different color. (d) Drawing of the same double strand with the central four base pairs selected and depicted with different color and thickness.

structural motif (e.g., the DNA double helix) using continuous ribbon shaped lines; again color, size, shading and textures may be combined to convey several layer of information, in combination with all-atoms conventions.

- Isosurfaces and slicing planes: the property under investigation (e.g., Van der Waals surfaces, atomic orbitals, electrostatic potentials) is drawn using a particular isosurface to show, at a constant value, its distribution in three dimensional space or the average along one axis for a slicing plane. The featured property may convey further data employing color gradients.

With molecular graphics (see Figure 5.10), a sequence of different structures and related properties may be visualized watching the system as it passes from a starting condition to a final one. These "frames" may originate from some type of atomistic simulation thus there is a well-defined temporal relation between all frames; but they can also originate from experiments, e.g., they can be models in which coordinates have

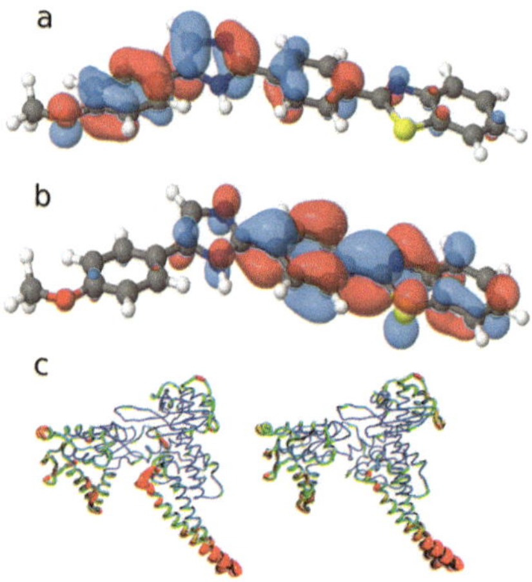

FIGURE 5.10 Different ways of representing different molecular structure *and properties.* (a-b) Highest occupied (HOMO, panel a) and lowest unoccupied (LUMO, panel b) molecular orbitals of the OMe-Phenylene bis imidazole molecule. Orbital surface and phase is represented using transparent isosurfaces of different color. (c) Altered ribbons used to represent the structural disorder (β factor) as calculated from MD simulations from Ref. [60]. The figure compares the changed induced on the b factor upon interaction of protein–DNA complex (the nucleic acid is not shown) with the topotecan drug. Ribbon color (from blue to red) and size are proportional to increasing b factor.

been fitted with respect to some kind of spectroscopic data, and thus represent alternative possible conformations of a molecular system; or they can represent related but different systems, e.g., point mutations of the same protein.

5.3.3 DATA MANAGEMENT AND INTEGRATION

Modern research is more than ever dealing with data management. Powerful computers and complex software analysis produce a very large amount of data, heterogeneous in terms of formats and meanings, but often correlated and fundamental for the success of the research. Management and integration are not trivial tasks, in the everyday life of a modern researcher, since data sets may be intrinsically different from each other and applications must deal with the a large variety of data. In particular, according on how they are organized, we can categorize the data as follows:

- Simple structured data: these data can be organized into simple tables that are structured based on business rules.
- Complex structured data: this kind of data is complex in nature and is suited for the object-relational features of the Oracle database such as collections, references, and user defined types.
- Semi-structured data: this kind of data has a logical structure that is not typically interpreted by the database. For example, an output file of a quantum mechanics program document that is processed by your application or an external service, can be thought of as semi-structured data.
- Unstructured data: this kind of data is not broken down into smaller logical structures and is not typically interpreted by the database or your application. A photographic image stored as a binary file is an example of unstructured data.

In order to create specialized search engines, data mining tools, and data visualization tools, the data will need to be annotated with relevant "metadata" giving information as to provenance, content, conditions, and so on [17]. This is a very important process in the integration phase, because it offers to other users a simple way for the comprehension of the data-sets. Metadata are, of course, specific in different fields and studies, but they are the key for sharing and analyzing data across different domains, designing software with custom functions that operate on data, but driven by metadata. As an example, we can consider what usually happens in many Computational Chemistry laboratories, where most of the researchers write their own scripts for extracting and handling the information suitable for their work. However, these scripts are often too tailored to specific needs to be employed in a wider community of users. In fact, Alsberg et al. [61] proposed a more general approach to manage quantum chemistry results where extracted data tables can be further processed and analyzed. Chemical data contained in output files from different quantum chemical programs are automatically extracted and incorporated into a relational database (PostgreSQL, http://www.postgresql.org). This type of data management is particularly suited for projects involving a large number of molecules.

Another significant aspect related to the datasets, is the Data Integration, which studies how to combine data from heterogeneous data sources in a single context. The researchers work with different instruments, producing a large amount of heterogeneous formats, including relational databases, text files, XML files, spreadsheets and images. This information is then stored in storage systems, with their own indexing and data access methods, often very specific for that particular research activity.

In 2003, Lincoln Stein in his Integrating Biological Databases paper [62] wrote that "life would be much simpler if there was a single biological database, but this would be a poor solution. The diverse databases reflect the expertise and interests of the groups that maintain them. A single database would reflect a series of compromises that would ultimately impoverish the information resources that are available to the scientific community. A better solution would maintain the scientific and political independence of the databases, but allow the information that they contain to be easily integrated to enable cross-database queries. Unfortunately, this is not trivial".

Various standardized data access (APIs ODBC, JDBC, XQJ, OLE DB and ADO.NET) have been developed to offer a specific set of commands to retrieve and modify data from a generic data source. There are also standard formats for representing data within a file that are very important to information integration. The best-known of these is XML, which has emerged as a standard universal representation format. The National Science Foundation initiatives Sustainable Digital Data Preservation and

Access Network Partner (DataNet) are intended to make data integration easier for scientists by providing cyberinfrastructures and setting standards. The first two steps are going to be *Data Conservancy* and *DataONE*.

Data Conservancy focuses on connection of systems into infrastructures through a program informed by user-centered design and research, sustained through a portfolio of funding streams, and managed through a shared, coordinated governance structure. Built on existing exemplar scientific projects, communities and virtual organizations that have deep engagement with scientists and extensive experience with large-scale, distributed system development. The goal of Data Conservancy is to support new forms of inquiry and learning that address important research challenges. Data Conservancy will also accomplish this goal through creation, implementation and sustained management of an integrated and comprehensive data curation strategy.

DataONE will link together existing CIs to provide a distributed framework, sound management, and robust technologies that enable long-term preservation of diverse multi-scale, multi-discipline, and multinational observational data. The distributed framework will be composed of Coordinating Nodes currently located at the Oak Ridge Campus, University of California Santa Barbara, and University of New Mexico, and many Member Nodes, located globally. DataONE will also provide an Investigator Tool Kit that will provide the DataONE users community with tools for accessing and using DataONE efficiently.

Advances in semantic expression, interoperability and integration, reasoning and validation, multilingualism and copyright issues, provided by powerful, automatic, and easy to use tools may lead to change in cultural data management and help the core of the community embrace semantic technologies.

5.4 FORGING SOFTWARE TOOLS INTO A CYBERINFRASTRUCTURE FOR MOLECULAR SCIENCES

5.4.1 SPACE AND HARDWARE DISTRIBUTION

We talked about what an CI could do, what can be made of and which fields can take advantages from such a structure. Something that we still miss in the discussion is where a CI could be placed in a research center.

We are not going to talk about this in details but a short mention on these topics is due, which concern of high speed networks part of a CI (see Section 5.1.2) before presenting the type of CI for molecular sciences we envision. Considering the complexity of the system and the variety of the hardware involved, we can think the infrastructure as a huge and interconnected network of pieces of hardware, with power consumption and high-speed network needs. Cluster machines may be placed in a single location, for a better management, but this may not the most common case. In fact, most of the institution fights everyday with the lack of space, not considering the costs of maintenance for big data centers. A logical solution could be a distributed architecture, not only from the software point of view, but also for the machines themselves. Different research centers could have part of the cyberinfrastructure, sharing calculation power and costs in a more efficient way, using optical fibers and secure network protocols. This, in our opinion, is an important issue, because the problem of the space could be a limitation in developing proper infrastructures, limiting the possibilities for researcher to extend their facilities. Sharing could be the solution, opening also to close collaborations not only from the research point of view, but also in the management of the resources.

5.4.2 SOFTWARE TOOLS

This paragraph illustrates some of the software tools which constitute the computational and data fruition aspects of a cyberinfrastructure for the creation, analysis and sharing of data related to molecular sciences, which are divided in four categories: computational tools, data visualization, data management and science gateways.

5.4.2.1 Computational Tools

This module is composed by JOYCE [63] and VMS-Comp applications. JOYCE is a tool for the parameterization of intramolecular force fields (FFs) from quantum mechanical data. It reads a trial FF file in which both all the selected (redundant) internal coordinates (RIC) and the associated model functions which define the intramolecular potential are specified.

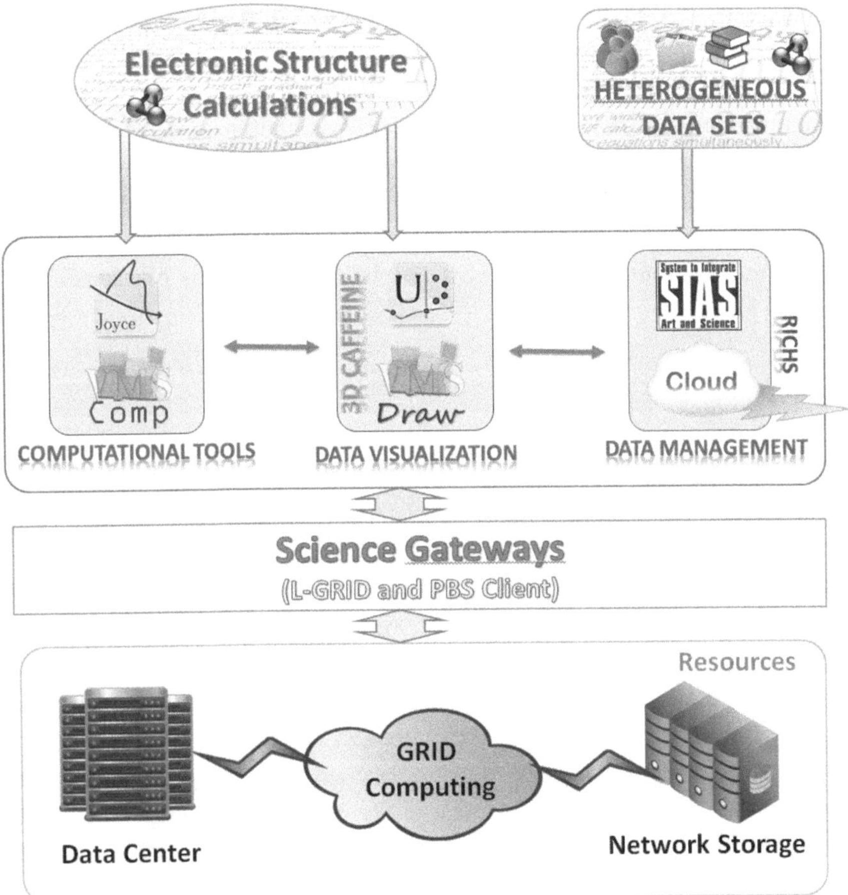

FIGURE 5.11 Our Cyberinfrastructure for analysis and visualization of chemical data.

The equilibrium values of the selected internal coordinates are read by the JOYCE program from a formatted checkpoint file (.fchk) produced by the GAUSSIAN 09 package. Finally, the force constants are computed by the JOYCE procedure from the first and second derivatives again read from the '.fchk' file.

The description of the peculiar chemical traits specifying the system under study encoded in the adopted force field allowing an adequate exploration of the phase space of such large systems can only be achieved with simulation methods based on classical mechanics. The selection of the most appropriate FF for the investigated system is therefore crucial for the reliability of the simulated results.

VMS-Comp is able to read data issuing from different Electronic Structure Codes (ESC, such as Gaussian09 or Dalton [64]) and to compute vibrational, electronic, electronic spin resonance (ESR), or microwave (MW) spectra. It simulates these spectra by links of the Gaussian 09 code or by specific codes for the interpretation of spectroscopies not yet included in Gaussian (e.g., ESR and MW). For details please refer to Section 5.3.

5.4.2.2 Data Visualization

The Data Visualization module contains two powerful multiplatform graphical tools, Ulysses [65] and VMS-Draw [66] able to analyze several types of numerical inputs and direct outputs from the Computational Tools component and to create, model, and save molecular structures, loading data from common file formats.

Ulysses includes an automated wizard, though several choices by the user are necessary to obtain the best FF, starting from the definition of the internal coordinates (IC) to be considered in the process (bonds, angles, dihedrals between either bonded or nonbonded quadruplets, and interacting nonbonded couples), that makes up the molecular 'topology.' Having collected all the QM data (energy, gradient, and Hessian matrix) and the topology interactively defined, Ulysses is able to call the Joyce routine to parameterize the intramolecular FF. When the parameterization procedure ends, the user can export the developed topology in formats (.top and .com) suitable for MD simulations with the Gromacs 4.5 software [35], or for MM/MD studies with Gaussian 09, and validate the FF as he prefers. Ulysses's workflow is described in Figure 5.12.

The ESCs and VMS-Comp results can be directly operated by VMS-DRAW (see Figure 5.13). It guides the user to analyze a number of different spectroscopic observables, and mix and compare them with experimental findings. VMS-Draw is a new all-in-one GUI with the goal of standardizing results, increasing the productivity of both computationally and experimentally oriented researchers, and allowing an easier and faster sharing of results, in all the significant ranges of the electromagnetic field.

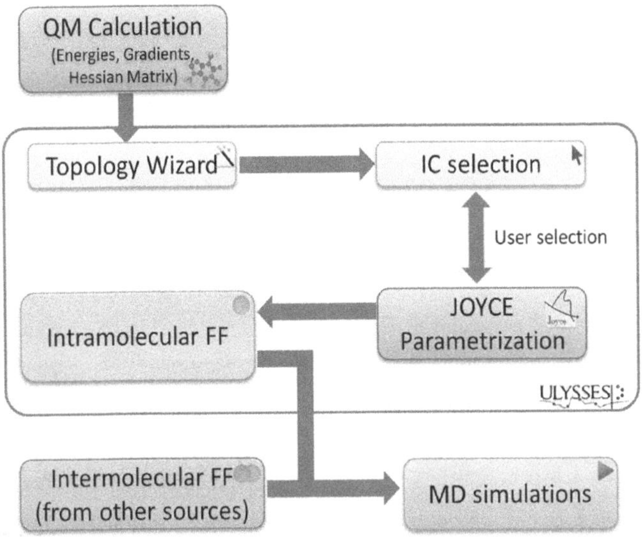

FIGURE 5.12 The user is only asked to provide a Gaussian 09 file containing the energy, gradient, and Hessian matrix of a minimum energy configuration of the selected molecule. Ulysses includes an automated wizard to create the molecular 'topology', the user has to choose the internal coordinates (IC) to be parameterized. Ulysses calls the Joyce routine to parameterize the intramolecular FF. When the parameterization procedure ends, the user can complete the topology file directly Ulysses adding Intermolecular FF which can be easily extracted from online libraries or literature and use it for MD simulations.

Among features not yet available in literature, we can mention the possibility of plotting anharmonic vibrational spectra, vibrationally resolved electronic spectra, Resonance Raman spectra, as well as of normalization, conversion and other manipulations of several spectra at the same time. VMS-Draw provides an integrated environment for direct comparison between different types of theoretical and experimental spectra, allowing several manipulations like shifting and rescaling. Furthermore, several computed spectra can be combined in a weighted mixture to better reproduce the experimental conditions. Special attention is paid to ease of use, generality and robustness for a panel of spectroscopic techniques and quantum mechanical approaches. Depending on the kind of data to be analyzed, VMS-Draw produces different types of graphical representations, including 2D or 3D plots, bar charts, or heat maps. In the current version of VMS Draw, parsers for Gaussian output files are supported,

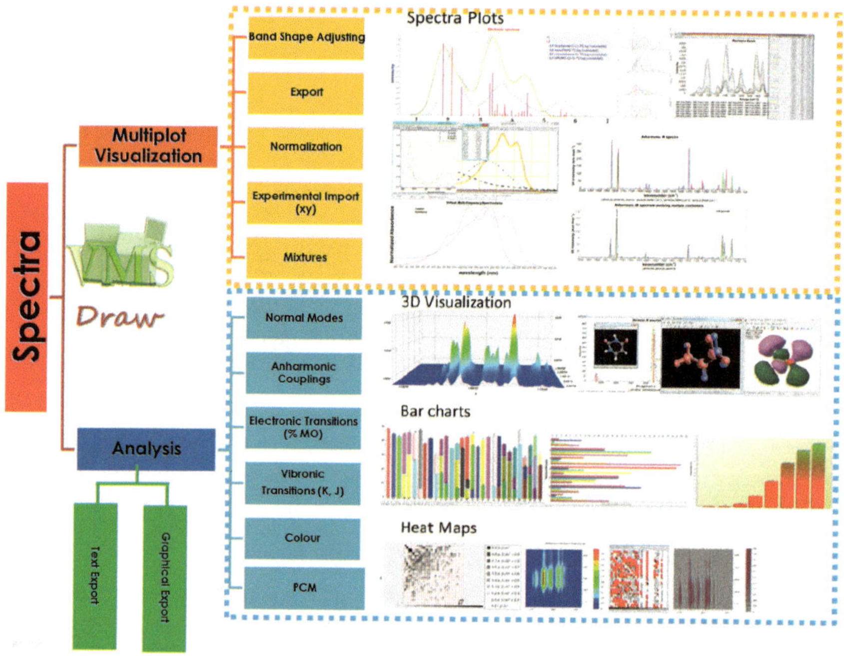

FIGURE 5.13 VMS can be used for two main purposes: for creating high-quality graphics or for analyzing the results of Gaussian 09 QM program through the interactive tools.

however a new format file was designed for use with a generic data source, which is independent of any QM code. Among other integrated features, one may quote the convolution of stick spectra in order to obtain realistic line-shapes or the deconvolution to obtain the stick spectrum from a convoluted one. It is also possible to analyze and visualize, together with the structure, the molecular orbitals, Polarizable Continuum Model (PCM) cavity and/or the vibrational motions of molecular systems thanks to 3D interactive tools.

Another software tool of our CI for molecular sciences is *Caffeine*. The aim of Caffeine project is to develop an integrated system for computational chemistry that will take maximum advantage from VR technologies in order to visualize and model complex molecular structures in a natural and effective way. Although initially designed for immersive VR systems, Caffeine is available for various environments, ranging from standard desktop systems to low cost VR platforms (e.g., the Oculus Rift). Caffeine, in particular, will provide innovative features to assist users along a typical computational path and it will provide state-of-the-art technologies for molecular visualization and user interaction. Currently, only the molecular viewer component is actively developed, for both desktop and immersive Virtual Reality systems [59]. In future developments, it will also allow users to create, model, and save molecular structures, loading data from the most diffused file formats or dedicated DBs. It will also be possible to run a simulation directly from the editor: Caffeine will take care of creating the corresponding job and submitting it to high-performance computing clusters. The results of these simulations, stored in files or in a chemical DB, will then be retrieved into Caffeine, which will provide the tools necessary for a complete analysis of the results. These features are summarized in Figure 5.14.

Caffeine is implemented in C++, and exploits some of the more popular and robust open-source cross platform toolkits, like OpenSceneGraph

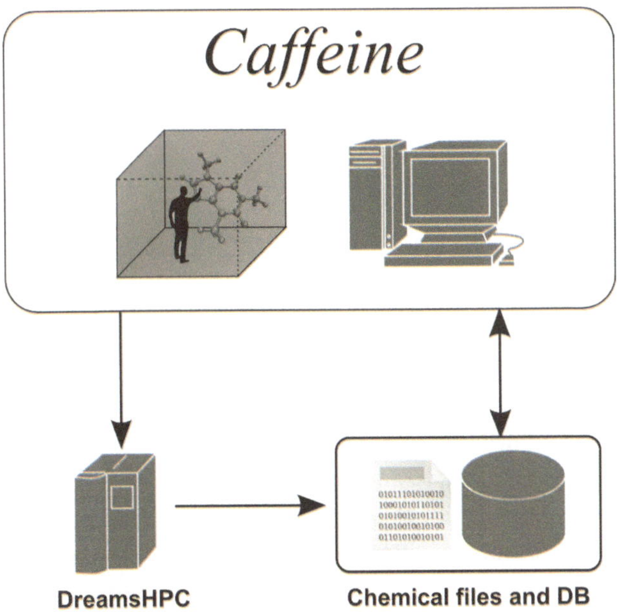

FIGURE 5.14 The aim of the Caffeine project, an integrated system for computational chemistry.

(http://www.openscenegraph.org), Open Babel (http://openbabel.org), Qt (http://qt-project.org) and the ENet Reliable UDP networking library (http://enet.bespin.org). In particular, the rendering module is based on OpenSceneGraph, a high-performance 3-D engine used in fields such as visual simulation, games, VR, and scientific visualization. On top of these toolkits, a set of dedicated software modules have been developed, providing features like:

- A distributed scene graph and modules dedicated to head-tracked stereoscopy and keystoning, which simplify the development of applications for cluster-based immersive virtual reality systems (like CAVE systems).
- GPU-accelerated rendering of spheres and cylinders (employed in the implementation of "all-atoms" representations of molecular structures, as shown in Figure 5.9a).
- Generation of three-dimensional "pathline" geometries (employed in the implementation of ribbon representations, as shown in Figure 5.9b–d).

Altogether, these modules form a library called "Moka" [59]. Moka has been developed as part of the Caffeine project, but it is designed with generality in mind, in the hope of make use of it also in others projects and application fields.

5.4.2.3 Data Management

The Data Management module aims to improve the sharing and the enrichment of heterogeneous data and expertise in a collaborative way through user-friendly tools. It allows a continuous exchange of information between the other three components of CI and it is used to simplify the creation of the complicated input files or to manage complex computational results in the effective and flexible way.

It promotes more efficient use and reuse of information; reduces data redundancy and promotes better data management. Most of computational results are semi-structured and unstructured data, often in large text file. Some databases support the Large Object (LOB) datatypes (http://docs.oracle.com/cd/E11882_01/appdev.112/e18294/toc.htm) which can hold up to 128 TB for single entry and are thus well suited to deal with this type of data. For this reason we have exploited LOBs as implemented in PostgreSQL (http://www.postgresql.org).

Storing data in LOBs enables us to access and manipulate the data efficiently in our application. With the growth of the internet and content-rich applications, it has become imperative that the database support a datatype that:

- can store unstructured and semi-structured data in an efficient manner;
- is optimized for large amounts of data;
- provides a uniform way of accessing data stored within the database or outside the database.

The design of this block is crucial for both effectiveness and reliability of the entire architecture. First, we need to build a high-speed access with fail-safe capabilities on top of the data stack. This is achieved by using an enterprise NAS providing an high-speed storing capability to our system. Then, we designed a framework aiming at managing and enriching what has been produced in the first step. RICHS [65] is a framework aims at the production and use of scientific content in a heterogeneous environment of increasing integration between different disciplines. It is easy to access and understand for various categories of users, offering the possibility of direct interaction with various data. It relies on a middleware Virtual Reality Modeling (VRM), which handles remote input and output infrastructure, converting different sets of information in a unified virtual view can be used by a wide range of different devices and applications. The database contains heterogeneous information such as molecular structures, chemical analysis, computational and experimental results, but also information to assist in the conservation and safeguard of cultural heritage. The data interchange to/from Data Visualization module is provided by standard formats and protocols, according to the paradigm of web-services.

5.4.2.4 Science Gateways

By definition, a Science Gateway is a "community-development set of tools, applications, and data that is integrated via a portal or a suite of applications, usually in a graphical user interface, that is further customized to meet the needs of a specific community". We have developed a pre-processing module in order to support the user in the process of creating complex input to HPC simulations and to interface efficiently with the module Computational Tools. This module supports the Portable Batch System (or simply PBS) allowing users to manage a job's life-cycle on spot and on-demand. Our client PBS, based on SSH protocol and Jamberoo program (Vasilyev, Jamberoo-Cross-PlatformMolecularEditor/BuilderHome http://sf.anu.edu.au/~vvv900/cct/appl/jmoleditor), allows users to transfer files from the local system to the queue system and vice versa. In addition, user can submit a job on selected queue, monitor job's status and retrieve results. The results will be analyzed by our data visualization module as shown in Figure 5.15.

This module include the L-GRID portal [82], a light gateway to access distributed European Grid resources. It provides the control over the complete lifecycle of a Grid Job, from its submission and status monitoring, to the output retrieval. The system, based on the Globus Grid middleware, is implemented as client-server architecture. There is no need of user registration on the server side, and the user needs only his own X.509 personal certificate. It allows to reduce the time spent for the job submission, granting at the same time a higher efficiency and a better security level in proxy delegation and management.

In the grid world it is often necessary for a remote service to act on a user's behalf, e.g., a job running on a remote site needs to be able to talk to other servers to transfer files, and it therefore needs to prove that it is entitled

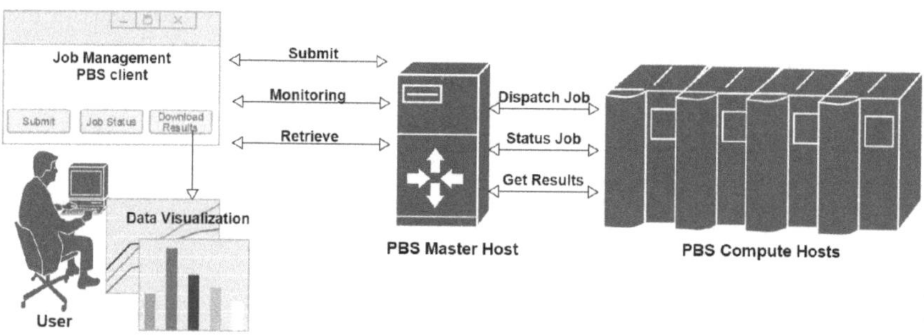

FIGURE 5.15 The user-friendly tool to manage a job's life-cycle on-demand.

to use the user's identity (this is known as delegation). The implementation of a mechanism for delegation dynamic server-side L-GRID allowed us to extend the concept of Grid to the connected clients too. The user client becomes itself part of the Grid infrastructure. The end user needs a valid X.509 personal certificate and an access to the Internet in order to benefit from large amounts of resources. The X.509 personal certificate does not get out from the local machine, strictly compliant to the security policies.

L-GRID provides access to the European Grid Infrastructure (EGI, http://www.egi.eu), which is a publicly funded e-infrastructure put together to give scientists access to more than 370,000 logical CPUs, 170 PB of disk capacity to drive research and innovation in Europe. Resources are provided by about 350 resource centers who are distributed across 56 countries in Europe, the Asia-Pacific region, Canada and Latin America (Figure 5.16).

5.4.3 *A CASE STUDY IN THE DEVELOPMENT OF A CYBERINFRASTUCTURE FOR MOLECULAR SCIENCES*

Our CI is mainly devoted to computational spectroscopy and includes both general utilities for experimentally-oriented scientists and advanced tools for theoreticians and developers. The infrastructure can offer an invaluable aid in pre-organizing and presenting in a more direct way the information produced by measurements and/or computations focusing attention on the underlying physical-chemical features without being concerned with

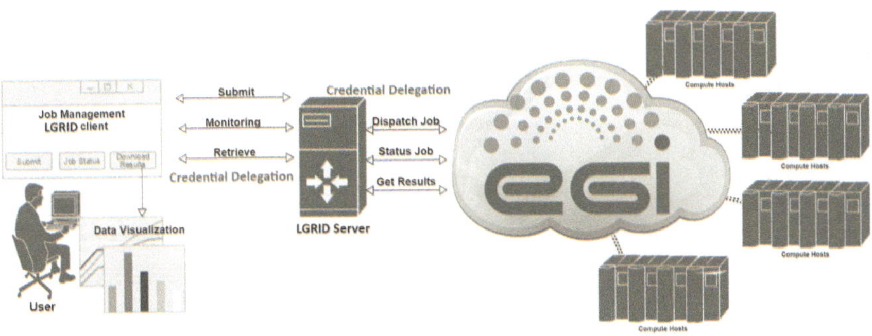

FIGURE 5.16 LGRID's interaction with the EGI Grid infrastructure.

technical details. It supports users in various tasks of research: complex input generation, job management, access to the latest computational tools for spectroscopy, data storage, data analysis, data presentation, and availability and reliability by using cloud-based resources.

We consider mandatory providing user-friendly access to the latest developments of computational spectroscopy also to non-specialists. User-friendly graphical tools are developed to better facilitate creating the complex text-based input files for ESCs, as shown in Figure 5.17. Different ESCs were implemented to simulate the various types of spectroscopies which are analyzed by our data visualization module.

An important consideration for all modern ESCs is their parallel scalability on high-performance computer (HPC) architectures that open up the possibility of simulating very large physical systems entirely *ab initio*. Therefore, it becomes very important to provide transparent access to parallel and distributed computing systems. Thanks to our science gateways, users can access to resources and services of the European Grid Infrastructure or of a workload manager such as the Portable Batch System (PBS) (Figures 5.18 and 5.19).

The computational results are saved in a cloud storage platform that facilitates the reuse and reduces the redundancy of Information.

Concerning the post-treatment of spectroscopic simulations, to the best of our knowledge, none of the available all-in-one packages is flexible

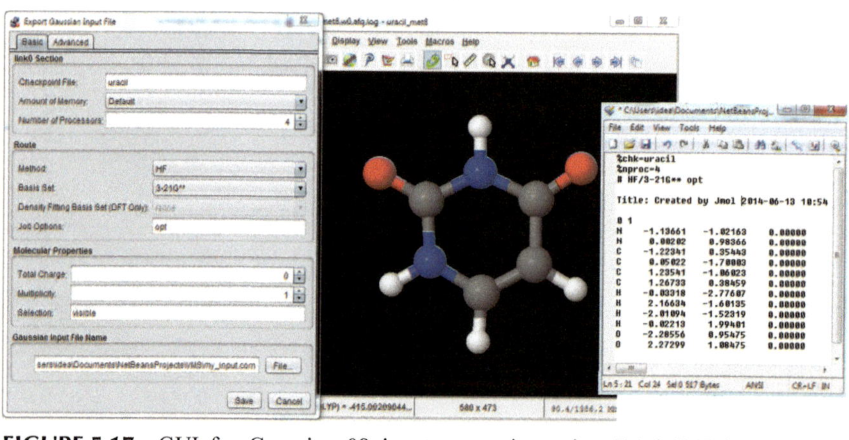

FIGURE 5.17 GUI for Gaussian 09 input generation using JMol [25] integrated in VMS Draw.

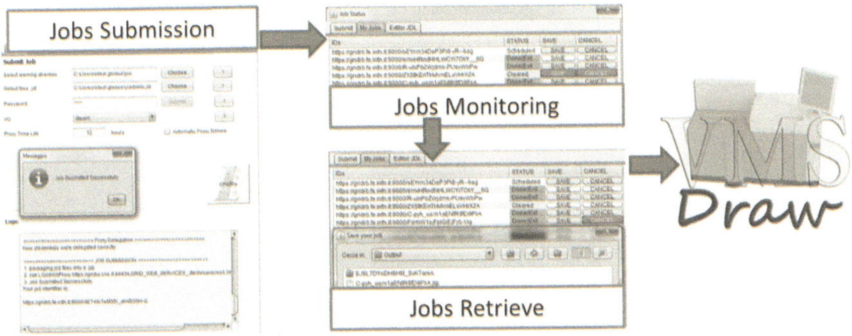

FIGURE 5.18 Science gateways can hide the complexities of distributed infrastructures from researchers.

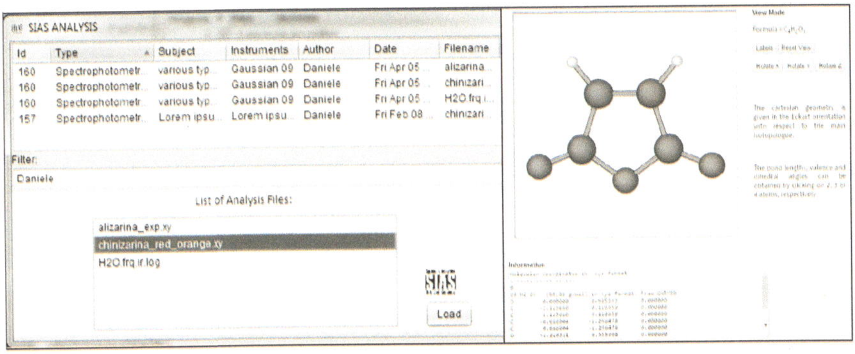

FIGURE 5.19 Users query the storage platform and display the results through visualization tools or web portal. VMS-Draw can retrieve, analyze, and compare the computational results and experimental data stored on our Cloud Storage Platform. In addition, the stored data can be accessed through our web portal.

enough to process the rich information produced by ESCs for excited electronic states. For this reason VMS-Draw has been designed to include several other post-processing tools, including, of course, the plot the contributions of different electronic transitions to the different bands of absorption of emission spectrum (Figure 5.20).

Whenever a Gaussian output file is opened, VMS-Draw is able to compare a reference vibrational spectrum with its computed harmonic and/or anharmonic counterparts. Figure 5.21 shows, as an example, the comparison between the experimental, harmonic, and anharmonic IR spectra of nicotine in chloroform solution. The stick-spectra are superposed to the

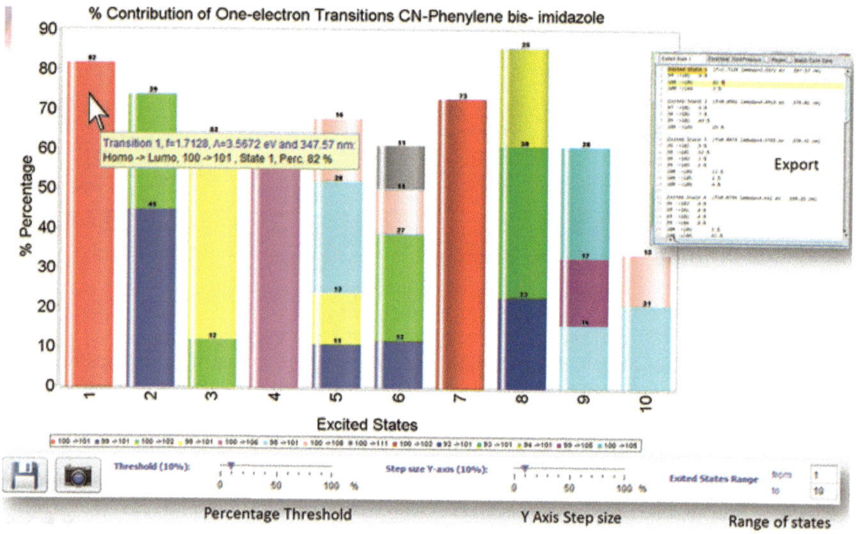

FIGURE 5.20 Contribution of One-Electron Transitions (Percentages) to the one photon absorption of CN-Phenylene bis-imidazole.

convoluted ones, allowing for an easier assignment of the experimental bands since clicking on a peak shows the nature of each excitation.

The Resonance Raman (RR) spectrum of a system can be recorded (and computed) for different incident frequencies, so that a 3D plot showing the dependence of the absorption on both the Raman shift (on the X axis) and the incident frequency (on the Y axis) could be more informative. VMS-Draw is able to perform the convolution of each stick spectrum and also the interpolation along the incident frequency axis producing either a 2D heath map or a 3D plot. An example of 3D spectrum is shown in Figure 5.22.

In anharmonic computations, one of the most effective approaches is based on the perturbative treatment of the molecular hamiltonian (the VPT2 approach), which takes into account the contribution of the third and fourth derivatives of the total energy in a perturbative way. An analysis of the K_{iij} force constants is helpful in order to perform the correct choice of the normal modes which need to be computed and of those which can be safely neglected (Figure 5.23).

Several tools are available to analyze in detail the outcome of anharmonic computations which are described in full detail in Ref [66].

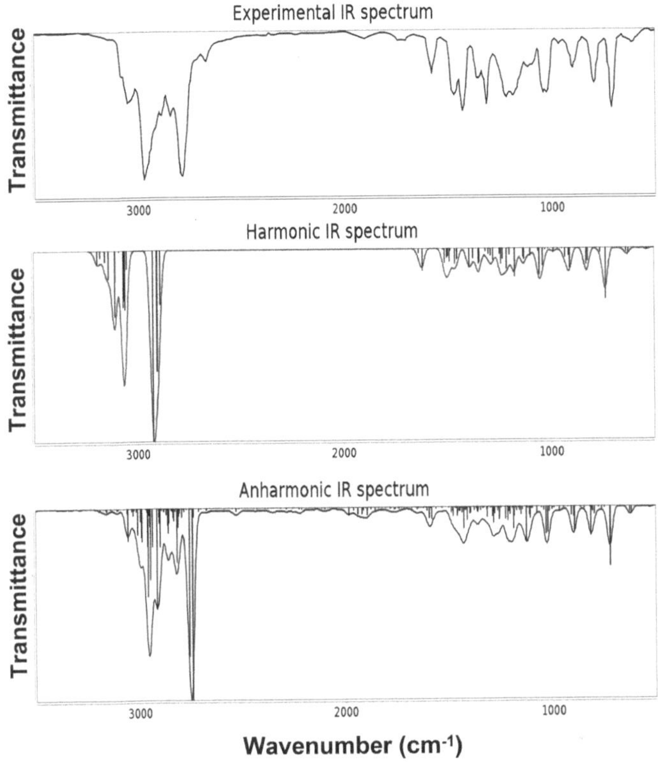

FIGURE 5.21 Experimental and Computed (Harmonic and Anharmonic) IR Spectra of Nicotine in chloroform solution.

One photon absorption and emission involve transitions between vibrational energy levels of two different electronic states. In order to visualize geometry changes accompanying electronic excitations, a tool has been set up, which superposes simplified stick and ball models of the equilibrium geometries of ground and excited states drawn in different colors. Further information concerning the Vibronic contributions can be obtained, including Duschinsky matrix, Shift Vector (in terms of the normal modes of the initial state) and Spectrum Convergence (Figure 5.24).

VMS Draw provides an integrated environment for direct comparison between different types of theoretical and experimental spectra. It offers runtime manipulation of spectra positions and intensities in

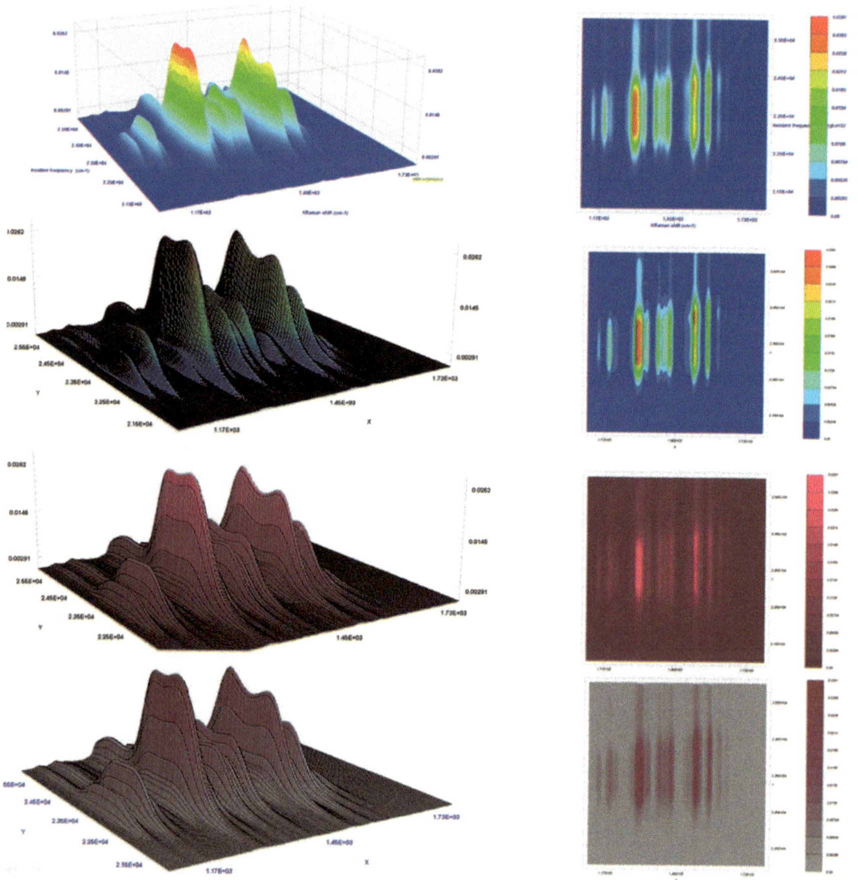

FIGURE 5.22 Resonance Raman spectrum of chlorophyll a1 as a function of both Raman shift (X axis) and incident frequency (Y axis) cm^{-1}.

order to compare more properly experimental and theoretical results. Furthermore, users can perform the deconvolution of an experimental spectrum.

Complex molecular mixtures can be evaluated varying the contributions of single-components. VMS Draw has integrated an algorithm for evaluating the color of a spectrum in the visible region. Finally, it is able to visualize the molecular orbitals and the Polarizable Continuum Model (PCM) cavity of molecular systems thanks to 3D interactive tools and change the iso-surface properties (Figure 5.25).

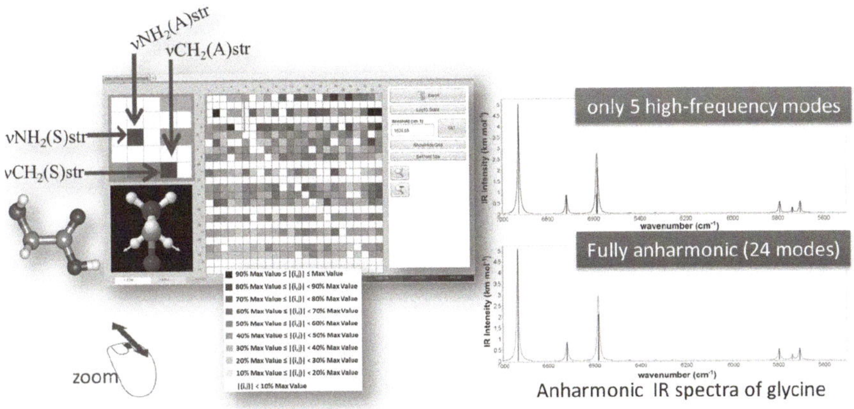

FIGURE 5.23 Heat diagram describing couplings between normal modes (from X matrix) of glycine. The strongly coupled NH2 symmetric and CH2 asymmetric stretching are highlighted by arrows. Fully anharmonic (24 modes) and reduced dimensionality (only 5 high-frequency modes) computed spectra are also shown.

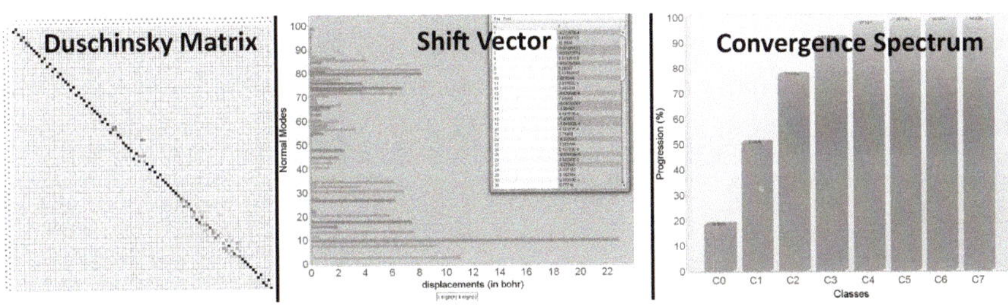

FIGURE 5.24 Analysis of vibrationally resolved electronic spectra.

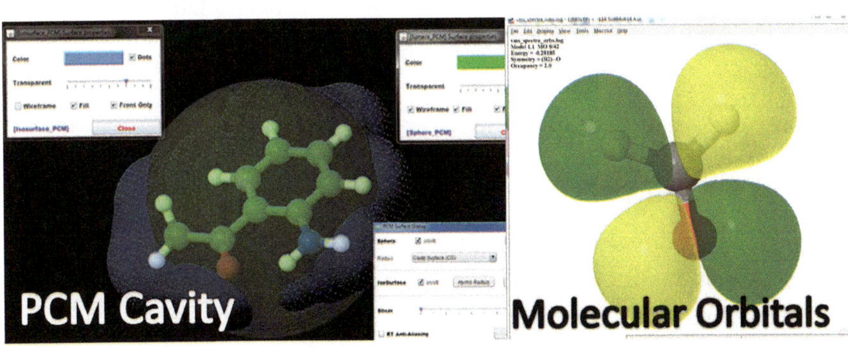

FIGURE 5.25 3D interactive tools.

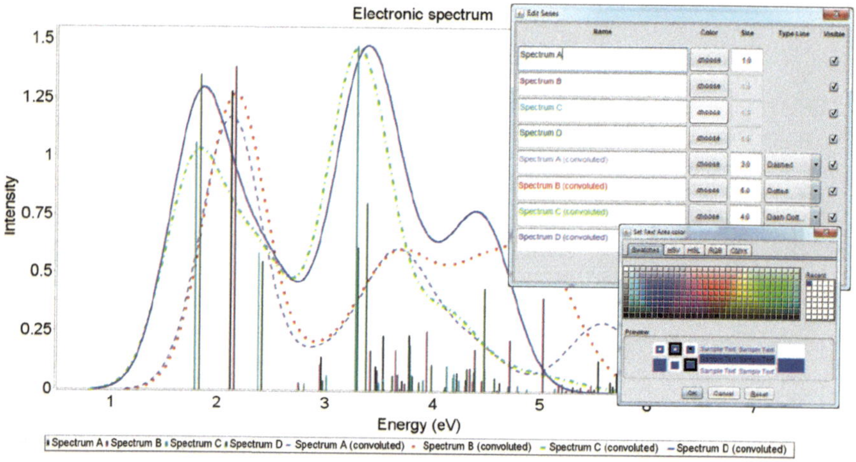

FIGURE 5.26 Highly customizable plotting functionality.

VMS Draw allows users a highly customizable plotting functionality via user-modifiable settings in order to produce publication quality graphs and to communicate complex ideas with clarity, precision, and efficiency. VMS DRAW "visually displays measured quantities by means of the combined use of points, lines, a coordinate system, numbers, symbols, words, shading, and color" [73] (Figure 5.26).

ACKNOWLEDGMENTS

The research leading to these results has received funding from the European Research Council under the European Union's Seventh Framework Program (FP/2007–2013)/ERC Grant Agreement no. [320951].

KEYWORDS

- **computational spectroscopy**
- **cyberinfrastructure**
- **data management**

- **eScience**
- **grid computing**
- **high-performance computers**
- **molecular modeling**
- **multiscale methods**
- **science gateways**
- **visualization systems**

REFERENCES

1. Metropolis, N., Rosenbluth, A. W., Rosenbluth, M. N., Teller, A. H., & Teller, E. (1953). *J. Chem. Phys. 21*, 1087.
2. Schaefer, H. F. *Quantum Chemistry: The Development of ab initio Methods in Molecular Electronic Structure Theory*; Dover Publications: Mineola, NY, 2004.
3. Barnett, M. P. (1963). *Rev. Mod. Phys. 35*, 571.
4. Clementi, E. Computational Chemistry, the "Launch Pad" to Scientific Computational Models: A Personal Account with a Forecast for Future Applications within the Global Simulation Approach. http://www.tacc2012.org/documents/Clementi_AIP.pdf.
5. Gaussian 09, Revision D.01, Frisch, M. J., Trucks, G. W., Schlegel, H. B., Scuseria, G. E., Robb, M. A., Cheeseman, J. R., et al., Gaussian, Inc., Wallingford CT, 2009.
6. ORTEP-III http://web.ornl.gov/sci/ortep/ortep.html (accessed Feb 27, 2015).
7. Ferrin, T. E., Huang, C. C., Jarvis, L. E., & Langridge, R. (1988). *J. Mol. Graph. 6*, 13–27.
8. Journal of Computational Chemistry – Wiley Online Library http://onlinelibrary.wiley.com/journal/10.1002/ (ISSN)1096–987X (accessed Feb 27, 2015).
9. The Nobel Prize in Chemistry 1998 http://www.nobelprize.org/nobel_prizes/chemistry/laureates/1998/index.html (accessed Feb 27, 2015).
10. The Nobel Prize in Chemistry 2013 http://www.nobelprize.org/nobel_prizes/chemistry/laureates/2013/ (accessed Feb 27, 2015).
11. Hill M., Jouppi N., Sohi G., *Readings in Computer Architecture* (Morgan Kaufmann Publishers, San Francisco, ed. 1), 2000, 56.
12. Allen, M. P., & Tildesley, D. J. (1989). *Computer Simulation of Liquids*; Clarendon Press; Oxford University Press: Oxford [England]; New York.
13. Molpro quantum chemistry package http://www.molpro.net/ (accessed Feb 27, 2015).
14. Gordon Group/GAMESS Homepage http://www.msg.ameslab.gov/gamess/ (accessed Feb 27, 2015).
15. Chismar, W., Horan, T. A., Hesse, B. W., Feldman, S. S., & Shaikh, A. R. (2011). *Am. J. Prev. Med., 40*, S108–S114.
16. Marris, E. (2006). *Nature, 444*, 985–991.
17. Hey, T. (2005). *Science, 308*, 817–821.

18. Foster, I. (2005). *Science*, *308*, 814–817.
19. Hey, A. J. G., Tansley, S., & Tolle, K. M. (2009). *The Fourth Paradigm: Data-Intensive Scientific Discovery*; Microsoft Research: Redmond, Wash.
20. Welcome to the Worldwide LHC Computing Grid. WLCG http://wlcg.web.cern.ch/ (accessed Feb 27, 2015).
21. Ensembl Genome Browser http://www.ensembl.org/index.html (accessed Feb 27, 2015).
22. RCSB Protein Data Bank – RCSB PDB http://www.rcsb.org/pdb/home/home.do (accessed Feb 27, 2015).
23. CombeChem http://www.combechem.org/ (accessed Feb 27, 2015).
24. Computational Chemistry Grid https://www.gridchem.org/ (accessed Feb 27, 2015).
25. Atkins, D.E., Droegemeir K. K., Feldman S. I., Garcia-Molina H., Klein M. L., Messerschmitt D. G., Messina P., Ostriker J. P., Wright M. H., *Revolutionizing Science and Engineering Through Cyberinfrastructure*, 2003.
26. LeDuc, R., Vaughn, M., Fonner, J. M., Sullivan, M., Williams, J. G., Blood, P. D., Taylor, J., & Barnett, W. (2014). *J. Am. Med. Inform. Assoc.*, *21*, 195–199.
27. XSEDE, https://www.xsede.org/.
28. Open Science Grid, http://www.opensciencegrid.org/.
29. EarthCube, http://earthcube.org/ (accessed Feb 27, 2015).
30. National Research Council (U.S.). *Integrated Computational Materials Engineering: A Transformational Discipline for Improved Competitiveness and National Security*; National Academies Press: Washington, D.C., 2008.
31. nanoHUB.org – Home https://nanohub.org/ (accessed Feb 27, 2015).
32. vhub – Home https://vhub.org/ (accessed Feb 27, 2015).
33. Ocean Observatories Initiative http://oceanobservatories.org/ (accessed Feb 27, 2015).
34. GriPhyN – Grid Physics Network in ATLAS http://www.usatlas.bnl.gov/computing/grid/griphyn/ (accessed Feb 27, 2015).
35. Pronk, S., Pall, S., Schulz, R., Larsson, P., Bjelkmar, P., Apostolov, R., Shirts, M. R., Smith, J. C., Kasson, P. M., van der Spoel, D., Hess, B., & Lindahl, E. (2013). *Bioinformatics*, *29*, 845–854.
36. Neese, F. (2012). *Wiley Interdiscip. Rev. Comput. Mol. Sci.*, *2*, 73–78.
37. Foresman, J. B., Frisch, Ae., & Gaussian, I. (1996). *Exploring Chemistry with Electronic Structure Methods*; Gaussian, Inc.: Pittsburgh, PA.
38. Cramer, C. J. (2004). *Essentials of Computational Chemistry: Theories and Models*, 2nd ed., Wiley: Chichester, West Sussex, England; Hoboken, NJ.
39. Leach, A. R. (2001). *Molecular Modeling: Principles and Applications*, 2nd ed., Prentice Hall: Harlow, England; New York.
40. Szabo, A. (1996). *Modern Quantum Chemistry: Introduction to Advanced Electronic Structure Theory*; Dover Publications: Mineola, NY.
41. Frenkel, D. (2002). *Understanding Molecular Simulation: From Algorithms to Applications*, 2nd ed., Computational Science Series; Academic Press: San Diego.
42. Dapprich, S., Komáromi, I., Byun, K. S., Morokuma, K., & Frisch, M. J. (1999). *J. Mol. Struct. THEOCHEM*, *461–462*, 1–21.
43. Scientific Computing Department – ChemShell http://www.stfc.ac.uk/SCD/research/app/ccg/40495.aspx (accessed Feb 27, 2015).
44. Principles of Data Visualization – What We See in a Visual – White Paper. FusionCharts http://www.fusioncharts.com/whitepapers/principles-of-data-visualization/ (accessed Feb 27, 2015).

45. Tufte, E. R. *The Visual Display of Quantitative Information*, 2nd ed., Graphics Press: Cheshire, Conn, 2001.
46. Data, data everywhere. The Economist http://www.economist.com/node/15557443 (accessed Feb 27, 2015).
47. *Cyberinfrastructure Vision for 21st Century Discovery*; National Science Foundation, Cyberinfrastructure Council, 2007.
48. Kelleher, C., & Wagener, T. (2011). *Environ. Model. Softw.*, *26*, 822–827.
49. Fry, B. (2008). *Visualizing data*; O'Reilly Media, Inc: Sebastopol, CA.
50. Salvadori, A., Licari, D., Mancini, G., Brogni, A., De Mitri, N., & Barone, V. (2014). In: *Reference Module in Chemistry, Molecular Sciences and Chemical Engineering*; Elsevier.
51. Humphrey, W., Dalke, A., & Schulten, K. (1996). *J. Mol. Graph.*, *14*, 33–38, 27–28.
52. Pettersen, E. F., Goddard, T. D., Huang, C. C., Couch, G. S., Greenblatt, D. M., & Meng, E. C. (2004). *J. Comput. Chem.*, *25*, 1605–1612.
53. MOLDEN a visualization program of molecular and electronic structure http://www.cmbi.ru.nl/molden/ (accessed Feb 27, 2015).
54. GaussView 5 Brochure http://www.gaussian.com/g_prod/gv5b.htm (accessed Feb 27, 2015).
55. Sloot, P. M. A., Abramson, D., Bogdanov, A. V., Gorbachev, Y. E., Dongarra, J. J., & Zomaya, A. Y. (2003). *Computational Science—ICCS 2003 International Conference, Melbourne, Australia and St. Petersburg, Russia, June 2–4, 2003 Proceedings, Part IV*; Springer-Verlag Berlin Heidelberg : Springer e-books: Berlin, Heidelberg.
56. Kokalj, A. (1999). *J. Mol. Graph. Model*, *17*, 176–179.
57. Spek, A. L. (2005). *Acta Crystallogr. D Biol. Crystallogr.* 2009, *65*, 148–155.
58. Phillips, J. C., Braun, R., Wang, W., Gumbart, J., Tajkhorshid, E., & Villa, E. *J. Comput. Chem.* *26*, 1781–1802.
59. Salvadori, A., Brogni, A., Mancini, G., & Barone, V. (2014). In *Augmented and Virtual Reality*; De Paolis, L. T., & Mongelli, A., Eds., Springer International Publishing: Cham., Vol. 8853, pp. 333–350.
60. Mancini, G., D'Annessa, I., Coletta, A., Sanna, N., Chillemi, G., & Desideri, A. (2010). *PLoS ONE*, *5*, e10934.
61. Stein, L. D. (2003). *Nat. Rev. Genet.*, *4*, 337–345.
61. Alsberg, B. K., Bjerke, H., Navestad, G. M., & Åstrand, P.-O. (2005). *Comput. Phys. Commun.*, *171*, 133–153.
62. Stein, L. D. (2003). *Nat. Rev. Genet.*, *4*, 337–345.
63. Cacelli, I., & Prampolini, G. (2007). *J. Chem. Theory Comput.*, *3*, 1803–1817.
64. Aidas, K., Angeli, C., Bak, K. L., Bakken, V., Bast, R., Boman, L., et al. (2014). *Wiley Interdiscip. Rev. Comput. Mol. Sci.*, *4*, 269–284.
65. Barone, V., Cacelli, I., De Mitri, N., Licari, D., Monti, S., & Prampolini, G. (2013). *Phys. Chem. Chem. Phys.*, *15*, 3736.
66. Licari, D., Baiardi, A., Biczysko, M., Egidi, F., Latouche, C., & Barone, V. (2015). *J. Comput. Chem.*, *36*, 321–334.
67. Database SecureFiles and Large Objects Developer's Guide – Contents http://docs.oracle.com/cd/E11882_01/appdev.112/e18294/toc.htm (accessed Feb 27, 2015).
68. Jamberoo – Cross-Platform Molecular Editor/Builder Home http://sf.anu.edu.au/~vvv900/cct/appl/jmoleditor/ (accessed Feb 27, 2015).
69. Licari, D., & Calzolari, F. (2011). *J. Phys. Conf. Ser.* *331*, 072043.

70. Goff, S. A., Vaughn, M., McKay, S., Lyons, E., Stapleton, A. E., Gessler, D. et al., (2011). "The iPlant collaborative: cyberinfrastructure for plant biology," *Front. Plant Sci. 2*, p. 34.
71. *Our Cultural Commonwealth: The Report of the American Council of Learned Societies Commission on Cyberinfrastructure for the Humanities and Social Sciences*. American Council of Learned Societies, 2006.
72. "Principles of Data Visualization – What We See in a Visual – White Paper," [Online]. Available: http://www.fusioncharts.com/whitepapers/principles-of-data-visualization/. (Accessed: 26/Jan/2015).
73. Tufte, E. R. *The Visual Display of Quantitative Information*, 2nd edition. Cheshire, Conn: Graphics Pr, 2001.
74. "Database SecureFiles and Large Objects Developer's Guide." [Online]. Available: http://docs.oracle.com/cd/E11882_01/appdev.112/e18294/toc.htm. [Accessed: 26/Jan/2015].
75. Stein, L. D. (2003). "Integrating Biological Databases," *Nat Rev Genet, 4*(5), 337–345.
76. Vavliakis, K. N., Karagiannis, G. T., & Mitkas, P. A. (2012). "Semantic Web in Cultural Heritage After 2020".
77. Tenopir, C., Allard, S., Douglass, K., Aydinoglu, A. U., Wu, L., Read, E., Manoff, M., & Frame, M. (2011). "Data Sharing by Scientists: Practices and Perceptions," *PLoS One*, *6*(6).
78. Healey, C. G. (2007). "Perception in visualization," *Retrieved February*, *10*, 2008.
79. Wooley, J. C., & Lin, H. S. (2005) National Research Council Committee on Frontiers at the Interface on Computing and Biology, "Computational Tools", National Academies Press (US) Washington (DC).
80. Hamdaqa, M. & Tahvildari, L. (2012). "Cloud Computing Uncovered: A Research Landscape," in *Advances in Computers*, vol. 86, pp. 41–85.
81. Licari, D. & Calzolari, F. (2011). "The Anatomy of a Grid Portal," *J. Phys.: Conf. Ser. 331*(7), 72043.
82. Jmol: an open-source Java viewer for chemical structures in 3D with features for chemicals, crystals, materials and biomolecules (2009).
83. Vasilyev, V. *Jamberoo*, http://sf.anu.edu.au/ vvv900/cct/appl/jmoleditor/.

CHAPTER 6

APPLICATION OF COMPUTATIONAL METHODS TO THE RATIONAL DESIGN OF PHOTOACTIVE MATERIALS FOR SOLAR CELLS

NARGES MOHAMMADI and FENG WANG

Molecular Model Discovery Laboratory, Department of Chemistry and Biotechnology, School of Science, Faculty of Science, Engineering and Technology, Swinburne University of Technology, Hawthorn, Melbourne, Victoria, 3122, Australia,
Tel.: +61 3 9214 5065; Fax: +61-3-9214-5921;
E-mail: fwang@swin.edu.au

CONTENTS

ABSTRACT

Computational chemistry methods are fast becoming an inevitable tool in design of new drugs and materials. Their applications to the *in silico* design of new photoactive materials have been very attractive although rational design of such materials is still under development. This chapter discusses the use of state-of-the-art computational methods to rationally design, to study and to model photoactive compounds for applications in solar cells. As a result, the main focus is a low-cost photovoltaic device known as dye sensitized solar cells (DSSCs).

The unique architecture of DSSC provides numerous possibilities to alter its components, one of which is to design and synthesize more efficient dye photosensitizers for DSSCs. In this Chapter, we discuss how computers can help in the development of photoactive molecules based on rational in solar cells. The well-established TA-St-CA organic dye has been employed as the reference structure for new compounds.

6.1 INTRODUCTION

It is estimated that the global need for energy would be double by 2050 and triple by the end of this century. Currently, fossil fuels are the primary source for energy supply in the world. A serious consequence of the excessive use of fossil fuels is the environmental impact such as global warming. Another warning issue is the limited resources of fossil fuels. As a result of such problems associated with traditional fossil fuels, it has been a significant challenge for scientists to replace fossil fuels with clean, renewable and sustainable energy sources.

Solar energy is the largest source of clean energy readily available. However, it is not the major source of electricity power generation yet; mainly because of the high cost of the current conventional silicon-based solar cells. Dye sensitized solar cells (DSSCs) [1] are a newer generation of solar cells compared to the well-known silicon-based cells. They have gained considerable attention in the last two decades, as potentially an inexpensive alternative to conventional costly silicon cells. Nevertheless, the efficiency of DSSCs in comparison to traditional silicon based solar cells is still very low.

A DSSC is a complex device composed of several components. That is, the conversion of solar radiation into electrical energy in this device depends on the interplay of several key components. The main components of a DSSC are a dye photosensitizer (photoactive material), a semiconductor film (usually TiO_2), and an electrolyte composition (usually iodide/tri-iodide). In a DSSC, the light is absorbed by a dye photosensitizer, which is grafted onto the semiconductor surface through a suitable anchoring group. Incident photons, with enough energy to be absorbed, create an excited state of the dye. Upon excitation, an electron will be injected from the photosensitizer to the conduction band of the semiconductor (e.g., TiO_2). The dye is then transferred into the oxidization state and the injected electron reaches the anode by diffusion through the nanoparticles of the semiconductor. The oxidized dye sensitizer is regenerated by a redox couple dissolved in electrolyte by transferring electrons from the counter electrode.

The unique modular structure and the process of electron generation and charge separation in the DSSCs differentiate them from all other types of solar cells. Such modular architecture of DSSCs provides many possibility of altering its components. As a result, over the past twenty years significant and increasing amount of research efforts have been dedicated to the design and synthesis of new materials such as dye photosensitizers. Such considerable research effort has been put forward in the hope of improving DSSC's power conversion efficiency. Nevertheless, most of these research efforts have been based on the "bottle neck" of the dye synthesis procedures.

With regard to the development of new photosensitizer materials, it is difficult for synthetic chemists to generate high-performance dyes with desirable properties prior to the experiments on the assembled cell, without necessary theoretical support. Unfortunately, the structure and property relationship of the new dyes can't be obtained only from "chemical intuition" without the support of quantum mechanical calculations. In some cases, disappointing results from final stage testing of the synthesized dyes indicate an urgent need to understand the physical behavior of dyes at the molecular level. Such information needs to be obtained prior to experiments taking place. To overcome this bottleneck in the development of new DCCSs with better efficiencies, state-of-the-art computational methods need to be employed.

Today, powerful supercomputing facilities have become important research recourses. As a result, accurate first-principle based quantum chemical calculations are more accessible. Such calculations are a powerful tool to understand the already existing materials, as well as to design new materials *in silico*. Therefore, it is possible to screen, to predict and to design desirable properties of new materials prior to synthesis by means of simulations (or modeling) and computational studies.

Computational modeling has proven to be efficient and practical in many aspects of DSSC materials and processes studies. A large and growing body of literature has investigated the applications of computational methods to the study of DSSC components. The applications of first-principles computational simulations for DSSCs have been reviewed by Pastore and co-workers [2] and other authors such as De Angelis and Fantacci [3] and Labat et al. [4]. A most recent review in 2014 summaries the importance of the prediction and design power of model discoveries for a number of areas including DSSCs [5]. According to these references, quantum mechanical calculations based on density functional theory (DFT) and time dependent DFT (TD-DFT), are acceptable and suitable tools to computationally model and study the dye sensitized solar cells.

The computational means available can be categorized into two main areas, (a) the computational simulation of individual components of DSSCs in isolation (i.e., the dye sensitizer, the TiO_2 semiconductor and the redox couple), and (b) the computational probe of the interactions between two or more components. Simulations of interactions are usually more demanding compared to the studies of the individual components. The most studied interactions between two components of DSSCs are the properties related to the adsorption of dyes on the surfaces of the TiO_2 semiconductor, such as binding modes and aggregation of photosensitizers on the TiO_2 surface, the UV-Vis absorption spectra of the dye-TiO_2 system, as well as the electron injection process [2b]. In this regard, computational modeling studies have been conducted on the surface and/or nanoparticles of TiO_2 by employing either cluster-based approaches or periodic boundary conditions (PBC). The PBC model is capable of modeling an infinite periodic solid/crystal. This model is usually applied on a periodic slab of 3–4 layers of TiO_2 [6].

To investigate individual components such as dyes in isolation, a large number of studies have been performed on the ground and excited state properties of the dye photosensitizers by means of DFT and TD-DFT calculations. For example, the molecular structure (geometry), the shape and the energy levels of the frontier molecular orbitals, polarizability and hyperpolarizability, and UV-Vis absorption spectra of photosensitizers can be investigated by DFT and TD-DFT approaches [2b]. The rational chemical modification of a high performing dye structure to produce more efficient new dyes has been the target of research for years [7, 8]. As a result, this chapter will concentrate on our recent study in this direction.

6.2 RATIONAL COMPUTATIONAL DESIGN

The term 'rational design' is generally understood to mean a design strategy to obtain a well-defined goal or target [9]. This goal is usually achieving a desirable behavior for the object under design. In the context of drug design, chemical structures of an existing drug can be rationally changed in order to obtain desirable properties and functionalities such as drug potency as well as to reduce or to eliminate unwanted properties such as toxicity and side effects, etc. The new molecular structures are designed based on the key principle that states "structure determines function". As a result, new compounds such as new drugs are designed for obtaining (or eliminating) particular properties by rationally changing their chemical structures. Computer based model discoveries can contribute in this process before the syntheses processes are implemented. In the study of photoactive compounds for DSSC applications, a similar process to drug design can be adopted. In the following sections, details of our rational design for generating new organic photoactive materials (dye sensitizers) will be discussed.

The ultimate goal of the rational design presented here is to produce new photosensitizers (dyes) which can enhance the total efficiency of the DSSCs. One of the very well-known routes to enhance the total efficiency of DSSC is to increase the short-circuit current density, J_{sc}, of the cell [10].

Increase of the short-circuit current density can be achieved through the expansion of the absorption region of the dye sensitizer into the near infrared (NIR) region [11]. In other words, shifting the absorption spectra to longer wavelengths (i.e., a bathochromic shift or red-shift) can enhance the overall efficiency of DSSC. [12]. In order to make a bathochromic shift in the absorption spectra of the rationally designed compounds, their highest occupied molecular orbital (HOMO) and their lowest unoccupied molecular orbital (LUMO) energy levels need to be fine-tuned. As a result, one can rationally modify the chemical structure of the existing dye compounds to reduce the HOMO-LUMO energy gap, in order to red-shift the absorption spectra of the new dyes. To achieve the desirable outcome, one needs a thorough understanding of the existing (reference) dyes and their structure-property relationships.

Central to the structure of organic dye sensitizers is the concept of D-π-A configuration. As shown in Figure 6.1(a), 'D' stands for an electron donor group, 'π' for a π-conjugated bridge (also known as spacer or linker) and 'A' for an electron acceptor moiety [13]. Some particular chemical groups which are employed as D-, π- and A-groups reported in literature [14], are shown in Figure 6.1(b). The D-π-A dye structure is an effective and flexible approach, which allows chemical changes of the structures and therefore the properties of dye sensitizers, in order to achieve and to enhance the desirable properties. For example, chemical changes will impact on the valence orbitals such as the HOMO and the LUMO of a dye photosensitizer. For many D-π-A structured dyes, rational chemical modifications can be made on a particular fragment, e.g., D or π or A individually, in order to shift the energy levels of the valence orbitals of the new dyes towards the desirable directions [15]. Computer modeling provides the information on the property changes which are caused by the chemical structural changes, prior to synthesis of the new dyes.

The rational design is applied on the structure of already well-performing photosensitizers with D-π-A configurations. These photosensitizers are designated as "the reference," "the original" or "the parent" dyes throughout this chapter. The reference photosensitizers are selected based on their overall performance in an assembled working solar cell, from the literature. By modifying the chemical structure of the reference

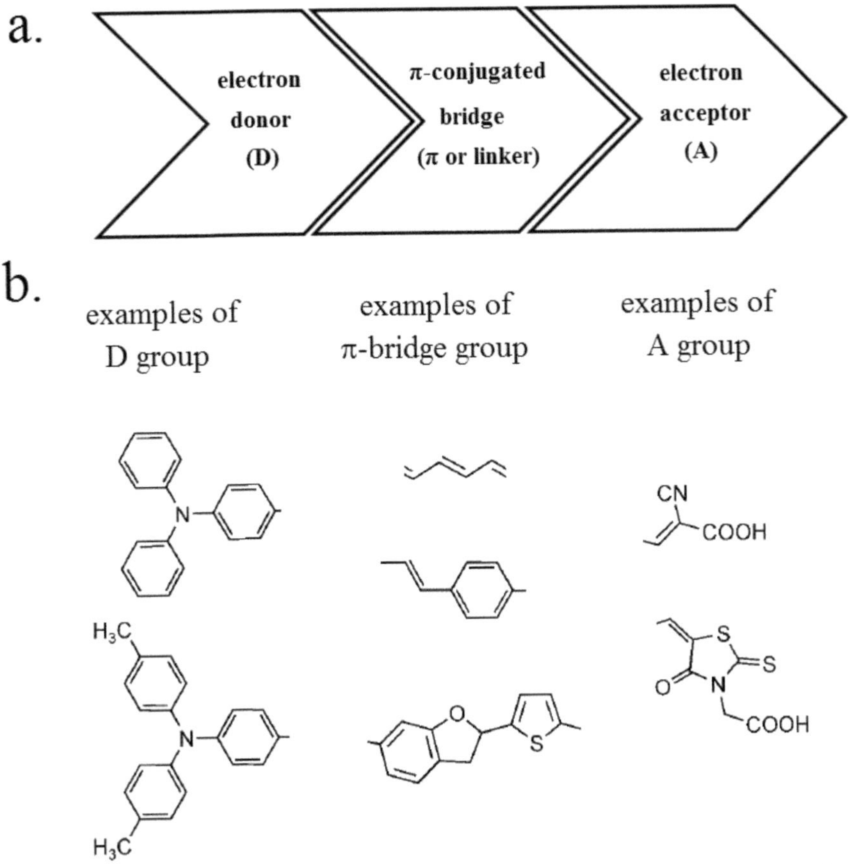

FIGURE 6.1 (a) A scheme of D-π-A dye configuration. (b) Some example of chemical groups employed for different moieties of the metal-free organic dye sensitizers.

dyes towards the desirable properties, new dye photosensitizers are designed. The main target is producing new dyes with reduced HOMO-LUMO energy gap and red-shifted absorption spectra in comparison to the parent (reference) dye.

A variety of approaches can be adapted to red-shift the absorption spectra of a particular D-π-A dye, such as: (i) making modifications into the structure of the π-conjugated bridge, (ii) employing stronger electron-donating groups (D), (iii) increasing the electron-withdrawing character of the acceptor group (A), and (iv) increasing the length of the π-conjugated bridge [16]. However, not all chemical changes would result in the desirable properties and would reduce the unwanted properties. That is, significant effort is needed to establish the structure-property relationship, so that the new dyes can be rationally designed.

As indicated before, three components of a reference D-π-A dye can be changed chemically for the desirable functionalities. For example, the structures of the π-conjugated bridges of two reference dyes, known as TA-St-CA [17] and Carbz-PAHTDDT [18] sensitizers, have been chemically changed to rationally design new dyes with improved properties [8]. These reference TA-St-CA and Carbz-PAHTDDT sensitizers are proven to be well performing dyes in experimental settings.

In this book chapter, a case study will be presented to give the details of our rational design to modify the donor (D) section of the TA-St-CA dye. It will be explained that how strong electron-donating groups can be employed (i.e., the second approach as listed above) to rationally design new photosensitizers with reduced HOMO-LUMO energy gap, as well as enhanced and red-shifted photo absorption spectra, compared with the parent TA-St-CA dye.

6.3 COMPUTATIONAL DETAILS

The new dyes are designed *in silico*. The rational of the chemical changes is based on our study of the structure-property relationship in recent years [8]. The previous studies suggest that modeling provides a very useful role in prediction of their properties before the expensive synthesis processes begin. In the present study, computational methods based on quantum mechanical calculations are employed to model and calculate properties of the new dyes for their energy levels such as the HOMOs and the LUMOs, as well as the UV-Vis absorption spectra. DFT and TD-DFT are employed in the calculations.

The DFT and TD-DFT models (here model means a combination of DFT functional and basis set) are validated based on the agreement of the calculations with available experimental data for the reference dye. Models which provide the best agreement with experiment for the reference dye are then selected to study the corresponding new dyes.

In the case study presented in this chapter, all quantum mechanical calculations are performed using the DFT-based PBE0 [19] hybrid density functional and polarized split-valence triple-zeta 6–311G(d) basis set. The geometries of the molecular systems are obtained without any geometric and symmetric constraints. The DFT-based PBE0 functional is a hybrid of PBE with 25% HF exchange term contribution and is shown to be a reliable functional to calculate the excitation energies of dye molecules [20]. As a result, it has been widely employed to study the colors of many industrial organic dyes. The PBE0/6–311G(d) model has been shown [8a] to provide a good agreement with the available experimental data for the reference TA-St-CA dye [17]. Therefore, the present study employs this model to study the new dye molecules derived from the reference TA-St-CA structure. All calculations are based on Gaussian 09 computational chemistry package [21].

The structures (geometries) of the TA-St-CA dye photosensitizer and its derivatives in three dimensional (3D) spaces were obtained through energy optimizations. To ensure that the optimized structure is a true minimum, it is necessary to ensure that no imaginary frequencies are found for the optimized structure. As a result, frequency calculations are performed on the optimized geometries of the reference dyes and the new dyes. Based on the optimized geometries, single point energy calculations are employed to obtain the properties of the dyes such as molecular energy levels and isodensity properties.

It is important to include the effects of solvation on the calculations of molecular properties, because many experimental measurements of the dyes take place in solutions. For example, the experimental measurements of the absorption spectra of the dye photosensitizers are usually measured in the presence of a solution. Ideally, solvent effects are to be calculated explicitly with the inclusion of solvent molecules. However, this approach is a computationally demanding task and is limited to very small solutes [22]. Alternatively, implicit solvation methods are employed in which the solute is placed into a cavity of the solvent reaction field. In implicit solvent models, the solvent molecules are treated as a structureless dielectric medium with surface tension at the solute-solvent boundary. Such an approach to the simulation of solvation is called “continuum” approximation.

In this case study, a continuum solvation method known as conductor-like polarizable continuum model (CPCM) [23] is employed to simulate the UV-Vis spectra in solutions. In order to make comparisons with the experimental UV-Vis spectrum of the reference TA-St-CA dye, which was measured in ethanol solution [17], the present study has simulated the UV-Vis spectra in ethanol solution. The theoretical UV-Vis spectra of the dyes are calculated using singlet-singlet transitions up to the 30th lowest spin-allowed excited state of each dye. These lowest-energy electronic transitions are then transformed into simulated UV-Vis spectra by GaussView 5 visualization software [24], using a Gaussian broadening function with the full width at half-widths maximum (FWHM) of 2500 cm^{-1}.

The optimized structure of the reference dye is modified by chemical substitutions to produce new dyes, which are then computationally studied by the same procedure of the reference dye to ensure the new properties are obtained as designed. That is, the property calculations and spectral simulations must be based on the geometries of the stable and true minimum structures of the new dyes. Results of the new dyes will be compared against the results of the reference dye to identify the improvement. It is apparent that to make a meaningful comparison, the same DFT and TD-DFT model should be employed for the simulations of the reference and the new dyes. Table 6.1 summarizes the key aspects of our rational design method.

6.4 A CASE STUDY: COMPUTATIONAL DESIGN OF NEW TA-ST-CA BASED PHOTOSENSITIZERS

The reference TA-St-CA dye is designed and synthesized by Hwang et al. [17]. This photoactive compound (see Figure 6.2) is a very successful photosensitizer. A DSSC based on this photosensitizer has shown the overall solar-to-energy conversion efficiency of 9.1% [17], which is very high for a solar cell with organic dye sensitizer. As a result, this push–pull dye is selected as the backbone reference structure to develop new dyes in the current case study.

The basic concept for the design of new photoactive materials in this case study is to reconstruct the donor (D) moiety of the reference

TABLE 6.1 A Summary of the Rational Design of Photosensitizers

Goal	Red-shifting the absorption spectra of rationally designed photosensitizers by reducing their HOMO-LUMO energy gap in compare to the reference compounds.
Reference	Already well-performing organic dye photosensitizers, exhibiting D-π-A structure.
Approach	Altering the chemical structure of the reference dye by adapting one of the following modifications: • Making modifications into the structure of the π-conjugated bridge. • Employing stronger electron-donating groups. • Increasing the electron-withdrawing character of the acceptor group. • Increasing the length of the π-conjugated bridge.
Quantum mechanical theory	DFT and TD-DFT.
Restrictions	• The LUMO energy level of the new dyes should be above the conduction band of the semiconductor employed in DSSC (usually TiO_2) for effective injection of the excited electron. • The HOMO energy level of the new dyes should be below the redox potential of the redox mediator (usually Iodide/Triiodide) for effective regeneration of the oxidized dye.

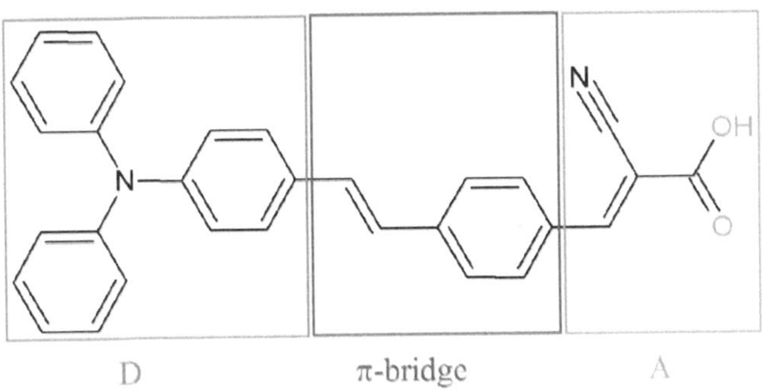

FIGURE 6.2 Molecular structure of the reference TA-St-CA dye.

compound. As shown in Figure 6.2, the D-moiety of the TA-St-CA reference dye consists of three *aromatic annulenes* which join at the center N atom. Annulenes, such as benzene, are conjugated monocyclic hydrocarbon rings without side chains. The general formula for an annulene is

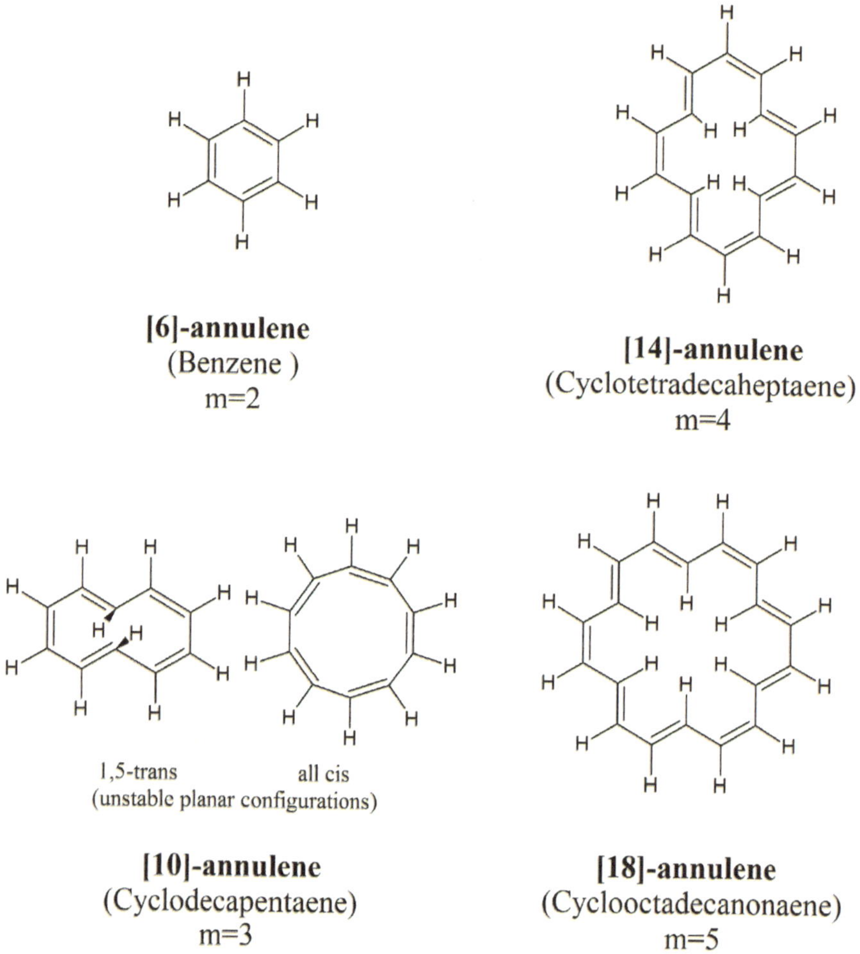

FIGURE 6.3 Molecular structures of different annulenes.

given as C_nH_n (if n is an even number) or C_nH_{n+1} (when n is an odd number) [25]. Figure 6.3 gives some examples of annulenes.

Aromaticity of annulenes was studied by the famous Hückel's rule of aromaticity in 1931 [26]. According to Hückel's rule, a *planer* (or almost planer) *cyclic* ring with a continuous system of π-orbitals is aromatic, if those π-orbitals are occupied by 4m+2 electrons (where m is a non-negative integer). The most well-known aromatic member of the annulene compounds is benzene (i.e., C_6H_6 or [6]-annulene, n=6, m=1).

There are exceptions to the 4m+2 rule due to other effects. For example, among the small annulenes, cyclodecapentaene or [10]-annulene (i.e., $C_{10}H_{10}$, n=10), which exhibits 10 π–electrons (m=2), and meets the 4m+2 requirements for aromaticity, does not exhibit sufficient aromaticity, due to a combination of steric strain and angular strain. The angular strain refers to the actual angle away from the angle formed by the planar sp^2 hybridized carbons of 120°. For cyclodecapentaene or [10]-annulene (see Figure 6.3), these angles are as large as 144°, a derivation of 24° strain! Therefore, the planar all cis-cyclodecapentaene is a less stable compound with little aromaticity. The other possible planar configuration for this substance is 1,5-trans cyclodecapentaene (i.e., the configuration in which two of the double bonds are trans, see Figure 6.3). This isomer has a minimum angle strain. However the repulsive force between the two flag hydrogen atoms (those ones that are forced together in the interior of the ring), again destabilizes this planar structure. As a result, this isomer is also relatively unstable with little aromaticity.

The molecular structures and aromaticity of the small [14]- and [18]-annulenes have been well studied [14]-annulene (also known as cyclotetradecaheptaene, i.e., $C_{14}H_{14}$, n=14, m=3) and [18]-annulene (i.e., $C_{18}H_{18}$, also known as cyclooctadecanonaene, n=18, m=3) exhibit certain aromaticity. For example, both of these ring structures (i.e. [14]-annulene and [18]-annulene) engage with conjugation and monocyclic hydrocarbons, which follows the Hückel's rule of aromaticity (see Figure 6.3). A number of studies, such as Wannere et al. [27], Jug et al. [28], Kennedy et al. [29] and Gellini et al. [30] indicated that both [14]- and [18]- annulenes indeed show aromatic characters. However, it should be noticed that [14]-annulene tends to be less planer, because of the steric interactions among the four inner-rings hydrogen atoms in its structure (see Figure 6.3). Such ring strain could destabilize this compound and makes it less stable.

Although annulenes have a number of applications, the current case study is the first attempt to employ the [14]-annulene in the structure of a dye sensitizer. As a result, rational design and computational study of new dyes based on [14]-annulene, will provide useful information in the design and development of new dyes for DSSCs applications.

Several properties of annulenes make them excellent candidates for the electron donor moieties of the new photosensitizers. The existence of n delocalized π-electrons in their rings (n refers to the number of carbon atoms in annulene formula) makes annulenes electron-rich compounds, which is ideal for the donor moiety of a push-pull dye. Moreover, structural conjugation of annulenes (i.e. [6]- [14]- and [18]- annulenes) enhances the chemical stability of the new dyes. Stability is an important requirement for the practically useful dye sensitizers. In addition, the conjugation of annulenes increases the electron delocalization, and therefore reduces the energy gap between the bonding (π) and anti-bonding orbitals (π*), which are often the HOMOs-LUMOs. As a result, by employing larger conjugated groups such as annulenes, the electronic structures of the dyes such as the HOMO-LUMO energy gap (should be reduced), and the maximum absorption (should move to longer wavelengths, i.e., red-shifting the absorption spectra). As a result, annulenes are employed to design new dyes in the present study.

In this case study, two new photosensitizers are rationally designed by altering the donor (D) moiety of the reference TA-St-CA dye. The new photoactive molecules inherit the same π-linker and acceptor (A) moieties of the parent TA-St-CA dye, but differ only in the donor (D) structures. Figure 6.4 shows the chemical structures of these new photoactive molecules. The D section of the parent photosensitizer consists of a triphenylamine (TPA) moiety. Three phenyl rings of the TPA moiety are labeled as R_1-R_3, respectively, as shown in Figure 6.4(a). The phenyl groups are derived from benzene ring (i.e., C_6H_6 or [6]-annulene). In the first new dye, AN–14, each phenyl ring (R_1, R_2 and R_3) is replaced with a [14]-annulene rings, respectively. The new dye AN–14 is given in Figure 6.4(b). In the second new dye, AN–18, two of the phenyl rings in the reference dye, TA-St-CA, i.e., R_1 and R_2 remain unchanged, while R_3 is substituted with [18]-annulene ring. Figure 6.4(c) shows structure of the new AN–18 dye. As seen in Figure 6.4, in the design of the new dyes, the "oligo-phenylenevinylene" group (π-linker) and the "cyanoacrylic acid" acceptor group remain unchanged, which are marked by the highlight box.

The present study calculates the properties such as electronic structures and adsorption spectra of the new photoactive molecular dyes using the DFT-based quantum calculations. The goal is to correlate the

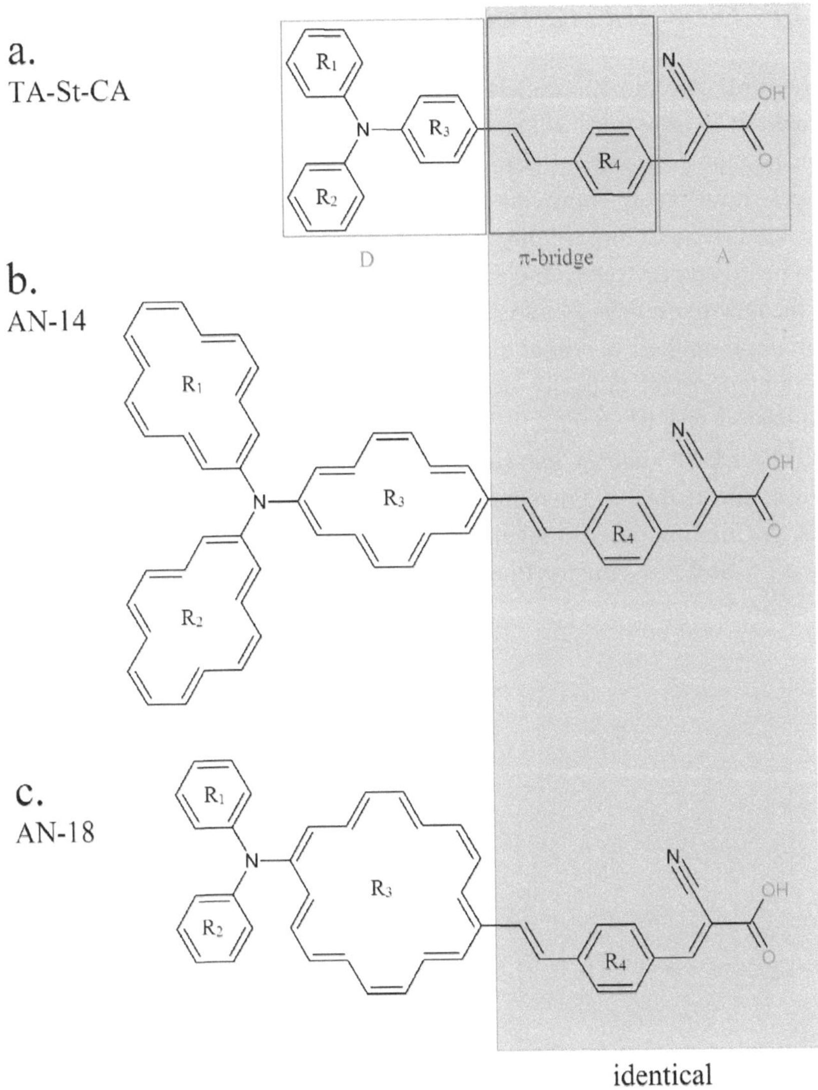

FIGURE 6.4 Molecular structure of the reference TA-St-CA sensitizer and new dyes AN–14 and AN–18.

structure-property relationships of the new AN–14 and AN–18 dyes. That is, to reveal the impact of the structure changes of the donor moiety on the important properties of the new dyes such as the HOMO-LUMO energy gap, as well as the UV-Vis absorption spectra.

6.4.1 FRONTIER MOLECULAR ORBITAL ANALYSIS

Structure determines properties and therefore, functionalities of molecules. It is important to know (i.e., to rational) if the structural changes enhance the desirable properties of the new dyes, rather than the opposite; if yes, how much improvement can be achieved? Apparently, quantum mechanical modeling is "the game in town". Following the computational details given previously, the PBE0/6–311G(d) model determines that both of the new dyes, AN–14 and AN–18 are indeed stable molecules with true minimum on their potential energy surfaces.

An important target of designing new photosensitizers in this rational design was to change the electronic structures such as to reduce the HOMO-LUMO energy gap (in comparison with to the reference dye). To investigate if this target is met, Figure 6.5 compares the frontier orbital energy (binding energy) levels of the new dyes with respect to the reference TA-St-CA dye, using the same DFT model, in their ground electronic structures.

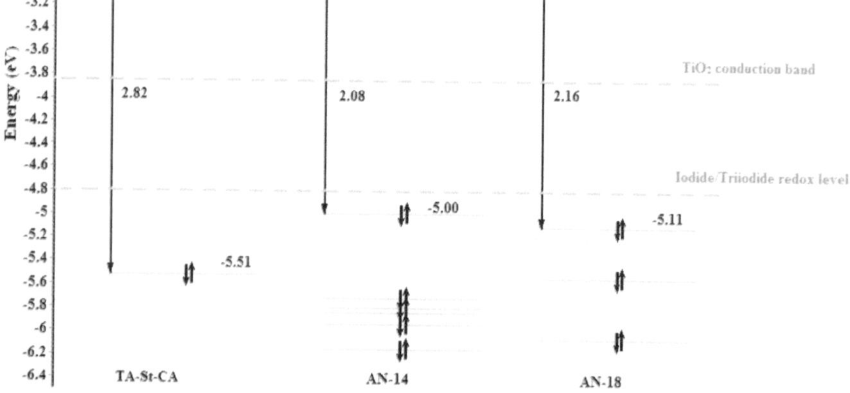

FIGURE 6.5 The calculated frontier MO energy levels using PBE0/6–31G* model in vacuum.

The orbital energies of the HOMO and the LUMO of the original TA-St-CA photosensitizer in isolation are calculated at –5.51 eV and –2.69 eV, respectively, corresponding to 2.82 eV energy gap between the HOMO and the LUMO of this dye. The TiO_2 conduction band and the Iodide/Triiodine redox level are also given in the figure. This figure indicates that the D-modified new dyes indeed improve the properties of the reference dye with respect to the HOMO-LUMO energy gap, as designed. For example, the new dyes AN–14 and AN–18 lead to reductions of their HOMO-LUMO gaps by apparently lifting up the HOMO energies and shifting down the LUMO energies. For example, the HOMO and LUMO energy levels of the new dye AN–14 are calculated at –5.00 eV and –2.92 eV, respectively, leading to the corresponding HOMO-LUMO gap of 2.08 eV, i.e., an energy reduction of 0.74 eV. In a similar way, the new dye AN–18 produced by the variation of the donor moiety of the reference dye, can achieve an increased HOMO energy (–5.11 eV) from –5.51 eV (TA-St-CA) and decreased LUMO energy (–2.95 eV) from –2.95 eV (TA-St-CA), resulting in the corresponding energy gap of 2.16 eV, a reduction of 0.66 eV.

From Figure 6.5, it can also be seen that the LUMO energies of the new dyes are very similar, whereas the HOMO energy of AN–14 is about 0.11 eV higher than that of the AN–18. These findings suggest that in general, the mechanism of the HOMO and LUMO energy changes of the molecular orbitals of AN–14 and AN–18 are very similar. In other words, chemical modifications of the D-section of the reference dye indeed achieve the desirable property changes in the frontier orbitals such as HOMO and LUMO towards to the wanted direction.

Figure 6.5 also gives positions of the conduction band of the TiO_2 semiconductor, as well as the redox potential of the iodide/triiodide redox mediator. As seen in the figure, the HOMO energy levels of both new dyes are well located below the redox energy level of the iodide/triiodide redox couple. This implies that sufficient driving force is available for the regeneration of the oxidized AN–14 and AN–18 by the iodide/triiodide redox mediator. In fact, the HOMO energy level of the new AN–14 sensitizer is very close to that of the well documented N3 sensitizer. The N3 (i.e., Ru(4,4-dicarboxylate–2,2-bipyridine)$_2$-(NCS)$_2$)) dye sensitizer [31], is believed to be one of the the best performing dyes from the ruthenium–polypyridyl sensitizer family. The HOMO energy level of the N3 dye sensitizer is caluclated at –5.08 eV using B3LYP/LANL2DZ level of theory [32]. It is suggested that "the sensitizer candidates with HOMO level close to that of the N3 dye would be promising for the regeneration since the N3 dye can be regenerated very well" [32]. As a result, the increased HOMO energies of the new dyes suggest that AN–14 and AN–18 are more efficient than the reference TA-St-CA dye.

Figure 6.5 also compares the LUMO energy levels of the new dyes with the conduction band edge of the semiconductor located at –4.0 eV for anatase TiO_2 (dotted line) [33]. Although the new dyes exhibit a down-shifted LUMO energy levels compared to that of the reference dye, their LUMO energies are still well above the CBE of TiO_2. That is, the LUMO levels of both AN–14 and AN–18 are higher than the conduction band edge of TiO_2 by at least 1.0 eV. Such energy pattern suggests that upon excitation, the photoexcited electrons possess sufficient driving force to be rapidly injected to the conduction band of the semiconductor. As a result, these two new dye senzitizers, i.e., AN–14 and AN–18, can operate functionally in working DSSCs. However, the influences of the donor modifications are more profound on the electronic structure of the AN–14 dye.

6.4.2 SIMULATION OF THE UV-VIS ABSORPTION SPECTRA

An important purpose in designing of the new dyes, AN–14 and AN–18, was to broaden and to red-shift the absorption spectra of the reference TA-St-CA dye by the variation of the structures of the dye donor moiety. To investigate the light absorption properties of the new photoactive compounds, their electronic spectra are simulated in Figure 6.6.

Figure 6.6 also gives the simulated UV-Vis spectrum of the parent TA-St-CA dye for comparison. Energetically, the UV-Vis spectrum of a dye is the transitions among the occupied molecular orbitals and the unoccupied virtual orbitals of the molecule, subject to selection rules. The electronic and optical data which are calculated using the TD-DFT-based PBE0/6–311G(d) model are given in Table 6.2. The results in Table 6.2 are compared with the related properties of the reference dye produced earlier [8a] where it has been shown that this DFT model closely reproduces the experimental UV-Vis spectra of the reference TA-St-CA photosensitizer.

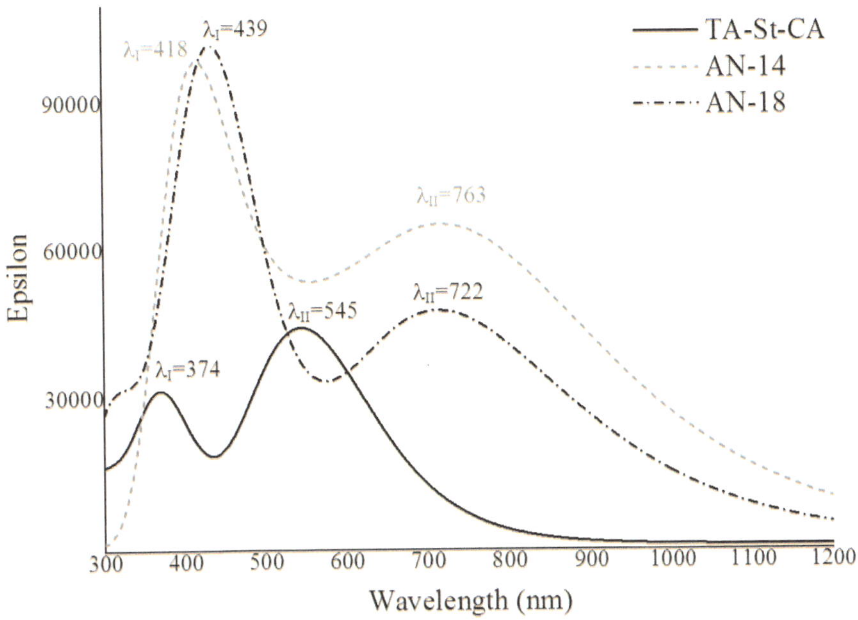

FIGURE 6.6 The simulated UV-Vis absorption spectra of the TA-ST-CA, AN–14 and AN–18 in ethanol solution using the (PBE0/6–311G*) TD-DFT calculations.

Figure 6.6 also compares the simulated UV-Vis spectra of the new dyes against the reference TA-St-Ca dye. Significant enhancement in the UV-Vis spectra is achieved in the new AN–14 and AN–18 sensitizers over the reference dye. Figure 6.6 shows that the new dyes indeed deliver the desirable property of bathochromic shift (i.e., to the longer wavelength or red-shift) of the spectral peaks. For example, the two most intensive spectral peaks of the reference dye are calculated at $\lambda_I = 374$ nm and $\lambda_{II} = 545$ nm. In the new dye "AN–14" (the dash spectrum), these peaks are significantly red-shifted to $\lambda_I = 418$ nm and $\lambda_{II} = 763$, respectively. For λ_I, a relatively small red shift of ca. 44 nm is obtained, whereas a very significant red-shift of ca. 218 nm is seen on the position of λ_{II} in AN–14, when compared to the reference compound. In addition, the intensities of the spectral peaks are also enhanced apparently, indicating that the new dyes will be able to cover substantially larger areas of the absorption band.

The effects of structural modifications on the electronic absorption spectra of AN–18 (the dash-dot spectrum in Figure 6.6) are similar to

TABLE 6.2 Absorption Properties of the TA-ST-CA Dye and the New Dyes, Calculated[a] in Ethanol Solution

Peak I			Peak II		
$\lambda_{(nm)}$	*f*	Assignment	$\lambda_{(nm)}$	*f*	Assignment
TA-St-CA					
374	0.77	H–1→L (91%) H→L+1 (7%)	545	1.22	H→L (99%)
AN–14					
418	0.61	H–2→L+1 (35%) H–2→L+2 (23%) H–1→L+1 (19%) H–1→L+2 (14%)	763	1.53	H→L (96%)
AN–18					
439	2.10	H–1→L+1 (48%) H–3→L (17%) H→L+2 (13%) H–1→L+2 (7%) H–2→L (5%) H–2→L+2 (5%)	722	1.27	H→L (95%) H–1→L+1 (3%)

[a]Calculated using TD-DFT-based PBE0/6–311G(d) model in ethanol (CPCM).

those of AN–14, except the λ_{II} peak of the AN–18 dye is less shifted with less intensity. For instance, in AN–18, the positions of both Peaks I at λ_I and Peaks II at λ_{II} are shifted to longer wavelengths, compared to those of the reference photosensitizer. The simulations on AN–18 produce a sharp intense peak at λ_I = 439 nm and a broader peak at λ_{II} = 722 nm. As a result, the rational donor design of AN–18 leads to a bathochromic shift of ca. 65 nm and ca. 177 nm on the positions of the spectral peaks I and II, compared to those of the reference dye, respectively.

As listed in Table 6.2, new dyes exhibit enhancement in the oscillator strengths (*f*) of their absorption peak I and II, with respect to the reference dye. For example, the peak II of the reference dye exhibits an oscillator strength of 1.22, which is increased to 1.53 and 1.27 in AN–14 and AN–18, respectively. For this peak the oscillator strengths follow a trend of $f_{\text{TA-St-CA}} < f_{\text{AN-18}} < f_{\text{AN-14}}$. For the other main absorption peak, the trend is quite different, that is, $f_{\text{AN-14}} < f_{\text{TA-St-CA}} < f_{\text{AN-18}}$.

The expansion of the absorption spectra of AN–14 and AN–18 dye photosensitizers with respect to the same spectrum of the reference dye can be attributed to the electronic structural differences of the dyes. For example, the HOMO-LUMO energy gaps of these dyes are quite different; the new dyes possess reduced energy gaps. However, the most significant contribution to the spectral changes of the new dyes is the substantially different electronic structures of the new dyes with respect to the reference dye. The spectra in Figure 6.6 are the result of the singlet-singlet transitions from the occupied molecular orbitals to all the possible virtual orbitals up to the 30th virtual orbital. As a result, the valence orbitals as well as the virtual orbitals will significantly impact on the UV-Vis spectra of the new dyes.

In Table 6.2, peak II is an important charge-transfer (CT) band, which is dominated by the HOMO→LUMO transition. For example, the dominant contribution of the the HOMO-LUMO transition is 99% in the spectra of the reference dye. The same transition in the new dye AN–14 becomes 96%, while it becomes 95% in in the new AN–18. The HOMO and the LUMO energy levels are brought closer to each other in new dyes AN–14 and AN–18, compared to the reference dye; therefore, this CT band (i.e., peak II) is red-shifted in both new candidates compared to TA-St-CA.

Similar justification can also be considered for peak I. However, as shown by Figure 6.5, within the energy brakects of [–6.4 eV, –1.6 eV], the distribution of the orbital energy levels in the three dyes are significantly different. Only the HOMO and LOMO of the reference dye position in this window. However, about half dozen valence orbitals and half donzen of virtual orbitals of the AN–14 dye move into this energy window and a few occupied and a few virtual orbitals of the AN–18 dye also fall into this ebergy window. As a result, a number of more transitions among those orbitals of the new dyes significantly enhanced the peak I in the simulated UV-Vis spectra in Figure 6.6. Such spectral enhancement and shift of the new dyes with respect to the reference dyes are the direct results of the electronic structure changes of the new dyes.

The new dyes, AN–14 and AN–18, which are produced from the replacement of the [6]-annulene rings in the reference TA-St-CA dye in the D-moiety by the electron-rich [14]-annulene [18]-annulene rings, respectively, exhibit significant improvement of the absorption spectra. The absorption spectra of the new dyes are red-shifted, more intensive, and broadened compared to that of the reference TA-St-CA dye, in accordance with the rational design goal. In addition, the light absorption spectrum of the AN–14 sensitizer seems to be superior to that of the AN–18. The new AN–14 and AN–18 photosensitizers are potentially promising photoactive candidates for better DSSCs with enhanced efficiencies. The practical application of the new dyes in the DSSCs settings will depend on a number of other issues such as the synthesis processes and environmental issues.

6.4.3 MOLECULAR ORBITAL SPATIAL DISTRIBUTION

To further investigate the influence of modifications on the electronic properties of the new dyes, Figure 6.7 presents the electron density distributions of the frontier molecular orbitals of the molecules.

The HOMO and the LUMO of the reference TA-St-CA dye as well as the AN–14 and AN–18 sensitizers are compared in the figure. These molecular orbitals represent the dominant transitions contributing to the main band (i.e., peak II, see Table 6.2) of the UV-Vis spectra. As indicated in Table 6.2, this band is dominated by a strong transition from the HOMO in the ground electronic state (S_0) to the LUMO in the first excited electronic state (S_1) in the visible region of the spectrum.

The HOMOs and LUMOs of the D-π-A dyes exhibit apparently different characters. The HOMOs spread over the entire donor (D) moieties and extend into the conjugated π-bridge (spacer) of all three dyes. For example, the HOMO of the reference dye not only distributes in the D-section of the D–π–A dye, that is, all over the R_1, R_2 and R_3 rings, but also expands into the R_4 ring of the π-bridge of the dye. On the other hand, the LUMO of the reference dye concentrates on the cyanoacrylic acid group which works as electron acceptor/anchoring moiety (A-section), the π-bridge as well as the R_3-ring of the D-section.

The HOMOs and LUMOs of the new AN–14 and AN–18 dyes inherit the characters of the reference dye. For example, the LUMOs of AN–14 and AN–18 are singlet π* orbitals which largely populates the cyanoacrylic

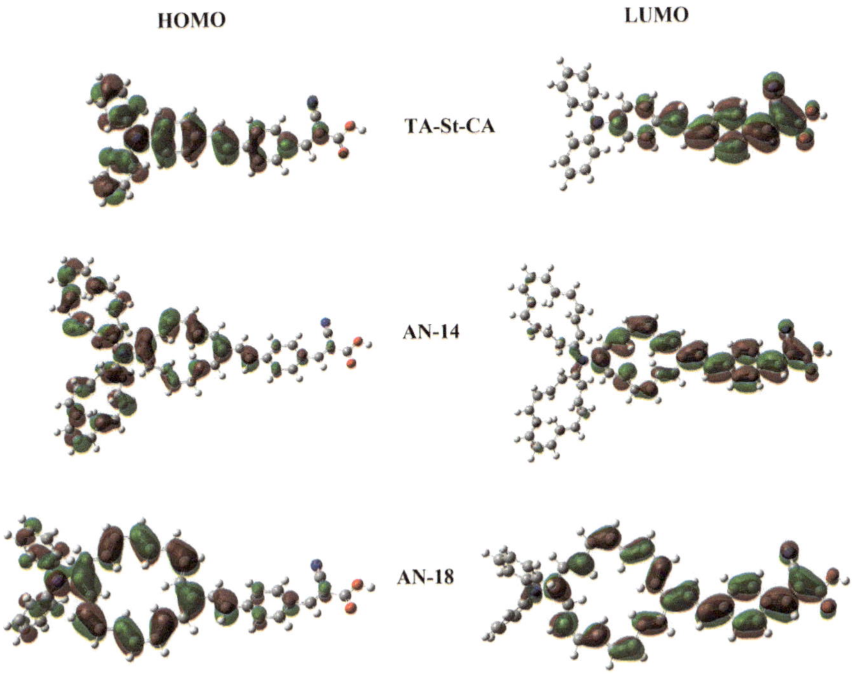

FIGURE 6.7 Comparison of the HOMOs (left) and LUMOs (right) of the new dye, AN–14 and AN–14 with respect to those of the reference TA-St-CA dye.

acid acceptor group and also spread into the π-conjugated bridge and into the R_3 ring in the D-section. Negligible contributions of the R_1 and R_2 rings in the D section to the LUMO are also observed for the dyes. The most noticeable difference between the HOMOs of the reference TA-St-CA dye and the new dyes is a reduced contribution from the R_4 ring of the π-conjugated bridge on the HOMO of AN–14 and AN–18. The enhanced HOMO-LUMO overlaps at R_3 of the D-section of the new dyes ensures the CT from HOMO to LUMO is easier in the new dyes, which is indicated by the shift of their orbital energy levels in Figure 6.5. Such distribution of the HOMOs and the LUMOs demonstrates that the HOMO-LUMO transition has an intra-molecular CT character. This distribution pattern for the HOMOs and the LUMOs is beneficial to a functional solar cell. This is because: (a) significant contribution of cyanoacrylic acid group (A-section) to the LUMO ensures a strong electronic coupling between the dye's excited state and conduction band of the semiconductor (TiO_2), which facilities the ultrafast electron injection, and (b) significant localization of the HOMO on the donor end (left-side of the dyes in the figure) minimizes the probability of charge recombination between the injected electrons and the resulting oxidized dye.

6.5 SUMMARY AND FURTHER PROSPECTIVE

The real potential of state-of-the-art computational methods to design new materials for solar cells has not been well recognized yet. In terms of computer-aided material studies, computational modeling and calculations have usually been performed for the understanding and interpretation of existing materials, rather than designing, predicting properties and screening new compounds. This chapter has addressed this issue and focused on the computational modeling of photosensitive compounds for DSSCs prior to the existences of the materials.

This study has focused on the photoactive component of the DSSCs, i.e., the dye photosensitizer, and has given an account for the computer-aided rational design of new photoactive materials. The rational design strategy employed in this work has been to modify the structure of the already well-performing organic dyes (or reference dyes) exhibiting donor, π-conjugated linker, acceptor structure (D-π-A), to produce new dyes with reduced HOMO-LUMO energy gaps and red-shifted absorption spectra. The rationale behind such modifications has been to produce new dyes with enhanced absorption spectra, as a route to enhance the overall efficiency of the DSSC. DFT has been exploited to study the ground state properties, while TD-DFT has been adopted for the study of the excited states of the compounds. Considering the size and the complexity of the photosensitizers, DFT and TD-DFT provide good balance between accuracy and performance. The theoretical models (i.e., functional and basis set) employed in the DFT and TD-DFT calculations of the new dyes, have been validated based on the agreement with available experimental data for the reference dyes available from the literature.

This study suggests that modifications made on the donor moiety of the D-π-A TA-St-CA reference photosensitizer, are able to produce new dyes with significantly appealing properties for DSSC applications. For example, the AN–14 dye and the AN–18 dyes, which have been designed by replacing the [6]-annulene rings of the reference dye with stronger donor groups, are both superior to their parent compound in their desirable properties. Results of this study indicate that both new photosensitizers exhibit reduced HOMO-LUMO gap and expanded absorption spectra. Furthermore, the rationally designed AN–14 sensitizer seems to possess a HOMO-LUMO energy gap which is very similar to that of the recognized benchmark N3 dye. With regard to the capability of collection of as much sunlight as possible, the new AN–14 and the AN–18 dyes are both superior to the reference TA-St-CA dye.

The present study adds to the growing body of literature on the computer-aided rational design of new materials for DSSCs. More studies in this direction are needed to understand the interaction of the dye molecule with other component of the cell. For example, future studies may address the adsorption of the new dyes designed in this study on the semiconductor (e.g., TiO_2) surface. Experimental studies are also essential to determine the efficiency of the new rationally designed dyes, as well as their stability, in real working cells.

It is a useful systematic method to modify the already-well performing reference dyes with D-π-A structures to produce new photosensitizers for DSSCs. In addition, the designed targets are well defined. That is, new photosensitizers need to exhibit a reduced HOMO-LUMO energy gap and red-shifted absorption spectra compared to their parent molecule. A few constraints should also be considered. For example, the LUMO energy level of the new dyes should be above the conduction band of the semiconductor employed in DSSC, and the HOMO energy levels of the new dyes should be below the redox potential of the redox mediator. The photosensitizer should carry a suitable anchoring group in its acceptor end to firmly graft it into the surface of the semiconductor. The later constraint is usually met by default, unless modifications are made on the acceptor section of the reference dye. As a result, for certain systematic improvements, it is possible to design a software tool for computer-aided rational design of new dye sensitizers, such as the score-docking software used in computational drug design. The input for such software can be the results of searching the database of the structures of the well performing reference dyes (such as the protein database (PDB) in bioinformatics).

The outputs are the structures of the new dyes, which will be deposited into the repository of useful dyes for DSSCs. Another direction of rational design is to develop a database of building blocks/fragments, which can be employed in the D, π and A sections of the D-π-A dye, to produce new photoactive molecules with enhanced properties. Therefore, it is recommended that further efforts be undertaken to design and code such a software tool and databases for fast and mass generation and screening of new photosensitizers.

ACKNOWLEDGMENTS

NM acknowledges Vice-Chancellors' Postgraduate Research Award at Swinburne University. This work has been made possible by the Swinburne supercomputing facilities and the Victorian Partnership for Advanced Computing (VPAC).

KEYWORDS

- **annulene**
- **charge-transfer band**
- **dye sensitized solar cells**
- **HOMO-LUMO**
- **photosensitizers**
- **TD-DFT**

REFERENCES

1. O'Regan, B., & Gratzel, M., (1991). A low-cost, high-efficiency solar cell based on dye-sensitized colloidal TiO2 films. *Nature 353*(6346), 737–740.
2. (a) Pastore, M., Mosconi, E., Fantacci, S., & De Angelis, F., (2012). Computational Investigations on Organic Sensitizers for Dye-Sensitized Solar Cells. *Curr Org Synth 9*(2), 215–232; (b) Pastore, M., & Angelis, F., Modeling Materials and Processes in Dye-Sensitized Solar Cells: Understanding the Mechanism, Improving the Efficiency. Springer Berlin Heidelberg: 2013; pp 1–86.

3. De Angelis, F., Fantacci, S., Selloni, A., Grätzel, M., & Nazeeruddin, M. K., (2007). Influence of the Sensitizer Adsorption Mode on the Open-Circuit Potential of Dye-Sensitized Solar Cells. *Nano Lett 7*(10), 3189–3195.
4. Labat, F., Le Bahers, T., Ciofini, I., & Adamo, C., (2012). First-Principles Modeling of Dye-Sensitized Solar Cells: Challenges and Perspectives. *Accounts of Chemical Research 45*(8), 1268–1277.
5. Wang, F., & Ahmed, M., (2014) Sitting above the Maze: Recent Model Discoveries in Molecular Science. *Molecular Simulation, 41, 205-229.*
6. Kalyanasundaram, K., *Dye-sensitized Solar Cells*. EFPL Press: 2010.
7. (a) Sanchez-de-Armas, R., San-Miguel, M. A., Oviedo, J., & Sanz, J. F., (2012). Molecular modification of coumarin dyes for more efficient dye sensitized solar cells. *The Journal of Chemical Physics 136*(19), 194702–7; (b) Gu, X., Zhou, L., Li, Y., Sun, Q., & Jena, P., (2012). Design of new metal-free dyes for dye-sensitized solar cells: A first-principles study. *Physics Letters A 376*(38–39), 2595–2599; (c) Fan, W., Tan, D., & Deng, W.-Q., (2012). Acene-Modified Triphenylamine Dyes for Dye-Sensitized Solar Cells: A Computational Study. *Chemphyschem 13*(8), 2051–2060; (d) Wang, J., Gong, S., Wen, S. Z., Yan, L. K., & Su, Z. M., (2013). A Rational Design for Dye Sensitizer: Density Functional Theory Study on the Electronic Absorption Spectra of Organoimido-Substituted Hexamolybdates. *J Phys Chem C 117*(5), 2245–2251.
8. (a) Mohammadi, N., Mahon, P. J., & Wang, F., (2013). Toward rational design of organic dye sensitized solar cells (DSSCs): an application to the TA-St-CA dye. *Journal of Molecular Graphics and Modelling 40*, 64–71; (b) Mohammadi, N., & Wang, F., (2014). First-principles study of Carbz-PAHTDDT dye sensitizer and two Carbz-derived dyes for dye sensitized solar cells. *J Mol Model 20*(3), 1–9.
9. Parnas, D. L., & Clements, P. C., (1986). A Rational Design Process – How and Why to Fake It. *Ieee T Software Eng 12*(2), 251–257.
10. Hamann, T. W., Jensen, R. A., Martinson, A. B. F., Van Ryswyk, H., & Hupp, J. T., (2008). Advancing beyond current generation dye-sensitized solar cells. *Energ Environ Sci 1*(1), 66–78.
11. Kim, B. G., Chung, K., & Kim, J., (2013). Molecular Design Principle of All-organic Dyes for Dye-Sensitized Solar Cells. *Chem-Eur J 19*(17), 5220–5230.
12. Geiger, T., Kuster, S., Yum, J.-H., Moon, S.-J., Nazeeruddin, M. K., Grätzel, M., & Nüesch, F., (2009). Molecular Design of Unsymmetrical Squaraine Dyes for High Efficiency Conversion of Low Energy Photons into Electrons Using TiO_2 Nanocrystalline Films. *Adv Funct Mater 19*(17), 2720–2727.
13. Mishra, A., Fischer, M. K. R., & Bauerle, P., (2009). Metal-Free Organic Dyes for Dye-Sensitized Solar Cells: From Structure: Property Relationships to Design Rules. *Angew Chem Int Edit 48*(14), 2474–2499.
14. Hagfeldt, A., Boschloo, G., Sun, L., Kloo, L., & Pettersson, H., (2010). Dye-Sensitized Solar Cells. *Chemical Reviews110*(11), 6595–6663.
15. Hagberg, D. P., Marinado, T., Karlsson, K. M., Nonomura, K., Qin, P., Boschloo, G., Brinck, T., Hagfeldt, A., & Sun, L., (2007). Tuning the HOMO and LUMO energy levels of organic chromophores for dye sensitized solar cells. *J Org Chem 72*(25), 9550–9556.
16. Peters, A. T., & Freeman, H. S., *Modern Colorants: Synthesis and Structures*. Blackie Academic & Professional: 1995.

17. Hwang, S., Lee, J. H., Park, C., Lee, H., Kim, C., Park, C., Lee, M.-H., Lee, W., Park, J., Kim, K., Park, N.-G., & Kim, C., (2007). A highly efficient organic sensitizer for dye-sensitized solar cells. *Chemical Communications 46*, 4887–4889.
18. Daeneke, T., Kwon, T. H., Holmes, A. B., Duffy, N. W., Bach, U., & Spiccia, L., (2011). High-efficiency dye-sensitized solar cells with ferrocene-based electrolytes. *Nature Chemistry 3*(3), 211–215.
19. Adamo, C., & Barone, V., (1999). Toward reliable density functional methods without adjustable parameters: The PBE0 model. *Chinese Journal of Chemical Physics 110*(13), 6158–6170.
20. (a) Tao, J., Tretiak, S., & Zhu, J.-X., (2008). Absorption Spectra of Blue-Light-Emitting Oligoquinolines from Time-Dependent Density Functional Theory. *The Journal of Physical Chemistry B 112*(44), 13701–13710; (b) Adamo, C., & Barone, V., (1999). Accurate excitation energies from time-dependent density functional theory: assessing the PBE0 model for organic free radicals. *Chemical Physics Letters 314*(1–2), 152–157.
21. Frisch, M. J., Trucks, G. W., Schlegel, H. B., Scuseria, G. E., Robb, M. A., Cheeseman, J. R. et al. Fox *Gaussian 09*, Gaussian Inc.: Wallingford CT, 2009.
22. (a) Pavone, M., Cimino, P., De Angelis, F., & Barone, V., (2006). Interplay of Stereoelectronic and Enviromental Effects in Tuning the Structural and Magnetic Properties of a Prototypical Spin Probe: Further Insights from a First Principle Dynamical Approach. *Journal of the American Chemical Society 128*(13), 4338–4347; (b) Crescenzi, O., Pavone, M., De Angelis, F., & Barone, V., (2004). Solvent Effects on the UV ($n \rightarrow \pi^*$) and NMR (13C and 17O) Spectra of Acetone in Aqueous Solution. An Integrated Car–Parrinello and DFT/PCM Approach. *The Journal of Physical Chemistry B 109*(1), 445–453.
23. Cossi, M., Rega, N., Scalmani, G., & Barone, V., (2003). Energies, structures, and electronic properties of molecules in solution with the C-PCM solvation model. *J Comput Chem 24*(6), 669–681.
24. Roy Dennington, Todd Keith, & Millam, J. *GaussView, Version 5*, Semichem Inc. Shawnee Mission KS: 2009.
25. Minkin, V. I., (1999). Glossary of terms used in theoretical organic chemistry (IUPAC recommendations 1999). *Pure Appl Chem 71*(10), 1919–1981.
26. Hückel, E., (1931). Quantentheoretische Beiträge zum Benzolproblem. *Z. Physik 70*(3–4), 204–286.
27. (a) Wannere, C. S., & Schleyer, P. V. R., (2003). How Aromatic Are Large $(4n + 2)\pi$ Annulenes? *Organic Letters 5*(6), 865–868; (b) Wannere, C. S., Sattelmeyer, K. W., Schaefer, H. F., Schleyer, P. V. R., (2004). Aromaticity: The Alternating C_C Bond Length Structures of [14]- [18]-, and [22]Annulene. *Angewandte Chemie 116*(32), 4296–4302.
28. Jug, K., & Fasold, E., (1987). Structure and aromaticity of 14-annulene and 18-annulene. *Journal of the American Chemical Society 109*(8), 2263–2265.
29. Kennedy, R. D., Lloyd, D., & McNab, H., (2002). Annulenes, 1980–2000. *Journal of the Chemical Society, Perkin Transactions 1, 14*, 1601–1621.
30. Gellini, C., & Salvi, P. R., (2010). Structures of Annulenes and Model Annulene Systems in the Ground and Lowest Excited States. *Symmetry 2*(4), 1846–1924.

31. Nazeeruddin, M. K., Kay, A., Rodicio, I., Humphry-Baker, R., Mueller, E., Liska, P., Vlachopoulos, N., & Graetzel, M., (1993). Conversion of light to electricity by cis-X2bis(2,2'-bipyridyl-4,4'-dicarboxylate)ruthenium(II) charge-transfer sensitizers (X = Cl-, Br-, I-, CN-, and SCN-) on nanocrystalline titanium dioxide electrodes. *Journal of the American Chemical Society 115*(14), 6382–6390.
32. Yang, L., Guo, L., Chen, Q., Sun, H., Liu, J., Zhang, X., Pan, X., & Dai, S., (2012). Theoretical design and screening of panchromatic phthalocyanine sensitizers derived from TT1 for dye-sensitized solar cells. *Journal of Molecular Graphics and Modelling 34*(1), 1–9.
33. Kavan, L., Gratzel, M., Gilbert, S. E., Klemenz, C., & Scheel, H. J., (1996). Electrochemical and photoelectrochemical investigation of single-crystal anatase. *Journal of the American Chemical Society 118*(28), 6716–6723.

CHAPTER 7

THEORETICAL STUDIES ON ADSORPTION OF ORGANIC MOLECULES ON METAL SURFACE

G. SARANYA and K. SENTHILKUMAR

Department of Physics, Bharathiar University, Coimbatore–641046, India, Fax: +91-422-2422387, E-mail: ksenthil@buc.edu.in

CONTENTS

ABSTRACT

Over the past two decades the science and engineering of organic semiconducting materials have improved very rapidly, leading to the optimization and demonstration of a range of organic-based solid-state devices. A key to design high performance molecular electronic device is the understanding of function and characteristics of organic materials while they are in-contact with metal. Hence, a systematic theoretical investigation on the structure and electronic properties of the organic molecule/metal interfaces is required to find the ideal material sets with improved properties for use in next generation organic electronic devices. Finding the adsorption geometry of organic molecule on the metal surface and their electronic properties is a challenging task while optimizing the new electronic devices. In this chapter, the first principle based theoretical methods to predict the stable adsorption geometry of organic molecule on the metal surface is discussed. The electronic properties, such as change in work function, density of states (DOS), population analysis, frontier molecular orbital analysis and Schottky barrier height (SBH) are used to characterize the charge transport properties of organic molecule/metal interface. The interface between pentacene organic molecule and Pd(100) surface and the electronic properties are discussed.

7.1 INTRODUCTION

Generally metals and inorganic semiconductors have been widely used in electronic components, because of their excellent conductive properties, and appear ubiquitously in everyday life in the form of copper wires and silicon microchip technology. The less flexible and environmental hazard nature of these materials demand for new functional materials with improved properties. A major contribution comes from organic materials which can be used at molecular level as active and passive electronic components, thus allowing the development of nanoscale electronic devices.

The interest in this field of organic conductors has arisen from the discovery of conduction in doped polyacetylenes in 1977 [1]. Recently, the organic semiconducting materials have emerged as a new class of material for opto-electronic devices due to their remarkable properties [2–4]. The organic materials have mechanical flexibility, less weight, semiconducting and luminescent properties. As a result, successful applications have recently been realized for π-conjugated organic molecules. Recent years have witnessed advanced experimental and theoretical studies to find the suitable organic materials and organic/metal interfaces for molecular electronic applications [5–11]. The promising organic electronic devices are organic light emitting diodes (OLEDs), dye sensitized solar cells

(DSSC), organic field effect transistors (OFETs), organic photovoltaic cells (OPVs) [12–19]. Hence, the physical, chemical and conductive nature of these materials is still an active and important area of research and development.

The adsorption of organic molecules on metal and dielectric surfaces received considerable attention due to its promising role in molecular electronics [20, 21]. A key to design high performance molecular electronic device is understanding the function and characteristics of organic semiconducting materials while they are in-contact with metal or dielectric [20]. The interaction between organic semiconductors and metal electrodes determines the charge transfer across the interface which is crucial for the performance of organic electronic devices. The energy offset between the metal Fermi level and the HOMO/LUMO level of the organic films play an important role in controlling the hole/electron injection from the anode/cathode to the transport material. Device performance can be understood and improved if a detailed picture of the orbital alignment of the organic molecule with the Fermi level of the metal is known.

The adsorption of organic molecule on the metal surface is either physisorption or chemisorption and is always an exothermic process. In physisorption the attractive force between the substrate (metal) and the adsorbate (molecule) is due to van der Waals interactions and electrostatic interactions [22]. For physisorption, the interaction energy between the substrate and adsorbate is typically less than –20 kJ/mol [23] and the distance between adsorbate and substrate is higher than the length of covalent or ionic bonds. Chemisorption was first proposed by the American chemist Irving Langmuir in 1916. The adsorbate is bound to the substrate by covalent or ionic bond, much like those that occur between the bonded atoms of a molecule [22]. This is a slow process compared to physisorption. In chemisorption, a bond of the molecule may be broken and new chemical bond is formed between the adsorbate and the substrate. In chemisorption the interaction energy between the substrate and adsorbate is large, i.e., values above –40 kJ/mol [24]. In addition, the length of the substrate-adsorbate bond is comparable with the normal covalent or ionic bonds. Because the chemisorption involves the formation of chemical bond with the metal surface, only a single layer of molecule, or a monolayer, can chemisorb to the surface.

Adsorbed species on single crystal surface is frequently found to exhibit long-range ordering with well-defined over layered structure. Based on the contribution of the π-electrons and surface coverage of the organic molecule, there are three possible adsorption geometries. They are flat-lying, tilted and standing-up or vertical configurations and are shown in Figure 7.1. Of these different configurations and adsorption sites, the most stable configuration is found by calculating adsorption or binding energy.

7.2 DFT IMPLEMENTATION FOR SOLIDS

Density functional theory (DFT) has long been used for electronic structure calculations in solid-state physics and it is an extremely successful approach for the description of ground state properties of metals, semiconductors, and insulators. The success of DFT not only encompasses standard bulk materials but also complex materials such as proteins and carbon nanotubes. For the past three decades it became very popular method in quantum chemistry because of its accuracy and less computational cost.

The basic idea of DFT is that the energy of a system is expressed as a functional of the electron density function. This representation provides a

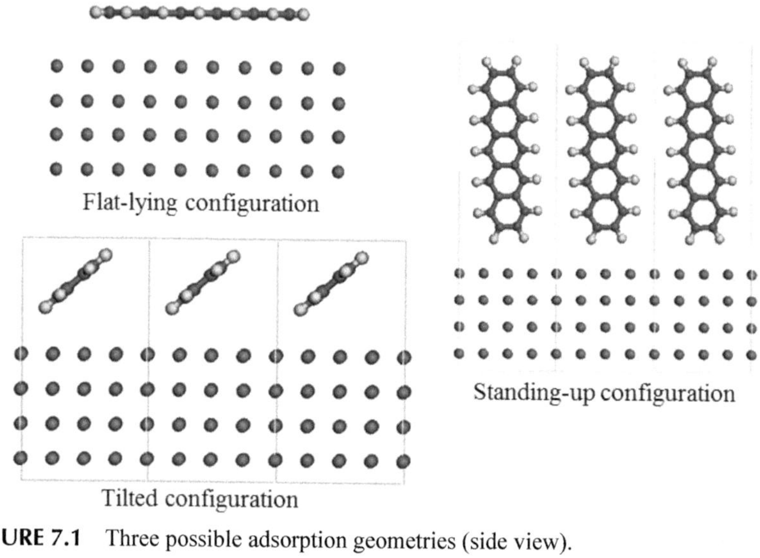

FIGURE 7.1 Three possible adsorption geometries (side view).

considerable simplification over the traditional electronic structure theory, in which the basic variable is the electronic wavefunction, ψ, which depends simultaneously on the coordinates of all the electrons (3N variables for N electrons). The electron density function $\rho(r)$ depends only on three variables.

The foundations of modern DFT are the Hohenberg-Kohn theorems [25], which show that the ground state density uniquely defines the system, and that the density can be derived from a variational procedure. In DFT, the energy functional can be written as,

$$E[\rho]=\int v(r)\rho(r)dr+J(\rho)+T_S(\rho)+E_{XC}(\rho) \tag{1}$$

The first term on the right-hand side of equation (1) represents the electrostatic interaction of the electron density with the 'external potential', this simply corresponds to the Coulombic interactions between electrons and nuclei. The second term is the Coulomb energy corresponding to the repulsion between the electrons. This term is non-zero even for one-electron systems because it includes electron self-interaction. The next term is an approximation to the electronic kinetic energy. Early attempts to describe the kinetic energy as a functional of the density (e.g., Thomas-Fermi theory) [26, 27] were not very successful. Instead in Kohn-Sham approach [28], the density is expanded in a set of orbitals, and the kinetic energy for this hypothetical system (in which the kinetic energy of the electrons is not affected by their interactions) is computed using the same expression as in traditional wavefunction based methods. The final term, the exchange-correlation functional corrects the first three terms. This term also corrects the spurious self-interaction introduced by the Coulomb energy and accounts the electron correlation.

The energy expression given in Eq. (1) leads to a set of equations defining the shape of the orbitals that are used to expand the density. These are the Kohn-Sham equations [28]. Solving these equations lead in principle to the exact energy and density of the target system. However, the degree of accuracy obtained depends on the form given to the exchange-correlation functional.

Commonly two different approaches have been used to solve the Kohn-Sham equations. While studying atoms and molecules, the Kohn-Sham orbitals are expressed as a linear combination of atomic centered basis functions,

$$\psi_i(r) = \sum_{\upsilon}^{k} C_{\upsilon i}\phi_{\upsilon} \tag{2}$$

where, $C_{\upsilon i}$ are the orbital co-efficients and ϕ_{υ} is the basis function. Another approach involves the use of plane waves and pseudopotentials. This approach is important particularly for the study of bulk systems such as metals and alloys [29].

7.2.1 BLOCH THEOREM AND PLANE WAVES

In a periodic system like crystals, the Hamiltonian appearing in the Kohn-Sham equation will have the lattice periodicity. According to Bloch's theorem [30], the wavefunction $\psi_{j,k}$ of an electron within a periodic potential is written as the product of a lattice part $u_j(r)$ and a wavelike part $e^{ik.r}$.

$$\psi_{j,k(r)} = u_j(r)\, e^{ik.r} \tag{3}$$

where the subscript j indicates the band index and k is a continuous wave vector that is confined to the first Brillouin zone of the reciprocal lattice. Since $u_j(r)$ has the same periodicity as the direct lattice, it can be expressed in terms of a discrete plane wave basis set with reciprocal lattice vectors of the crystal as the wave vector (G) i.e.,

$$u_j(r) = \sum_{G} C_{j,G}\, e^{iG.r} \tag{4}$$

where, $G.r = 2\pi m$ in which m is an integer, r is the crystal lattice vector and $C_{j,G}$ are the plane wave coefficients. Hence the electron wavefunction can be expanded in terms of a linear combination of plane waves,

$$\psi_{j,k} = \sum_{G} C_{j,k+G}\, e^{i(k+G).r} \tag{5}$$

In principle, an infinite plane wave basis is needed to expand the wavefunctions, however, for practical implementations the basis set must be finite. Typically, the plane waves with small kinetic energies is given by

$$E_k = \frac{\hbar^2 |k+G|^2}{2m} \quad (6)$$

The solutions with lower energies are more physically important than solutions with very high energies. The introduction of a plane wave energy cut off reduces the basis set to a finite size.

$$E_{cut} = \frac{\hbar^2}{2m} G_{cut}^2 \quad (7)$$

The infinite sum reduces to

$$\psi_{j,k} = \sum_{|G+k|<G_{cut}} C_{j,k+G}\, e^{i(k+G).r} \quad (8)$$

A good approximation to the total energy can therefore be achieved by only including plane waves that have kinetic energy less than cutoff energy. Using the plane waves, the Kohn-Sham equation can be written as [31],

$$\sum_G \left\{ \begin{array}{l} \frac{\hbar^2}{2m}|k+G|^2 \delta_{GG'} + V_{en}\left(G-G'\right) \\ +V_{ee}\left(G-G'\right) + V_{XC}\left(G-G'\right) \end{array} \right\} C_{j,k+G'} = C_{j,k+G'}\, \varepsilon_i \quad (9)$$

where, $V_{ee}(G–G)$, $V_{ve}(G–G')$ and $V_{XC}(G–G')$ are the Fourier transforms of electron-electron, electron-nuclei and exchange correlation potentials. Plane waves have a number of advantages because,

- It is easy to calculate all kinds of matrix elements through fast Fourier transform techniques.
- The size of basis set can be increased systematically in a simple way.
- The same basis set can be used for all atomic species.
- Forces acting on atoms are equal to Hellmann-Feynman forces.
- Convergence towards completeness can be tested.
- The plane waves do not depend on nuclear positions, so unlike localized basis sets, correlation terms are not needed for the calculation of forces.
- Plane wave basis sets are free from 'basis set superposition error' (BSSE).

The main disadvantage is that the basis sets become large. For finite systems such as clusters and model bulk systems, it is essential to construct a supercell in each dimension, in which the system is localized. In order to neglect any interactions with the images, the supercell should be large enough and a large number of plane waves should be used.

7.2.2 PSEUDOPOTENTIAL APPROXIMATION

With the standard Coulomb potential model, including the potential terms corresponding to the motion of the core electrons and the interactions of core electrons with the nucleus would lead to a very complicated plane-wave basis set. This in turn would take a huge amount of computational time, especially when dealing with large unit cells or atoms with many electrons. Most physical and chemical properties of crystals depend on the distribution of valence electrons rather than the tightly bound core electrons. The core electrons do not participate in the chemical bond. They are strongly localized around the nucleus, and their wavefunctions overlap is very little with the core electron wavefunctions of the neighboring atom. Therefore, the distribution of the core electrons basically does not change when the atoms are placed in the different chemical environment. Thus the core electrons are assumed to be "frozen" in the crystal environment. With this frozen core approximation, one can have less number of electrons and hence less eigenstates of the Kohn-Sham equation to be calculated. The second advantage is that the total energy scale is largely reduced when the core electrons are removed from the calculation which makes the calculation of the energy difference between atomic configurations numerically much more stable.

In order to make plane wave approaches to be of practical use and to obtain smoothly varying wavefunctions, pseudopotentials are introduced to replace the Coulomb potential of the electron-nucleus interaction. The pseudopotential (PP) approximation was first introduced by Hans Hellman in 1934. By introducing PPs, two goals can be achieved. First, the core electrons can be removed from the calculations. The contribution of the core electrons to the chemical bonding is negligible but they contribute mostly to the total energy of the system. The required accuracy for the total energy will be of much less important than all electrons energy. Second, by introducing PPs the true valence wavefunctions is replaced by the pseudo wavefunctions which match exactly with the true valence wavefunctions outside the ionic core region but are nodeless inside. These pseudo wavefunctions can be expanded using plane wave basis sets. A further advantage of pseudopotential is that relativistic effects can be incorporated easily into the potential by treating the valence electrons non-relativistically. The Figure 7.2 shows the difference between the Coulomb and pseudopotentials as well as the real and pseudo wavefunctions. The Coulomb potential is much steeper and deeper than the pseudopotential. As part of the requirements of a proper pseudopotential, the real and pseudo wavefunctions and potentials must match up at certain radius (the cut off radius, r_c). When using the PPs, only the valence electrons will contribute to bonding and only these electrons are explicitly dealt with, while the nucleus and core electrons are considered as an effective potential [32].

7.2.2.1 Norm-Conserving Pseudopotentials

The concept of "norm-conservation" has a special place in the development of ab initio pseudopotentials, it simplifies the application of the pseudopotentials and it makes them more accurate and transferable. Norm-conserving pseudofunctions $\psi^{PS}(r)$ are normalized and are solutions of a model potential chosen to reproduce the valence properties of

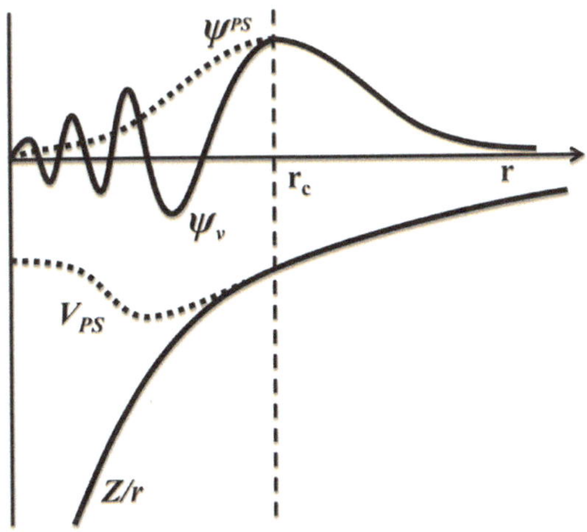

FIGURE 7.2 Schematic representation of pseudopotential.

an all electron calculations [33]. The norm-conserving pseudo potentials (NCPPs) are developed based on the following conditions proposed by Hamann, Schuter, and Chiang (HSC) [34].

1. All electron and pseudo valence eigenvalues agree for the chosen atomic reference configuration.
2. All electron and pseudo valence wavefunctions agree beyond the chosen core radius r_c.
3. The integrated charge inside r_c for all electron wavefunction and pseudo wavefunction are same.
4. The logarithmic derivatives of the all electron and pseudo wavefunctions should coincide at r_c.
5. The first energy derivative of the logarithmic derivatives of the all electron wavefunctions and pseudo wavefunctions should coincide at $r = r_c$.

The above points 1 and 2 infer that the NCPP is equal to the atomic potential outside the 'core region' of radius r_c, because the potential is uniquely determined by the wavefunction and the energy ε, that need not be an eigen energy. Inside r_c, the pseudopotential and radial pseudo-orbital $\psi_l^{PS}(r)$ differ from their all electron counterparts; however, the point 3 requires that the integrated charge,

$$Q_l = \int_0^{r_c} r^2 |\psi_l(r)|^2 dr = \int_0^{r_c} dr\, \phi_l(r)^2 \tag{10}$$

is same for all electron wavefunction and pseudo wavefunction. The conservation of Q_l ensures that, the total charge in the core region is correct and the normalized pseudo-orbital is equal to the true orbital at $r > r_c$.

According to the point 4, for any smooth potential the wavefunction $\psi_l(r)$ and its radial derivative $\psi_l'(r)$ are continuous at r_c. The dimensionless logarithmic derivative of pseudopotential D is defined as,

$$D_l(\varepsilon,r) \equiv \frac{r\psi_l'(\varepsilon,r)}{\psi_l(\varepsilon,r)} = r\frac{d}{dr}\ln\psi_l(\varepsilon,r) \tag{11}$$

The point 5 is a crucial step towards the goal of constructing a good pseudopotential. In a molecule or solid when bonding or reaction take place the wavefunctions and eigenvalues will change and the pseudopotential that satisfies the point 5 will reproduce the changes in the eigenvalues linearly with respect to change in the self-consistent potential.

7.2.2.2 Non-Local Pseudopotentials

In local pseudopotential each state of angular momentum l and m are treated independently except that the total potential is calculated self consistently within the given approximation for exchange and correlation and for the given configuration of the atom. The next step is identifying the valence states and generating the pseudopotentials $V_l(r)$ and pseudo orbitals $\psi_l^{PS}(r) = r\phi_l^{PS}(r)$. The procedure varies with different approaches, but common task is finding the total "screened" pseudopotential acting on the valence electrons in an atom [35]. The screened Hartree and exchange correlation pseudopotential, $V_{HXC}^{PS}(r)$ can be written as,

$$V_{HXC}^{PS}(r) = V_{Hartree}^{PS}(r) + V_{XC}^{PS}(r)$$

For few applications, it is sufficient to consider screened pseudopotentials. But, for many applications, it is necessary to consider unscreened pseudopotentials, which is obtained from the screened pseudopotential after subtracting the screening part of the potential from the total potential and is written as,

$$V_l(r) = V_{l,total}(r) - V_{HXC}^{PS}(r) \tag{12}$$

Here, $V_l(r)$ and $V_{l,total}(r)$ are depending on angular momentum, l and $V_{HXC}^{PS}(r)$ is defined for the valence electrons in their pseudo orbitals [33]. It is useful to define the ionic pseudopotential as a local (l independent) part of the potentials plus non local terms,

$$V_l(r) = V_{local}(r) + \delta V_l(r) \tag{13}$$

where $V_{local}(r)$ is the local pseudopotential. When $r > r_c$, since the eigenvalues and orbitals are required to be the same for the pseudo and all electron cases, the potential $V_l(r)$ is equal to the local (l independent) all electron potential, that is, $V_l(r) \to -\frac{Z_{ion}}{r}$ for $r \to \infty$. Thus $\delta V_l(r) = 0$ for $r > r_c$

and all the long range effects of the Coulomb potential are included in the local potential, $V_{local}(r)$. The semilocal operator can be written as,

$$\hat{V}_{SL} = V_{local}(r) + \sum_{lm} |Y_{lm}\rangle \delta V_l(r) \langle Y_{lm}| \tag{14}$$

This semilocal potential is computationally very expensive. The number of matrix elements which need to be calculated is equal to the square of the number of basis sets.

7.2.2.3 Kleinman-Bylander Pseudopotentials

To speed up the calculations with pseudopotentials in plane wave basis set methods, the fully separable form of the pseudopotential is introduced. Kleinman and Bylander (KB) [36] constructed a separable pseudopotential operator, i.e. $\delta V(r,r')$ which is written as a sum of products of the form $\sum_i f_i(r) g_i(r')$. KB showed that the potential $\delta V_l(r)$ in equation (13) can be replaced by a separable operator δV_{KB} so that the total pseudopotential has the form,

$$\hat{V}_{KB} = V_{local}(r) + \sum_{lm} \frac{|\psi_{lm}^{PS} \delta V_l\rangle \langle \delta V_l \psi_{lm}^{PS}|}{\langle \psi_{lm}^{PS} | \delta V_l | \psi_{lm}^{PS} \rangle} \tag{15}$$

where, $\langle \delta V_l \psi_{lm}^{PS}|$ is the projectors that operate upon the wavefunction. Unlike the semilocal operator, it is fully non local with respect to the angular variables θ, ϕ and radial variable r. When operating on the reference atomic states, ψ_{lm}^{PS} the $\delta\hat{V}_{KB}(r,r')$ acts the same as $\delta V_l(r)$, and it is an excellent approximation for the operation of the pseudopotential on the valence states in a molecule or solid [33].

The Kleinman-Bylander scheme is usually good. Nevertheless, for some cases an inappropriate cutoff might lead to an un-physical ghost states which leads unphysical situations, like the energy of the lowest ‘p’ states is lower than the energy of ‘s’ state Therefore, checking of ghost states is essential for Kleinman-Bylander pseudopotential generation [36].

7.2.2.4 Ultrasoft Pseudopotentials

The norm-conservation constraint is the main factor that determining the smoothness of pseudopotential, especially for 'p' states in first row elements and 'd' states in second-row transition metals. For these states, there is no core state of the same angular momentum to which they have to be orthogonal. Therefore, the wavefunction corresponding to core electrons is nodeless and quite compressed compared to the other valence states. Therefore, it requires a large number of plane waves, which often makes calculations for such elements prohibitively expensive. Generally, the goal of pseudopotential is to create pseudofunctions that are as smooth as possible and yet are accurate. A different approach known as ultrasoft pseudopotentials reaches the goal of accurate calculations by a transformation that re-expresses the problem in terms of a smooth function and an auxiliary function around each ion core that represents the rapidly varying part of the density [33].

Blöchl [37] and Vanderbilt [38] proposed a transformation for non-local potential $\delta\hat{V}_{NL} = \sum_{lm}\left[\sum_{s,s'} B_{s,s'}|\beta_s\rangle\langle\beta_{s'}|\right]_{lm}$ by introducing a smooth function $\tilde{\phi} = r\tilde{\psi}$ that is not norm-conserving. The difference between norm conserving integrated charge given in Eq. (10) and the charges corresponding to norm-conserving function $\phi = r\psi$ (either for all electron function or for pseudofunction) is given by,

$$\Delta Q_{s,s'} = \int_0^{r_c} \Delta Q_{s,s'}(r)\,dr \tag{16}$$

where

$$\Delta Q_{s,s'}(r) = \phi_s^*(r)\phi_{s'}(r) - \tilde{\phi}_s^*(r)\tilde{\phi}_{s'}(r) \tag{17}$$

A new non-local potential that operates on the $\tilde{\psi}_{s'}$ can be defined as

$$\delta\hat{V}_{NL}^{US} = \sum_{s,s'} D_{s,s'}|\beta_s\rangle\langle\beta_{s'}| \tag{18}$$

where

$$D_{s,s'} = B_{s,s'} + \varepsilon_{s'} \Delta Q_{s,s'} \tag{19}$$

For each reference atomic states s, it is straightforward to show that the smooth functions $\tilde{\psi}_s$ are the solutions of the generalized eigenvalue problem,

$$[\hat{H} - \varepsilon_s \hat{S}]\tilde{\psi}_s = 0 \tag{20}$$

with $\hat{H} = -\frac{1}{2}\nabla^2 + V_{local} + \delta\hat{V}_{NL}^{US}$ and $\hat{S}$ is an overlap operator,

$$\hat{S} = 1 + \sum_{s,s'} \Delta Q_{s,s'} |\beta_s\rangle\langle\beta_{s'}| \tag{21}$$

which is different from unity only inside the core radius. The eigenvalues ε_s agree with the all electron calculation at as much energy as desired. The full density can be constructed from the functions $\Delta Q_{s,s'}(r)$ which can be replaced by a smooth version of the all electron density.

The advantage of relaxing the norm-conservation condition, $\Delta Q_{s,s'} = 0$ is that each smooth pseudofunction $\tilde{\psi}_s$ can be formed independently, with the constraint of matching the value of the functions $\tilde{\psi}_s(r_c) = \psi_s(r_c)$ at the radius r_c. Thus, it becomes possible to choose r_c much larger than for a norm-conserving pseudopotential, while maintaining the desired accuracy by adding the auxiliary functions $\Delta Q_{s,s'}(r)$ and an overlap operator $\hat{S}$.

7.2.3 PROJECTOR AUGMENTED WAVE METHOD

The projector augmented wave (PAW) method is another general approach to solve the electronic structure problem. Like the ultrasoft pseudopotential method, it introduces projectors and auxiliary functions. The PAW approach defines a functional for the total energy that involves auxiliary functions and it uses advances in algorithms for efficient solution of the generalized eigenvalue problem. However, the PAW approach keeps the full all electron wavefunction. Since the full wavefunction varies rapidly near the nucleus, all integrals are evaluated as a combination of integrals of smooth functions extending throughout the space plus localized contributions evaluated by radial integration over muffin-tin spheres [33].

Let us consider the basic ideas of the definition of the PAW method for an atom. We can define a smooth part of a valence wavefunction $\tilde{\psi}_i^{\upsilon}(r)$ and a linear transformation $\psi^{\upsilon} = \mathrm{T}\tilde{\psi}^{\upsilon}$ that relates the set of all electron valence functions $\psi_i^{\upsilon}(r)$ to the smooth functions $\tilde{\psi}_i^{\upsilon}(r)$. The transformation is assumed to be unity except with a sphere centered on the nucleus

$$T = 1 + T_0 \tag{22}$$

For simplicity, we omit the superscript υ, assuming that the ψ's are valence states. Adopting the Dirac notation, the expansion of each smooth function $|\tilde{\psi}\rangle$ in partial waves m within each sphere can be written as

$$|\tilde{y}\rangle = \sum_m C_m |\tilde{y}_m\rangle \tag{23}$$

with the corresponding all electron function,

$$|\psi\rangle = T|\tilde{\psi}\rangle = \sum_m C_m |\psi_m\rangle \tag{24}$$

Hence the full wavefunction in all space can be written as,

$$|\psi\rangle = |\tilde{\psi}\rangle + \sum_m C_m \{|\psi_m|\tilde{\psi}\rangle\} \tag{25}$$

If the transformation T is required to be linear, then the coefficients must be given by a projection in each sphere

$$C_m = \langle \tilde{p}_m | \tilde{\psi} \rangle \tag{26}$$

for some set of projection operators $\tilde{p}$. If the projection operators satisfy the biorthogonality condition,

$$\langle \tilde{p}_m | \tilde{\psi}_{m'} \rangle = \delta_{mm'}$$

then the one center expansion $\sum_m |\tilde{\psi}_m\rangle\langle \tilde{p}_m|\tilde{\psi}\rangle$ of the smooth function $\tilde{\psi}$ equals $\tilde{\psi}_m$. Just as for pseudopotentials, there are many possible choices for the projector. The difference between pseudopotentials and PAW is that the transformation T involves the full all electron wavefunction

$$T = 1 + \sum_{m} \{ |\psi_m\rangle - |\tilde{\psi}_m\rangle \} \langle \tilde{p}_m | \tag{27}$$

Furthermore, the expressions apply equally well to core and valence states so that we can derive all electron results by applying the expressions to all electron states.

The general form of the PAW equations can be written in terms of transformation equation (27). For any operator $\hat{A}$ in the original all electron problem, one can introduce a transformed operator $\tilde{A}$ that operates on the smooth part of the wavefunctions,

$$\tilde{A} = T\hat{A}T = \hat{A} + \sum_{mm'} |\tilde{p}_m\rangle \{ \langle \psi_m | \hat{A} | \psi_{m'} \rangle - \langle \tilde{\psi}_m | \hat{A} | \tilde{\psi}_{m'} \rangle \} \langle \tilde{p}_{m'} | \tag{28}$$

One can add the right hand side of equation (27) to any operator of the form,

$$\hat{B} - \sum_{mm'} |\tilde{p}_m\rangle \langle \tilde{\psi}_m | \hat{B} | \tilde{\psi}_{m'} \rangle \langle \tilde{p}_{m'} |$$

without changing its expectation value. The expressions for physical quantities in the PAW approach follow from Eqs. (27) and (28). The density is given by,

$$n(r) = \tilde{n}(r) + n^1(r) - \tilde{n}^1(r)$$

where the first term is a smooth part arising from the density of pseudo (PS) wavefunctions, while the last two terms are depend on the true all electron and PS partial waves respectively, within each augmentation region. A similar decomposition can also be done for the total energy. By effectively separating the problem into different parts, the PAW method allows for an efficient treatment of separated problems, while at the same time it offers the possibility for implementation of the PAW method in plane wave pseudopotential codes. Hence, the PAW method has become a very useful method, with the ability of performing accurate calculations with a limited basis set size [39].

7.3 EXAMPLE: STRUCTURE AND ELECTRONIC PROPERTIES OF PENTACENE/Pd(100)* INTERFACE

Pentacene (PEN) is an aromatic molecule composed of five linearly fused benzene rings, which is the most intensively studied p-type organic semiconductor and serves as a good point of reference. PEN has recently been identified as one of the most promising candidates for organic thin film transistors (OTFTs) [40], OFETs [41], OLEDs [42] and photovoltaic devices [43] due to its high hole mobility up to 5.5 cm^2/Vs in OFETs [44]. Both single crystal and thin films of PEN have been used for fabrication of transistor devices. Previously, adsorption of PEN on the metal surfaces Cu(111) [45], Cu(100) [46], Ag(111) [47], Ag(110) [48], Al(100) [49–51], Fe(100) [52], Au(111) [53] and Au(001) [54] were studied using DFT method.

In the present work, we have studied the adsorption of PEN molecule on palladium (100) surface. Palladium is frequently used in various chemical and electronic applications due to its chemical stability and excellent electrical conductivity. Palladium is used as an alternative to gold in various electronic components, such as hybrid integrated circuits (HIC), lead frames, soldering materials, multilayer ceramic capacitors (MLCC) [55] and palladium-silver alloys are used as electrodes [56].

The adsorption characteristics of single PEN molecule on Pd(100) surface of c(5x5) unit cell was studied by using first-principle DFT methods with van der Waals correction. For the stable adsorption geometry, the electronic properties like population analysis, density of states (DOS), partial density of states (PDOS), band structure and Schottky barrier heights (SBH) are calculated to study the charge transfer along PEN/Pd(100) interface. Further, change in work function of the Pd(100) surface due to the adsorption of PEN also calculated.

7.3.1 COMPUTATIONAL DETAILS

The electronic structure calculations are performed within the framework of DFT using Vienna Ab initio Simulated Package (VASP) code [57–59]. In the VASP calculations, plane wave basis sets are employed to expand the electronic wave functions. The exchange and correlation functional has been parameterized according to the Perdew-Burke Ernzerhof gradient corrected exchange functional [60] within the generalized gradient approximation (GGA). Electron-ion interactions were described by the projected augmented wave (PAW) method of Kresse, Joubert and Blöchl [61]. While solving the Kohn-Sham equations, the electronic wavefunctions are expanded by plane waves up to a kinetic energy of 400 eV. The occupation of electronic states is calculated using a Methfessel-Paxton smearing of 0.2 eV. The Brillouin-zone sampling was carried out according to the Monkhorst-Pack [62] scheme with 2 × 1 × 1 k-points meshes. The Pd(100) surface is constructed by using a periodically repeated slab approach with super cell geometry of c(5 × 5) surface unit cell. This slab consists of four layers with a vacuum region of 10 Å. A single PEN molecule is deposited on the upper surface of the slab with the aromatic rings parallel to the surface. We used a conjugate-gradient algorithm in all the calculations based on the reduction in the Hellman-Feynman forces on each constituent atom to less than 10 meV.

Since DFT calculations are unable to describe the van der Waals (vdW) interactions between molecules and substrate or between molecules, an empirical dispersion-corrected DFT-D2 method proposed by Grimme et al. [63, 64] has also been used in the DFT calculations. Within this method, the total energy (E_D) of the system is defined as a sum of the self-consistent Kohn-Sham energy (E_{KS}) and a semiempirical dispersion correction (E_{disp}). That is, in DFT-D2 method the total energy can be expressed as,

$$E_{DFT-D} = E_{KS-DFT} + E_{disp} \tag{29}$$

Since the vdW interaction is basically a long-range dynamical dipole-dipole interaction, the dispersion correction term (E_{disp}) can be evaluated using semi-empirical pair potentials, i.e.,

$$E_{disp} = -s_6 \sum_{i=1}^{i=N-1} \sum_{j=i+1}^{j=N} \frac{C_6^{ij}}{R_{ij}^6} f_{damp}\left(R_{ij}\right) \tag{30}$$

where, s_6 is a scaling factor, N is the total number of atoms, R_{ij} is the distance between the atoms i and j and the parameter C_6^{ij} describes the strength of the dispersion interaction between atoms i and j. The parameter C_6^{ij} can be evaluated as $\left(C_6^i C_6^j\right)^{1/2}$, where C_6^i is a semi-empirical parameter optimized for each type of atom. The damping function $f_{damp}(R_{ij})$ eliminates the dispersion correction at short distance where the DFT energy works well. Recent studies show that the DFT-D2 method successfully describes the long range interactions and the calculated adsorption energy of organic molecule on the metal surface is in good agreement with the experimental values [65–67].

7.3.2 ADSORPTION GEOMETRY

The geometry of isolated PEN and clean Pd(100) surface was optimized. The PEN molecule has a rod-like planar structure and it is shown in Figure 7.3. The optimized bond lengths and bond angles of the PEN molecule are in good agreement with the experimental [68] and previous theoretical results [54]. The lattice constant of the relaxed Pd(100) surface is found to be 3.88 Å, which is in agreement with the previous theoretical [69] and experimental [70] value of 3.98 and 3.89 Å, respectively.

As we discussed earlier, the PEN molecule may have different orientations and the position on the metal surface. In the present work, parallel (flat-lying) orientation of the PEN on Pd(100) surface is considered rather than the perpendicular orientation. Because, previous experimental and theoretical studies show that polycyclic aromatic molecules tend to adsorb on a metal surface in a flat lying geometry [71, 72]. Three adsorption geometries with respect to center of the molecule namely, top, hollow and bridge sites on Pd(100) surface are shown in Figure 7.4.

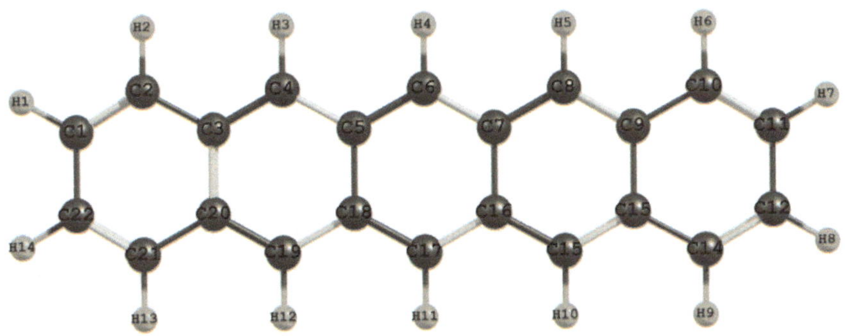

FIGURE 7.3 The optimized structure of pentacene molecule.

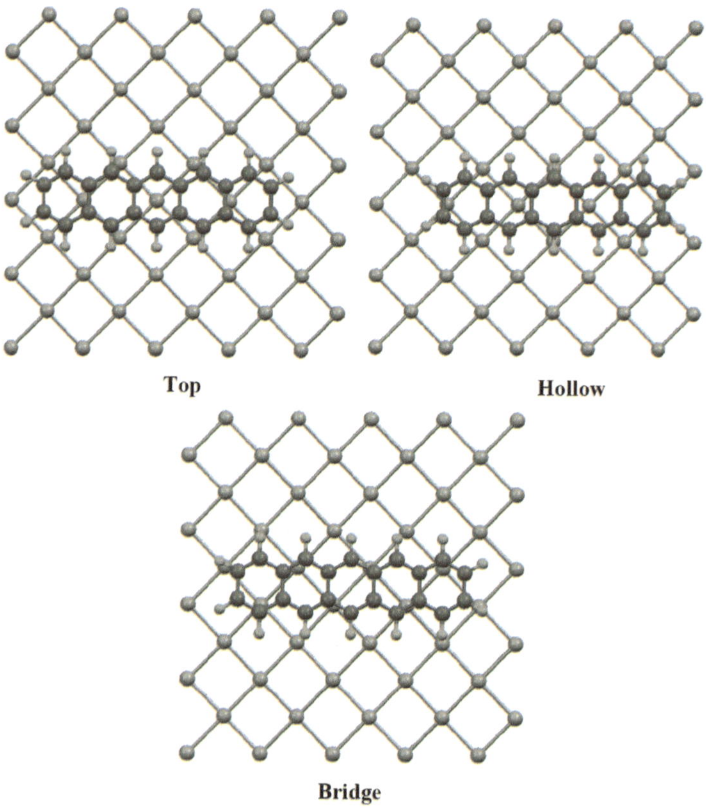

FIGURE 7.4 Orientation of the pentacene molecule on Pd(100) surface. (Only the top most of Pd atoms are shown for clarity). Palladium, carbon and hydrogen atoms are shown in blue, violet and green colors, respectively.

The adsorption energy (E_{ads}) of PEN on Pd(100) surface was calculated by subtracting the total energies of isolated PEN molecule (E_{PEN}) and Pd(100) surface ($E_{Pd(100)}$) from the total energy of the adsorbed system ($E_{Pd\text{-}PEN}$). i.e.,

$$E_{ads} = E_{Pd\text{-}PEN} - E_{Pd(100)} - E_{PEN}$$

The calculated E_{ads} values are –0.067, –0.93 and –1.55 eV, which corresponds to the top, hollow and bridge sites, respectively. These results show that the bridge site orientation is the stable orientation than the top and hollow sites. Previous theoretical study on adsorption of PEN molecule on the Ag(111), [47] Al(100) [50] and Au(001) [54] surface also reported that the bridge site orientation is the stable orientation.

Note that the above mentioned adsorption energy values were calculated with the vertical distance of 2.2 Å at 0° of rotational angle. To find the stable adsorption geometry with respect to the vertical distance between the PEN and Pd(100) surface, the adsorption energy calculations were performed for PEN/Pd(100) interface with different vertical distances in the bridge site configuration using GGA and DFT-D2 functionals. The adsorption energy as a function of vertical distances are shown in Figure 7.5(a). The maximum adsorption energy of –1.19 and –2.11 eV is found at a distance of 2.6 and 2.4 Å at GGA and DFT-D2 methods, respectively. With these vertical distances, the PEN molecule is rotated on the Pd(100) surface from 0° to 90° in the steps of 15° and the adsorption energy is calculated and is shown in Figure 7.5(b).

From Figure 7.5b, it has been observed that the adsorption energy is maximum at 45° of rotational angle. At 45° of rotational angle, the calculated adsorption energy is –2.10 eV and–3.12 eV at GGA and DFT-D2 methods. The most stable geometry of PEN on Pd(100) surface is shown in Figure 7.6.

As shown in Figure 7.6, in the most stable adsorption geometry, the chemical bonding is observed between the carbon atoms of PEN molecule and the palladium atoms in the upper surface of the slab. That is, the interaction between the PEN and Pd(100) surface is chemisorption. It has been studied in earlier studies that the adsorption of PEN [50] and perfluoropentacene [51] on Al (100) surface is physisorption with the adsorption energy of –0.36 eV and –0.27 eV. As expected, the GGA-PBE functional slightly underestimates the adsorption energy because of its poor description for dispersion interactions. Previous studies on the interaction between organic molecule and metal surface show that in comparison with the GGA functionals, the results obtained with DFT-D2 methods are in good agreement with the experimental results [66, 67]. Hence, DFT-D2 method is used to study the electronics properties for the most stable PEN/Pd(100) interface.

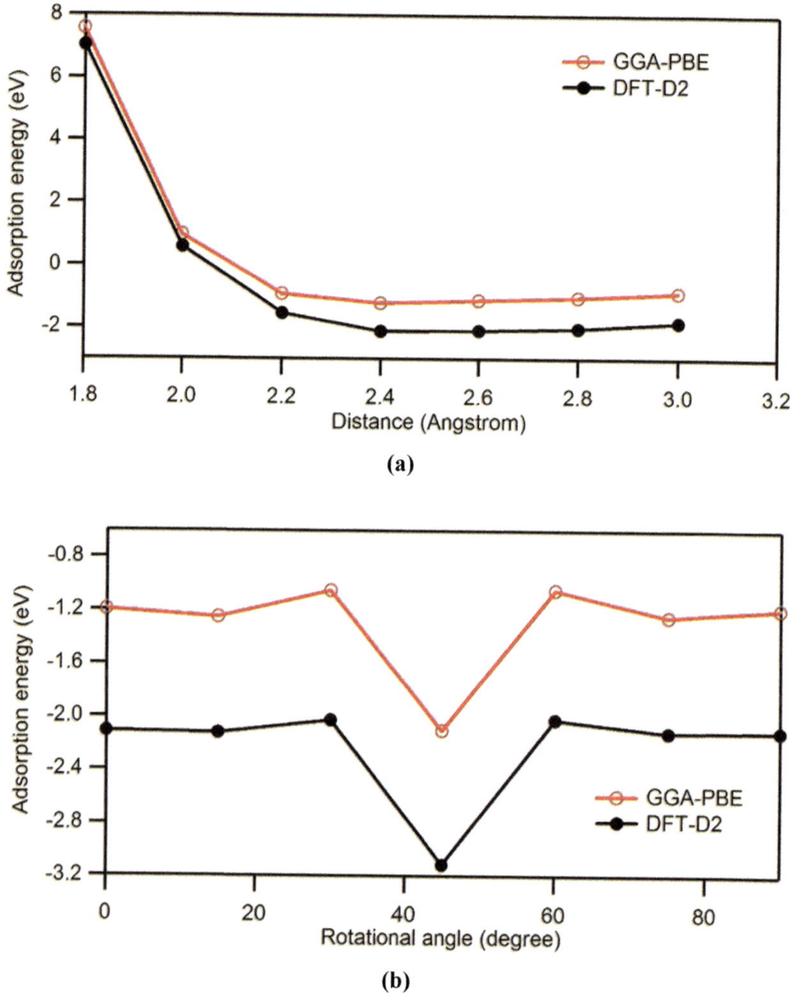

FIGURE 7.5 The calculated adsorption energy values with respect to (a) the vertical distance between the pentacene molecule and the Pd(100) suface and (b) the rotational angle of the pentacene molecule on the Pd(100) surface using GGA-PBE and DFT-D2 methods.

7.3.3 CHANGE IN WORK FUNCTION

The work function of the clean Pd(100) surface ($\phi_{Pd(100)}$) was calculated as the energy difference between the Fermi energy (E_F) of the surface and the

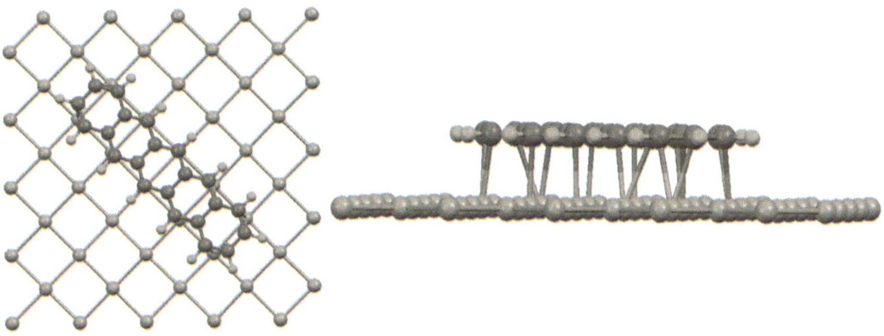

Top View **Side iew**

FIGURE 7.6 The most stable geometry of pentacene molecule on Pd(100) surface (fisrt layer) with the rotational angle of 45° and vertical distance of 2.4 Å.

electrostatic potential energy (V_{vac}) of an electron in the vacuum region at the distance where the potential energy has reached its asymptotic value. The calculated work function of the Pd(100) surface is 5.34 eV, which is in agreement with the experimental value of 5.65 eV [73]. The work function of the PEN/Pd(100) interface was calculated as described for the clean Pd(100) surface. The work function of the adsorbed system is found to be 3.91 eV. That is, the adsorption of PEN decreases the work function of the palladium surface by 1.43 eV. Previously, Li et al. [53] studied the electronic structure of PEN - gold interface and reported that the adsorption of PEN molecule reduced the work function of the Au(111) surface by 0.96 eV. The above results clearly shows that the adsorption of PEN significantly decreases the work function of the Pd(100) surface.

7.3.4 POPULATION ANALYSIS

The charge transfer between the PEN molecule and the Pd(100) metal surface is studied by calculating the atomic charge on the individual systems and on the adsorbed system by using a Mulliken atomic charge analysis of bond population [74]. The calculated Mulliken charges in the single molecule (q_M) and in the adsorbed system (q_{ads}) are summarized in Table 7.1. The charges on selected carbon and hydrogen atoms of PEN molecule are presented, because the similar trend was observed for all the other carbon

TABLE 7.1 Mulliken Charges (in Electron Charge Units) in the Isolated Pentacene Molecule (q_M) and in the Adsorbed System (q_{ads})

Atom	q_M	q_{ads}
C1	–0.239	–0.293
C2	–0.257	–0.296
C3	–0.240	–0.290
C4	–0.033	–0.066
C5	–0.222	–0.292
C6	–0.035	–0.069
H1	0.257	0.241
H2	0.263	0.246
H3	0.252	0.235
H4	0.251	0.237

*For numbering of atoms see Figure 7.3.

and hydrogen atoms. It has been observed that the charge on all the atoms of PEN molecule is significantly modified by the adsorption. Particularly, the carbon and hydrogen atoms of the PEN molecule acquire charge from the metal surface. With respect to the isolated PEN molecule, the total Mulliken charge on the PEN molecule in the adsorbed system is increased by –0.22 electrons due to the chemisorption. This result shows that there is a net charge transfer from the Pd(100) surface to the PEN molecule and the charge transfer leads to a positively charged metal surface and a negatively charged PEN overlayer, suggesting strong electrostatic interaction between PEN and Pd(100) surface. Previous study on adsorption of PEN on Al(100) surface [50] shows that the total Mulliken charge on the PEN molecule in the adsorbed system is increased by –0.05 electrons with respect to isolated PEN molecule due to physisorption.

7.3.5 DENSITY OF STATES AND ENERGY LEVELS

In order to study the charge transfer and the contribution of a particular atomic orbital or atom to the energy levels of the adsorbed system, we have calculated the DOS and PDOS on the atoms of PEN molecule in the adsorbed system and are compared with the isolated PEN molecule. Figure 7.7 shows the total DOS for the isolated PEN, clean Pd(100) surface and PEN/Pd(100) interface. The Fermi level is set to zero. The calculated DOS of the overall system is approximately equal to the sum of the DOS of isolated systems.

The PDOS on carbon of the isolated PEN molecule, in the adsorbed system are shown in Figure 7.8. The PDOS for the specific carbon atoms C1-C6 (see Figure 7.1 for labeling of atoms) are considered and all the remaining atoms follow the similar trend due to symmetry. Figure 7.8 (a-f) shows the PDOS on the carbon atoms of isolated PEN molecule and on the adsorbed system. It has been observed that the PDOS on the carbon atoms of PEN molecule are significantly altered by the adsorption and in the adsorbed system PDOS are shifted towards the higher energies in comparison with that of the isolated PEN molecule. Figure 7.9 shows the PDOS for a specific Pd atom in the clean Pd(100) substrate and in the adsorbed system. This Pd atom is located at the bridge site, which is nearer to the C6 atom of the PEN molecule. It has been observed that the PDOS on the Pd atom decreases while interacting with PEN molecule and shifted

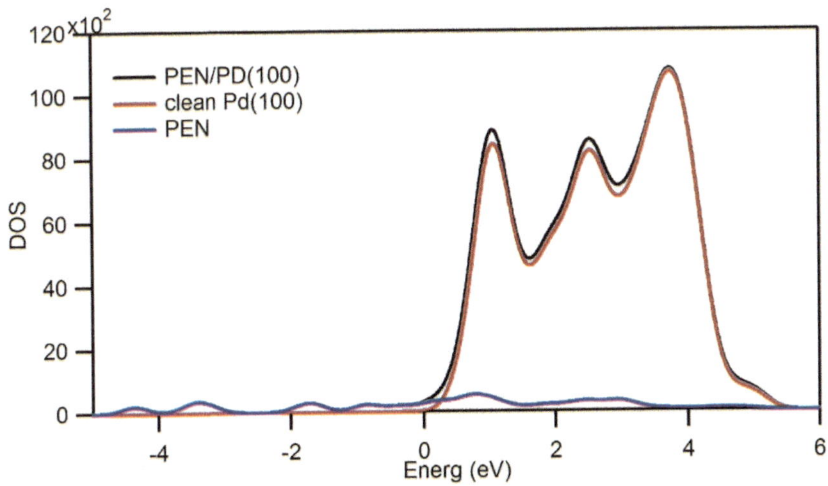

FIGURE 7.7 Total density of states of the isolated pentacene molecule, clean Pd(100) surface and pentacene/Pd(100) system. The Fermi level is set to zero.

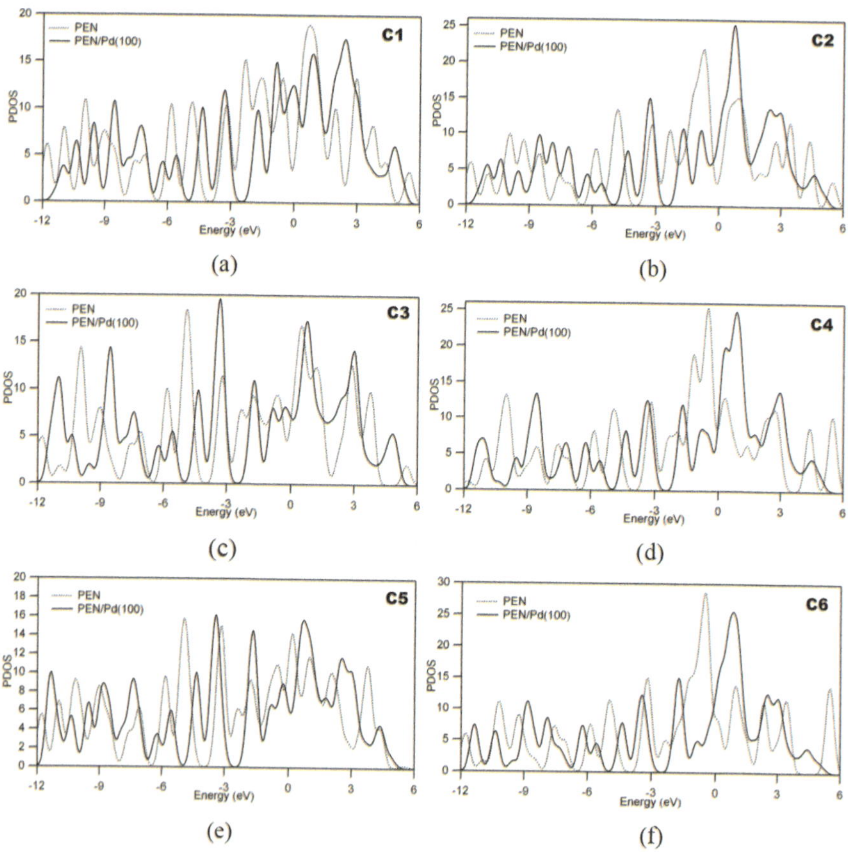

FIGURE 7.8a–f Partial density of states on the different carbon atoms in the isolated pentacene molecule and in the adsorbed system. The Fermi level is set to zero.

slightly to lower energies, indicating the transfer of charge from Pd substrate to the PEN molecule. This is in agreement with the results obtained from Mulliken population analysis and PDOS on atoms of PEN.

The density plot of the highest occupied molecular orbital (HOMO) and lowest unoccupied molecular orbital (LUMO) of PEN is calculated and is shown in Figure 7.10. It has been observed that HOMO as well as LUMO are delocalized over the entire molecule and are having π-orbital character. It has been found that the HOMO (H) and LUMO (L) levels of adsorbed system are shifted by 0.26 and 1.62 eV, respectively from the values of isolated PEN molecule. As shown in Figure 7.11, the H-L gap

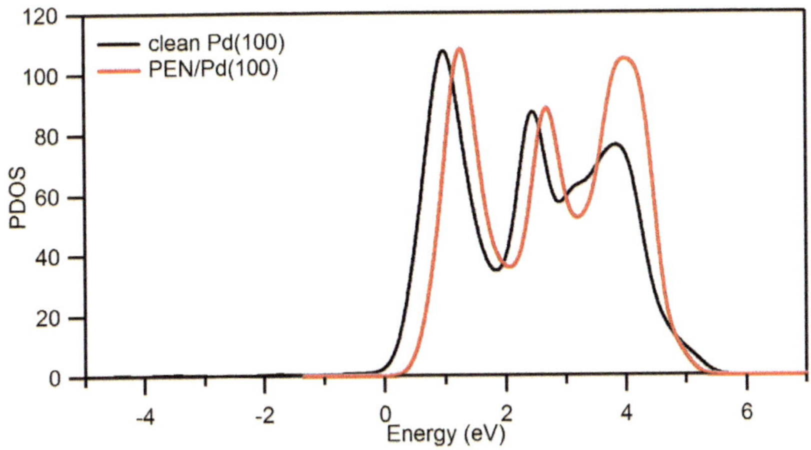

FIGURE 7.9 Partial density of states of the palladium atom in the clean Pd(100) surface and in the adsorbed system. The Fermi level is set to zero.

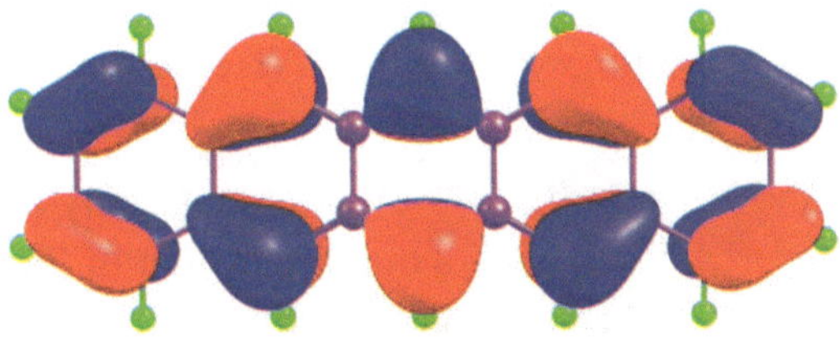

(a)

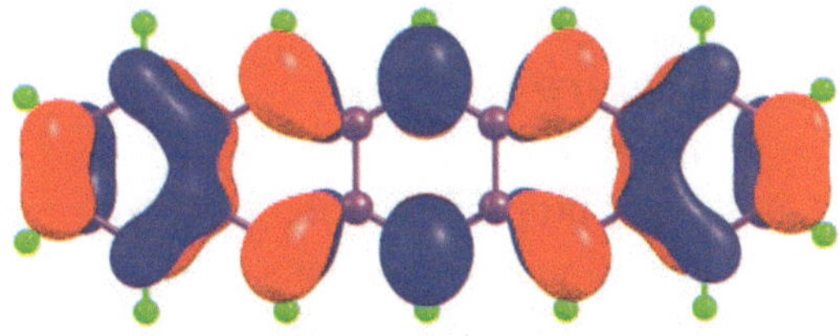

(b)

FIGURE 7.10 The density plot of (a) highest occupied molecular orbital (HOMO) and (b) lowest unoccupied molecular orbital (LUMO) of pentacene.

of isolated PEN molecule is 1.72 eV and in the adsorbed system the H-L gap is 0.31 eV. That is, the adsorption of PEN molecule on the Pd(100)

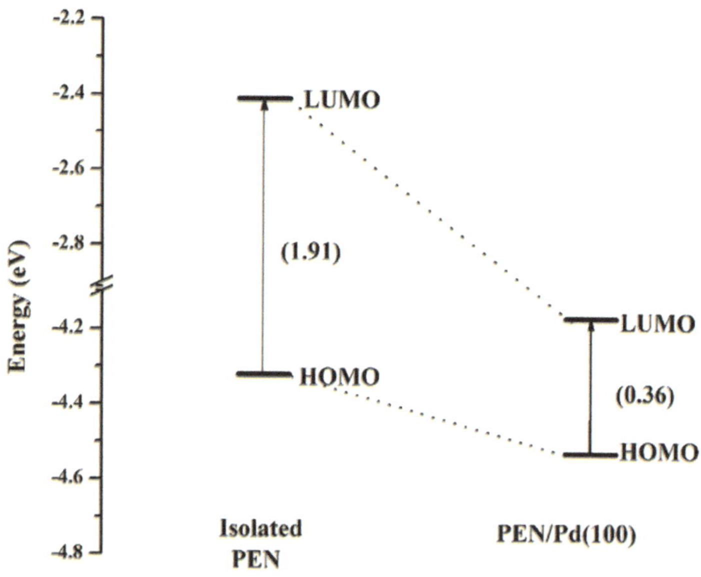

FIGURE 7.11 Energies of frontier molecular orbitals of pentacene molecule and pentacene/Pd(100) interface. Energy gap value (in eV) is given within the parenthesis.

surface reduces the H-L gap by 1.41 eV. Previous study on adsorption of PEN on Al(100) surface [50] shows that adsorption of PEN molecule on Al(100) surface reduces the H-L gap upto 1.33 eV. Recently, Hahn et al. [75] reported that the H-L gap of pyridine molecule decreases by about 1 eV while interacting with Ag(110) surface.

7.3.6 SCHOTTKY BARRIER HEIGHT

One of the most important properties in the study of organic molecule-metal interface is the Schottky barrier heights (SBHs). The SBH is calculated from first principle calculations taking into account both the charge rearrangement at the interface and bulk contributions of the separate constituents [76, 77] based on core level energy as reference energy [78, 79]. In the case of the metal/molecule interface, as shown in Figure 7.12, the p-type Schottky barrier height (ϕ_B^p) is calculated as,

$$\phi_B^p = E_F - E_{1s}^{\text{int}} + BE^{1s}$$

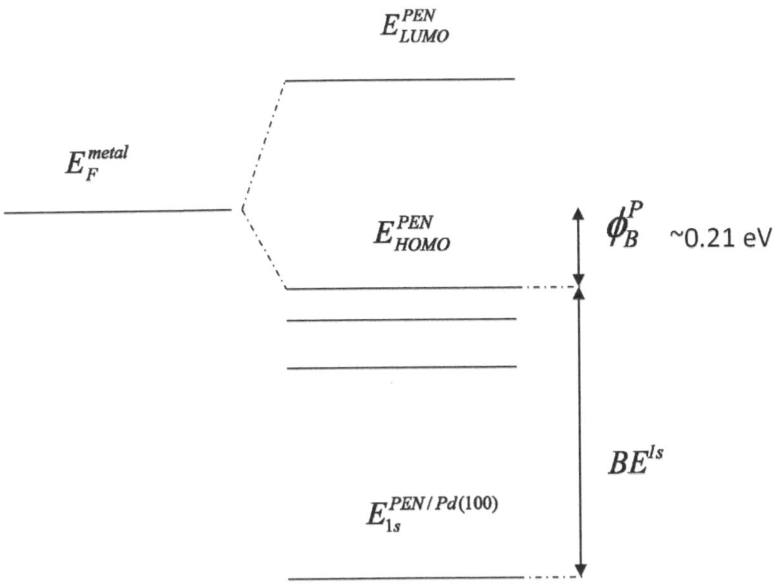

FIGURE 7.12 Schematic diagram showing calculation of Schottky barrier height.

The charge rearrangement at the interface is calculated as the difference in energy between the Fermi energy of the adsorbed system (E_F) and the 1s core level energy of carbon atom in the interface ($E^{\text{int}}{}_{1s}$). The bulk contribution is included interms of the binding energy of the same core level energy of the carbon atom within the isolated molecule, denoted as BE^{1s}. The bulk contribution of the metal surface is built-in with E_F, since it is expected that the Fermi level of the interface is very nearer to the Fermi level of clean Pd surface.

The calculated p-type SBH value for the PEN/Pd(100) interface is equal to 0.27 eV. The n-type SBH (ϕ_B^n) is calculated as,

$$\phi_B^n = E_{gap}^{PEN} - \phi_B^p$$

where E_{gap}^{PEN}, is the band gap of the PEN molecule. The calculated band gap for the isolated PEN molecule is about 1.72 eV, with this band gap value the calculated n-type SBH is 1.45 eV. The experimental band gap value of PEN reported by Sakamoto et al. is 2.09 eV [80]. The DFT method underestimates the band gap by about 0.4 eV. With the experimental band gap value, the calculated n-type SBH is 1.82 eV. The calculated SBH values show that PEN/Pd(100) interface is favorable for hole injection than the electron injection, because the SBH for electron is higher than that of hole. This result is in agreement with the previous study on adsorption of PEN on Al(100) surface, which show that the injection of hole is easier in the PEN/Al(100) interface than the electron injection [50]. These results clearly show that the PEN molecule can be used as a p-type organic semiconducting material in organic electronic components.

7.4 CONCLUSIONS AND OUTLOOK

The first principle calculations within the framework of DFT have been performed to study the adsorption of pentacene molecule on Pd(100) surface. The most stable adsorption geometry and the electronic properties of PEN/Pd(100) interface have been calculated. The most stable adsorption structure was found at bridge site orientation with 45° of orientation of PEN molecule on Pd(100) surface and vertical distance of 2.4 Å, and the adsorption is chemisorption with the adsorption energy of –3.12 eV. The analysis of electronic states and atomic charges shows that the charge is transfer from Pd(100) surface to pentacene molecule and PEN/Pd (100) interface is favorable for hole transport. The above results clearly shows that the DFT with plane wave and pseudopotential approach can be used to study the structure and electronic properties of organic molecule, metal surface and organic molecule/metal interfaces.

KEYWORDS

- **charge transport**
- **density functional theory**
- **density of states**
- **electronic properties**
- **pentacene**
- **Schottky barrier height**

REFERENCES

1. Shirakawa, H., Louis, E. J., MacDiarmid, A. G., Chiang, C. K., & Heeger, A. J. (1977). *Chem. Commun.* 578.
2. Troisi, A. (2010). *Adv. Polym. Sci. 223*, 259.
3. Mei, J., Diao, Y., Appleton, A. L., Fang, L., & Bao, Z. (2013). *J. Am. Chem. Soc. 135*, 6724.
4. Trang, N. T., Quynh, L. M., Nam, T. V., & Nhat, H. N. (2013). *Surf. Sci. 608*, 67.
5. Götzen, J., Schwalb, C. H., Schmidt, C., Mette, G., Marks, M., Höfer, U., & Witte, G. (2011). *Langmuir 27*, 993.
6. Coropceanu, V., Cornil, J., da Silva Filho, D. A., Olivier, Y., Silbey, R., & Bredas, J. L. (2007). *Chem. Rev. 107*, 925.
7. Track, A. M., Rissner, F., Heimel, G., Romaner, L., Kafer, D., Bashir, A., Rangger, G. M., Hofmann, O. T., Bucko, T., Witte, G., & Zojer, E. (2010). *J. Phys. Chem. C 114*, 2677.
8. Picozzi, S., Pecchia, A., Gheorghe, M., Carlo, A. D., Lugli, P., delley, B., & Elstner, (2004). M. *Surf. Sci.* 566–568, 628.
9. Glowatzki, H., Heimel, G., Vollmer, A., Wong, S. L., Huang, H., Chen, W., Wee, A. T. S., Rabe, J. P., & Koch, N. (2012). *J. Phys. Chem. C 116*, 7726.
10. Jager, C. M., Schmaltz, T., Novak, M., Khassanov, A., vorobiew, A., Hennemenn, M., Krause, A., Dietrich, H., Zahn, D., Hirsch, A., Halik, M., & Clark, T. (2013). *J. Am. Chem. Soc. 135*, 4893.
11. Iqbal, Z., Wu, W. Q., Zhang, H., Han, L., Fang, X., Wang, L., Kuang, D. B., Meier, H., & Cao, D. (2013). *Org. Elect. 14*, 2662.
12. Dimitrakopoulos, C. D., & Malenfant, P. R. L. (2002). *Adv. Mater. 14*, 99.
13. Duan, Y. A., Geng, Y., Li, H. B., Tang, X. D., Jin, J. L., & Su, Z. M. (2012). *Org. Elect. 13*, 1213.
14. Burrows, P. E., Gu, G., Bulovic, V., Shen, Z., Forrest, S. R., & Thompson, M. E. (1997). *IEEE Trans. Electron Devices 44*, 1188.
15. Dobbertin, T., Kroeger, M., Heithecker, D., Schneider, D., Metzdorf, D., Neuner, H., Becker, E., & Johannes, H.-H., *Appl.* (2002). *Phys. Lett. 82*, 284.
16. DRew, A. J. (2009). *Nat. Mater. 8*, 691.
17. Hoppe, H., & Sariciftci, N. S. (2004). *J. Mater. Res. 19*, 1924.
18. Hung, L. S., & Chen, C. H. (2002). *Mater. Sci. Eng. R. 39*, 143.
19. Inoue. Y; Sakamoto. Y; Suzuki. T; M, K., & Gao, Y. S, T. (2005). *Japan. J. Appl. Phys. 44*, 3663.
20. Belkada, R., Shirakawa, Y., Kohyama, M., Tanaka, S., & Hidaka, J. (2006). *Mat. Trans. 7*, 2701.
21. Zhou, H.-C., Long, J. R., & Yaghi, O. M. (2012). *Chem. Rev. 112*, 673.
22. Mc Quarrie, D. A., Simon, J. D., Eds., *Physical Chemistry: A Molecule Approach*; Viva books Pvt. Ltd., 1998.
23. Mete, E., Demiroglu, I., Danisman, M. F., & Ellialtioglu, S. (2010). *J. Chem. Phys. C 114*, 2724.
24. Shi, X. Q., Li, Y., Van Hove, M. A., & Zhang, R. Q. (2012). *J. Chem. Phys. C 116*, 23603.
25. Hohenberg, P., & Kohn, W. (1964). *Phys. Rev. 136*, b864.
26. Thomas, L. (1927). *Proc. Camb. Phil. Soc 23*, 542.
27. Fermi, E. (1927). *Rend. Accad. Lincei 6*, 602.

28. Kohn, W., & Sham, L. J. (1965). *Phys. Rev. 140*, A1133.
29. Leach, A. R., Ed., *Molecular Modelling Principles and Applications*; 2nd edn., Donling Kindersley Pvt. Ltd., 1996.
30. Ashcroft, N. W., Mermin, N. D., Eds., *Solid State Physics*; Harcourt College Publishers, 1976.
31. Payne, M. C., Teter, M. P., Allan, D. C., Arias, T. A., & Joannopoulos, J. D. (1992). *Rev. Mod. Phys. 199*, 1045.
32. Meyer, B., Grotendorst, J., Blügel, S., Marx, D., (eds , Bernd Meyer),The Pseudopotential Plane Wave Approach; (2006). NIC Series, 31, 71, ISBN 3-00-017350-1.
33. Martin, R. M., Ed., *Electronic Structure Basic Theory and Practical Methods*; Cambridge University Press, 2004.
34. Hamann, D. R., Schluter, M., & Chiang, C. (1979). *Phys. Rev. Lett. 43*, 1494.
35. Srivastava, G. P., Ed., *Theoretical Modelling of Semiconductor Surfaces: Microscopic Studies of Electrons and Phonons*; World Scientific Publishing co. Pvt. Ltd., 1999.
36. Kleinman, L., & Bylander, D. M. (1982). *Phys. Rev. Lett. 48*, 1425.
37. Blöchl, P. E. (1990). *Phys. Rev. B 41*, 5414.
38. Vanderbilt, D. (1990). *Phys. Rev. B 41*, 7892.
39. Kohenoff, J., Ed., *Electronic Structure Calculation for Solids and Molecules: Theoretical and computational methods*; Cambridge University Press, 2006.
40. Lin, Y.-L., Gundlach, D. J., Nelson, S. F., & Jackson, T. N. (1997). *IEEE Trans. Electron Devices 44*, 1325.
41. Butko, V. Y., Chi, X., Lang, D. V., & Ramirez, A. P. (2003). *Appl. Phys. Lett. 83*, 4773.
42. Kitamura, M., & Imada, T., Y., A. (2003). *Appl. Phys. Lett. 83*, 3410.
43. Mayer, A. C., Lloyd, M. T., Herman, D. J., Kasen, T. G., & Malliaras, G. G. (2004). *Appl. Phys. Lett. 85*, 6272.
44. Lee, S., Koo, B., Shin, J., Lee, E., Park, H., & Kim, H. (2006). *Appl. Phys. Lett. 88*, 162109.
45. Toyoda, K., Hamada, I., Lee, K., Yanagisawa, S., & Morikawa, Y. (2010). *J. Chem. Phys. 132*, 134703.
46. Ferretti, A., Baldacchini, C., Calzolari, A., Di Felice, R., Ruini, A., Molinari, E., & Betti, M. G. (2007). *Phys. Rev. Lett. 99*, 046802.
47. Mete, E., Demiroglu, I., Danisman, M. F., & Ellialtioglu, S. (2010). *J. Phys. Chem. C 114*, 2724.
48. Wang, Y. L., Ji, W., Shi, D. X., Du, S. X., Seidel, C., Ma, Y. G., Gao, H. J., Chi, L. F., & Fuchs, H. (2004). *Phys. Rev. B 69*, 075408.
49. Simeoni, M., Picozzi, S., & Delley, B. (2004). *Surf. Sci. 562*, 43.
50. Saranya, G., Shiny Nair; Natarajan, V., Kolandaivel, P., & Senthilkumar, K. (2012). *J. Mol. Graph. Mod. 38*, 334.
51. Saranya, G., Shiny Nair; Natarajan, V., Kolandaivel, P., & Senthilkumar, K. (2014). *Comp. Mater. Sci. 89*, 216.
52. Sun, X., Suzuki, T., Kurahashi, M., Wang, Z. P., & Entani, S. (2008). *Surf. Sci. 602*, 1191.
53. Li, H., Duan, Y., Coropceanu, V., & Bredas, J. L. (2009). *Org. Elect. 10*, 1571.
54. Lee, K., & Yu, J. Y. (2005). *Surf. Sci. 589*, 8.
55. Zuo , R., Li, L., Gui, Z., Ji, C., & Hu, X. (2001). *Mater. Sci. Eng. B 83*, 152.

56. Jeong, S.-J., Lee, D.-S., Song, J.-S., & Lee, H.-K. (2007). *J. Mater. Sci. 42*, 604.
57. Kresse, G., & Hafner, J. (1993). *Phys. Rev. B 47*, 558.
58. Kresse, G., Furthmuller, J. (1996). *Phys. Rev. B 54*, 11169.
59. Kresse, G., Marsman, M., & Furthmuller, J. *Vienna Ab-inition Simulation Package, A–1130 Wien, Austria,* 2009.
60. Perdew, J. P., Burke, K., & Ernzerhof, M. (1996). *Phys. Rev. Lett. 77*, 3865.
61. Kresse, G., & Joubert, D. (1999). *Phys. Rev. B 59*, 1758.
62. Monkhorst, H. J., & Pack, J. D. (1976). *Phys. Rev. B 13*, 5188.
63. Grimme, S. (2006). *J. Comp. Chem. 27*, 1787.
64. Grimme, S., Antony, J., Ehrlich, S., & Krief, H. (2010). *J. Chem. Phys. 132*, 154104.
65. Grimme, S. (2004). *J. Chem. Chem. 25*, 1463.
66. Toyoda, K., Hamada, I., Lee, K., Yanagisawa, S., & Morikawa, Y. (2011). *J. Phys. Chem. C 115*, 5767.
67. Wang, Y. L., Urban, C., Rodriguez-Fernandez, J., Gallego, J. M., Otero, R., Martin, N., Miranda, R., Alcami, M., & Martin, F. (2011). *J. Phys. Chem. C 115*, 13080.
68. Campbell, R. B., Robertson, J. M., & Trotter, J. (1961). *Acta Cryst. 14*, 705.
69. Singer-Miller, N. E., & Mazrzari, N. (2009). *Phys. Rev. B 80*, 235407.
70. Kittel, C., Ed., *Introduction to Solid State Physics*; 7th ed., JohnWiley & Sons: New York, 1996.
71. Tautz, F. S., Eremtchenko, M., Schaefer, J. A., Sokolowski, M., Shklover, V., Glocker, K., & Umbach, E. (2002). *Surf. Sci. 502–503*, 176.
72. Umbach, E., Sokolowski, M., & Fink, R. (1996). *Appl. Phys. A 63*, 565.
73. Hulse, J., Küppers, J., Wandelt, K., & Ertl, G. (1980). *Appl. Surf. Sci. 6*, 453.
74. Mulliken, R. S. (1955). *J. Chem. Phys. 23*, 1833.
75. Hahn, J. R., & Kang, H. S. (2010). *Surf. Sci. 604*, 258.
76. Picozzi, S., Continenza, A., Massidda, S., & Freeman, A. J. (1998). *Phys. Rev. B 57*, 4849.
77. Picozzi, S., Continenza, A., Satta, M., Massidda, S., & Freeman, A. J. (2000). *Phys. Rev. B 61*, 16736.
78. Wei, S. H., & Zunger, A. (1987). *Phys. Rev. B 59*, 144.
79. Massidda, S., Min, B. I., & Freeman, A. J. (1987). *Phys. Rev. B 35*, 9871.
80. Sakamoto, Y., Suzuki, T., Kobayashi, M., Gao, Y., Fukai, Y., Inoue, Y., & Sato, F. S. T. (2004). *J. Am. Chem. Soc. 126*, 8138.

CHAPTER 8

A COMPARATIVE THEORETICAL INVESTIGATION ON THE ACTIVATION OF C-H BOND IN METHANE ON MONO AND BIMETALLIC Pd AND Pt SUBNANOCLUSTERS

PAKIZA BEGUM and RAMESH C. DEKA

Department of Chemical Sciences, Tezpur University, Tezpur, Napaam, 784 028, Assam, India, E-mail: ramesh@tezu.ernet.in

CONTENTS

ABSTRACT

We report a comprehensive Density Functional Theory (DFT) exploration on the C-H bond activation of methane, an important step of the conversion reaction of methane to liquid fuels, over pristine and bimetallic Pt and Pd tetramers. Subnanoclusters exhibit great promise for catalytic activities, implicitly much greater than monolith. Understanding the C-H cleavage of methane is significant in designing suitable catalysts. Doping is known to be an excellent and simple way of catalyst design and bimetallic nanoparticles are more promising due to synergetic effect. The lowest energy barrier is calculated for $PdPt_3$ system with more active Pt-site than Pd-site. The reaction is found to exothermic when methane is bound to the Pt-site of the subnanometer clusters. For methane attached to the Pd-site, the reaction is mostly endothermic in nature. Our study demonstrates that bimetallic subnanometer clusters with active Pt-site, enhance the ability for methane C-H bond activation.

8.1 INTRODUCTION

Functionalization of alkane through C-H bond activation by transition metals has attracted enormous attention in the last several decades to secure the supply of chemicals, energy and fuels in the future [1–4]. The controlled activation of small, relatively inert molecules has been a ubiquitous subject matter in transition metal chemistry since its resurgence began in the late 1950s. To date, binding of methane to metal centers has resulted in altered and/or improved reactivity in a variety of molecules, through associated changes in the relative energies of their orbitals or their polarity. In this regard, many experimental works [5–8] and rather limited theoretical works [9–12] have been presented on C-H bond activation. These studies have been motivated by the availability of methane- it is the main constituent of natural and bio-gas and present in enormous quantity on our planet [8]. Methane chemistry experiences a bang not only because of its massive reserves and resources but also because of enhanced production technology.

But there are very few practical methods for direct conversion of alkanes to more valuable products due to chemical inertness. Reactions are possible at high temperatures, but desired products are not obtained as such reactions are not readily controllable and usually give economically unattractive products [13]. Gas phase reactions over solid catalysts constitute a significant fraction of large–scale industrial processes for energy conversion [14]. Mono or bimetallic nanoclusters have received much interest in scientific research and industrial applications due to their unique large surface to volume ratio and quantum size effects [15, 16]. Noble metals exhibit very good performance for a number of these processes, but usage can be sometimes too expensive. Therefore, highly selective, efficient and robust catalysts are therefore important to meet this challenge. One method for reducing the amount of noble metals required for a reaction is substitutional doping [17]. It has been demonstrated by various studies that metal-doped bimetallic clusters are more compelling due to the complex structure which offers more opportunity to enhance catalytic performance [18–22].

Binding of methane to metal centers has resulted in altered and/or improved reactivity in a variety of molecules, through associated changes in the relative energies or polarity. In this regard, many experimental works and rather limited theoretical works have been presented on C-H bond activation. Nizovtsevix [23] carried out activation of C–H bond in methane by Pd atom from the bonding evolution theory perspective. He found that redistribution of electron density occurred in the course of C-H bond activation in methane. In another study by Choudhary and Goodman [10], using Ni and Ru model catalysts, which illustrates the efficiency of model catalyst studies in providing a greater understanding of the methane activation process. Siegbahn et al. [24] studied Pt^{+}-catalyzed oxidation of methane in detail by extensive calculations and experiments to gain complete understanding of different elementary steps involved and to investigate the accuracy of computational methods for reactions involving transition metals. Their study found that calculations are quite successful in reproducing the experimental results. Zapol et al. [25] carried out methane bond activation by Pt and Pd subnanometer clusters in gas phase as well as supported on graphene and carbon nanotubes. Pt-clusters are found to have higher activity over Pd cluster for CH_4 dissociation.

For the supported clusters, size and chirality of the support tailor their catalytic activity resulting in lower barriers. Sun et al. [12] performed a theoretical study to explore the catalytic activation of C–H bonds in methane by the iron atom, Fe, and the iron dimer, Fe_2. Their findings indicate that the iron dimer Fe_2 has a stronger catalytic effect on the activation of methane than the iron atom. This study serves as the probe for the activation mechanism of methane on very small Fe clusters, providing an interesting contrast to the bulk solid. Although a number of studies on the C-H bond activation on transition metals have been performed. But, to date, only a few studies have been carried out on the bimetallic clusters on how methane will interact on the bimetallic environment. By dint of potential prominence and finite information, we carry out a detailed Density Functional calculation with the objective to prepare robust catalyst with higher activity and shed light on the fundamental step of methane conversion.

8.2 MATERIALS AND METHODS

We presented here a comparative theoretical study of the catalytic activity of pristine palladium and platinum tetramer nanoclusters and their bimetallic counterparts. All the density functional calculations were performed using double numerical plus polarization (DNP) basis set implemented in $DMol^3$ package [26, 27]. Geometry optimizations and frequency analysis were done by treating the exchange-correlation interaction with generalized gradient approximation (GGA) using Becke-Lee-Yang-Parr (BLYP) [28–30] exchange-correlation functionals. The valence electrons are described by double numerical basis set with polarization function and the core electrons are described with local pseudo-potential (VPSR) which accounts for the scalar relativistic effect expected to be significant for heavy metal elements in the periodic table [31, 32]. The DNP basis set has a computational precision quite comparable to split-valence basis set 6–31G** of Gaussian. We used inversion iterative subspace (DIIS) approach in order to speed up SCF convergence. To improve computational performance, a global orbital cutoff of 4.5 Å was employed. Self-consistent field (SCF) procedures are done with tolerances of the energy, gradient and displacement convergences: 1×10^{-5} Ha, 2×10^{-3} Ha. Å^{-1} and 5×10^{-3} Å, respectively.

The DFT calculations are performed under the GGA with exchange-correlation functional BLYP, which is combination of exchange functional developed by Becke with the gradient corrected functional of Lee-Yang-Parr [28]. Systematic density functional study has been performed on transition metal clusters and the results are in good agreement with experimental values [24]. Spin multiplicity plays an important role on the stability of metal complexes. Considering the spin polarization, every structure is optimized at two spin states (M = 1 and 3), and the structure with the lowest energy for each case is considered. For the superior systems obtained from the above considered spin states –the reaction has been studied for six spin multiplicities, M = 1, 2, 3, 4, 5 and 6, in order to have a better understanding. The potential energy profile (including geometries and energies of reactant, intermediates, transition states and products) for the activation of CH_4 from molecular form, passing over the transition state (TS) to the dissociative form by mono and bimetallic palladium and platinum tetramers investigated at the DFT level of theory. A schematic potential energy surface of the reaction is constructed to search the most dominant cluster to carry out the reaction suitably. For a bimetallic system, e.g., Pd_2Pt_2, the reaction starts with the adsorption of CH_4 on Pd_2Pt_2 forming the molecularly adsorbed species. However, there is the possibility of adsorbing CH_4 on Pd- or Pt-site. Therefore, we have considered both the pathways; Pd_2Pt_2 – where CH_4 is adsorbed on the Pt-site and Pt_2Pd_2 – where CH_4 is adsorbed on the Pd-site of the considered system. Thus, the possible adsorption sites have been considered for all the systems studied (Pd_3Pt, Pd_2Pt_2 and $PdPt_3$). Frequency calculations have been performed on the optimized geometries of the clusters and almost all the clusters and are found to be positive for all the structures confirming them to be at energy minima.

The most plausible pathways for the activation of CH_4 investigated during present study

$$\underset{\text{(Reactant)}}{Pt_nPd_m + CH_4} \rightarrow \underset{\text{(MA)}}{Pt_nPd_mCH_4} \rightarrow \underset{\text{(TS)}}{\text{transition state}} \rightarrow \underset{\text{(DA)}}{Pt_nPd_mHCH_3}$$

We have calculated the bond length, bond angle, vibrational frequency and binding energy of the adsorbed CH_4 molecule onto the metal clusters. The binding energy of methane molecule on Pt_mPd_n cluster is defined as:

$$BE = E_{(\text{cluster})} + E_{(CH4)} - E_{(\text{cluster} - CH4\ adduct)} \quad (1)$$

where, $E_{(\text{bare cluster})}$ is the total energy of the cluster, $E_{(CH4)}$ is the total energy of the methane molecule and $E_{(\text{cluster} - CH4\ adduct)}$ is the total energy of the combined system.

In order to verify the reliability of our calculation we have compared the results with one of the reported studies on structural stability of Pt_4CH_4 [24]. It was found that Pt-C, Pt-H and C-H bond lengths are in good agreement with the reported study.

8.3 RESULTS AND DISCUSSION

8.3.1 CH_4-Pt_4 AND CH_4-Pd_4 SYSTEMS

For the CH_4-Pt_4 system the calculations have been done in six different spin multiplicities, M = 1, 2, 3, 4, 5 and 6, without imposing any symmetry constraints. For the palladium tetramer two spin states are considered, M = 1 and 3. The reaction for the activation of methane from molecular form starts with the reactants (methane and the cluster) separated, passing over the C-H bond insertion TS to the dissociative adsorption form (DA). The relative energies for this reaction are reported in Table 8.1, the optimized

TABLE 8.1 Table for Barrier Height (kcal.mol^{-1} and eV)

System	kcal.mol^{-1}	eV
Pt_4	4.30	0.18
Pd_4	25.29	1.01
$PdPt_3$ (Pt-active site)	3.01	0.13
Pt_3Pd (Pd-active site)	21.13	0.91
Pd_2Pt_2 (Pt-active site)	6.20	0.26
Pt_2Pd_2 (Pd-active site)	24.00	1.03
Pd_3Pt (Pt-active site)	10.59	0.46
$PtPd_3$ (Pd-active site)	43.8	1.89

geometries (including bond lengths, bond angle and the Hirshfeld charge of reactant, intermediates, transition states and products) for the reaction are illustrated in Figures 8.1 and 8.2, energy profile (including energies of reactant, intermediates, transition states and products) of these reaction steps is given in Figures 8.3 and 8.4.

Our calculations show that the ground state of the Pt- tetramer is a quintet state and Pd- tetramer is stable at the triplet state (the total spin multiplicity: M = 2S+1, where S is the total spin). Of all the spin states taken into account, each structure having the most stable lowest energy structure has been taken into consideration. For the MA, the lowest energized structure is in spin quartet, and the ground state for TS and DA are found to be stable in their triplet electronic state for the platinum tetramic system. For the Pd_4 system the stable spin states for MA, TS and DA are singlet. It has been seen from the Hirshfeld charges in Figure 8.1(a) and 8.1(b), a small amount of electron density is transferred from cluster to RC and TS (with a difference of 0.10 and 0.20 |e| across the systems, respectively). The binding energy of methane with Pt_4 is obtained to be 24.67 kcal.mol^{-1}, which is similar to the one reported by Siegbahn et al.

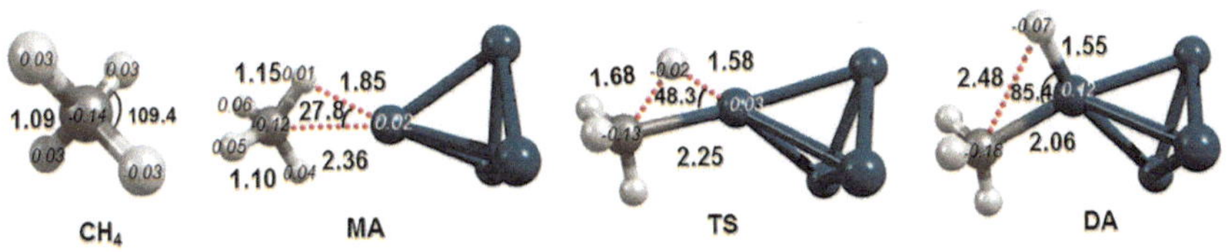

FIGURE 8.1 Optimized structures of transition state and intermediates for methane activated with Pt_4, for selected structural parameters (distances in Å, bond angle in degrees) and Hirshfeld charges (shown in italic, in |e|).

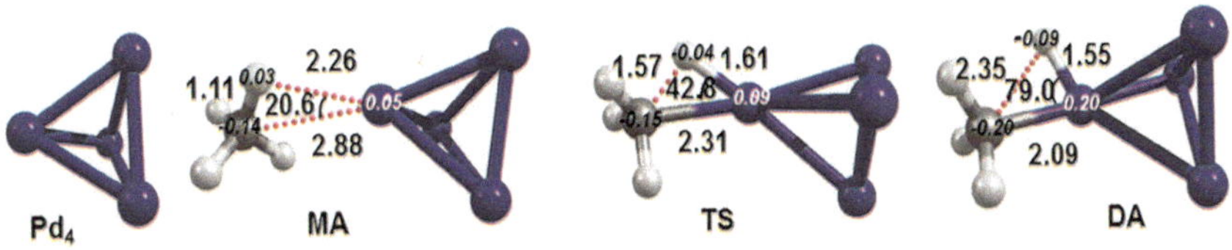

FIGURE 8.2 Optimized structures of the intermediates involved for methane activation over Pd_4, for selected structural parameters (distances in Å, bond angle in degrees) and Hirshfeld charges (shown in italic, in |e|).

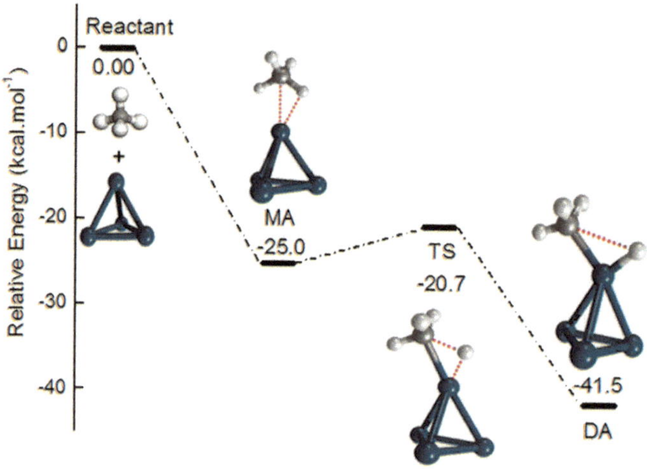

FIGURE 8.3 Calculated energy profile for the activation of methane over Pt_4. The total energy of the reactants is taken as zero.

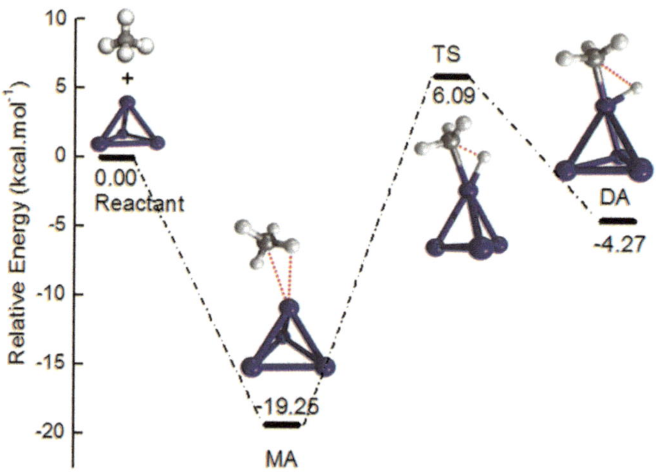

FIGURE 8.4 Calculated energy profile for the activation of methane over Pd_4. The total energy of the reactants is taken as zero.

[24] and that for Pd_4 is 19.25 kcal.mol^{-1}, indicating a higher affinity for binding methane by the platinum cluster. As the reaction proceeds, the angle between C-Pt-H increases and the bond Pt-C and Pt-H strengthens. The Pd-C and Pd-H bond is compressed by 0.30 Å each along the

reaction for the Pt_4 system, whereas, 0.79 Å and 0.71 Å, respectively for the Pd_4 system. During the cleavage of the C-H bond, the reactants go through the TS with an imaginary frequency of 625i cm^{-1} Pt_4 and 978.7i cm^{-1} for Pd_4 system, assigned to the vibration of C-H bond. The activation energy for C–H bond breaking reaction is exothermic on Pt_4 clusters as shown in Figure 8.1, but is endothermic on Pd_4 system as seen from Figure 8.4. In comparison, other studies [25] have found barriers for C–H bond breaking in methane to be 0.1 eV on a Pt_4 cluster, which is in good agreement to our findings. The barrier height of the TS is computed to be 4.3 kcal.mol^{-1} (0.18 eV), as shown in Table 8.1–8.4, and the dissociation is exothermic by 41.5 kcal.mol^{-1} (1.7 eV). For the Pd_4 system, a barrier height of 25.29 kcal.mol^{-1} (1.01 eV) is obtained, in agreement with previous calculation by Zapol et al. [25]. Thus, tetramer platinum cluster serves as a predominant system for methane activation as compared to the palladium counterpart.

8.3.2 $PdPt_3$ AND Pt_3Pd SYSTEMS

For a bimetallic system, the reaction starts with the adsorption of CH_4 on $PdPt_3$ forming an electrostatically bound methane complex. However, there is the possibility of adsorbing CH_4 on Pd- or Pt-site. Figures 8.5 and 8.6 shows the various intermediate states along with the geometrical parameters involved in the reaction over the dimeric $PdPt_3$ system. Therefore, we have considered both the pathways; $PdPt_3$-CH_4 (path I shown in Figure 8.5), where methane is adsorbed on the Pt-site and Pt_3Pd-CH_4 (path II shown in Figure 8.6) where methane is adsorbed on the Pd-site of system.

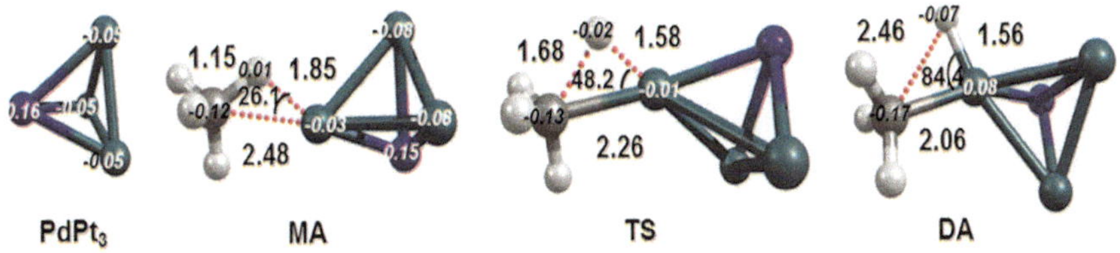

FIGURE 8.5 Optimized structures of transition and intermediate states for methane activated with $PdPt_3$- Pt site considered for the reaction, for selected structural parameters (distances in Å, bond angle in degrees) and Hirshfeld charges (shown in italic, in |e|).

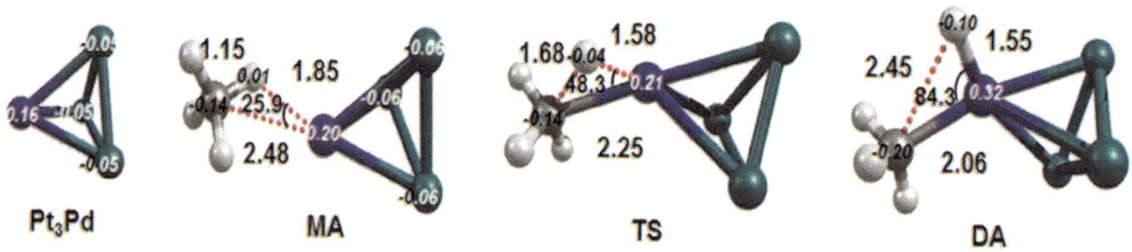

FIGURE 8.6 Optimized structures of transition state and intermediates for methane activated with $PdPt_3$-CH_4 adsorbed on Pd-site, for selected structural parameters (distances in Å, bond angle in degrees) and Hirshfeld charges (shown in italic, in |e|).

For the $PdPt_3$-CH_4 system the calculations have been done in six different spin multiplicities, M = 1, 2, 3, 4, 5 and 6, without imposing any symmetry constraints. For the considered system, triplet state is found to be the lowest energy states for cluster and MA, whereas, TS and DA are found to be stable at their doublet spin state. The system with Pd as the active site (Pt_3Pd) is calculated at singlet and triplet states. For this system all the structures are stabilized in the triplet state. The binding energy of methane for $PdPt_3$ and Pt_3Pd systems are 24.98 kcal.mol^{-1} and 21.28 kcal.mol^{-1}, respectively. Thus, CH_4 has a higher binding affinity for the Pt-site as compared to the Pd-site. Similar trend in the transfer of electron density has been observed with a difference of 0.13 and 0.16 electrons across the reaction for the Pt- and Pd-sites, respectively. For both the considered sites the C-M-H (M = Pt or Pd) bond angle increases and the M-C and M-H bond lengths decrease. The difference in M-C and M-H bond lengths from the reactant to the product is almost same for either sites, i.e. compressed by 0.32 Å and 0.29 Å, respectively. As seen from Figure 8.7, the reaction is exothermic for both sites, but the exothermicity is higher when methane is attached to the Pt-site of the $PdPt_3$ cluster. The TS is associated with an imaginary frequency of 840.9i cm^{-1} and 992.3i cm^{-1}, respectively for the pathway I and II, owing to the vibration of C-H bond. The C-H bond is dissociated by crossing TS involving energy barriers of 3.0 kcal.mol^{-1} (path I) and 21.12 kcal.mol^{-1} (path II) as summarized in Table 8.1. Thus, it is seen that energy barrier gap has a considerable difference when methane is adsorbed at the two different sites of a bimetallic system. This shows that Pt-site has a higher activity than Pd-site of the $PdPt_3$ system for C–H bond breaking of methane.

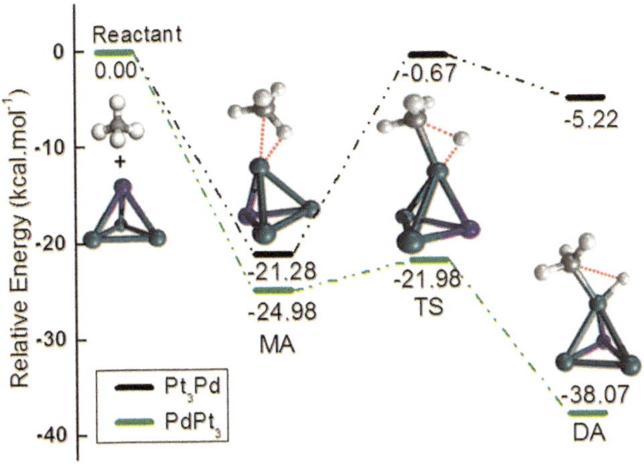

FIGURE 8.7 Potential energy profile for the activation of methane over $PdPt_3$ at the two active sites. The total energy of the reactants is taken as zero. (Green lines – Pt as the reactive site, black lines – Pd as the active site for carrying out the reaction).

8.3.3 *Pd_2Pt_2 AND Pt_2Pd_2 SYSTEMS*

As mentioned above two possible sites have been taken into consideration for the said reaction- Pd_2Pt_2-CH_4 (path I shown in Figure 8.8), where methane is adsorbed on the Pt-site and Pt_2Pd_2-CH_4 (path II shown in Figure 8.9) where methane is adsorbed on the Pd-site of the system. For pathway I reaction has been performed for six spin multiplicities (M = 1, 2, 3, 4, 5 and 6) and found that cluster is stable at triplet state, MA and TS at doublet, and DA is stabilized at singlet state. Singlet and triplet states have been considered for path II, where all the intermediates are found to have attained stability in their triplet states. The binding energy values for methane adsorption on Pt-site is 24.56 kcal.mol^{-1} and that for Pd-site is 21.34 kcal.mol^{-1}, showing a higher binding affinity of the Pt-site for the Pd_2Pt_2 system also. The difference in electron density is found to be 0.16 and 0.17 electrons across the reaction for the Pt- and Pd-site, respectively. The C-M-H bond angle (M = Pt or Pd) increases and the M-C and M-H bond lengths decrease. The difference in M-C and M-H bonds from the reactant to the product is almost same for either sites, i.e., shortened by 0.64 Å (Pt-C bond) and 0.54 Å (Pt-H bond) for path I, likewise, 0.64 Å (Pd-C bond) and 0.55 Å (Pd-H bond) for path II.

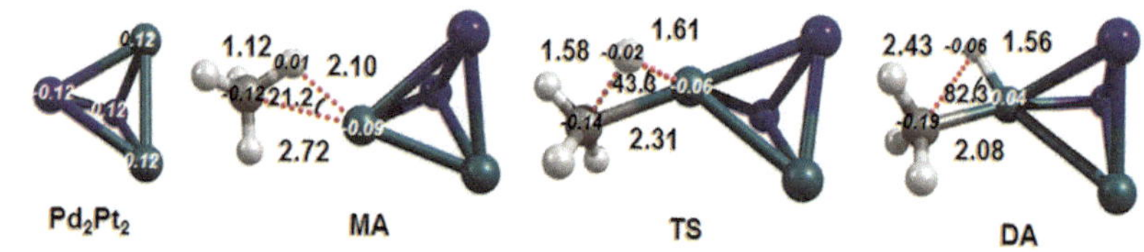

FIGURE 8.8 Optimized structures of the intermediates involved in the activation of methane on Pd_2Pt_2-Pt as active site, for selected structural parameters (distances in Å, bond angle in degrees) and Hirshfeld charges (shown in italic, in |e|).

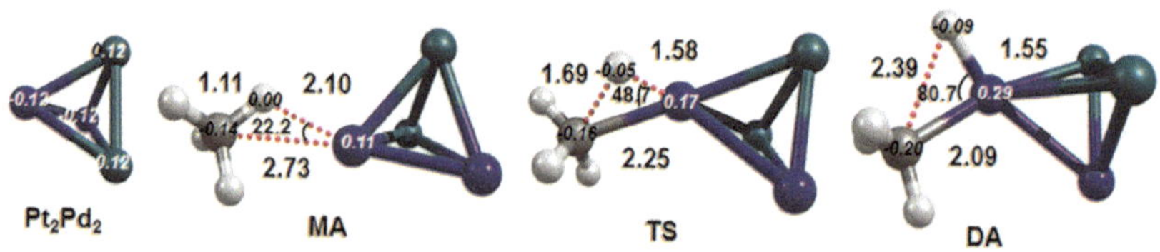

FIGURE 8.9 Optimized geometries of the transition and intermediate states involved in the activation of methane on Pt_2Pd_2-Pd as active site, for selected structural parameters (distances in Å, bond angle in degrees) and Hirshfeld charges (shown in italic, in |e|).

From Figure 8.10, it is seen that the reaction is exothermic when methane is adsorbed on the Pt-site, but slightly endothermic when the reaction is associated with Pd-site. The transition states are associated with an imaginary frequencies of 639.5i cm^{-1} and 871.1i cm^{-1}, respectively for the pathway I and II, corresponding to the vibration of C-H bond. The C-H bond is activated by crossing TS involving barrier heights of 6.2 kcal.mol^{-1} (path I) and 24.0 kcal.mol^{-1} (path II), summarized in Table 8.1. For the Pd_2Pt_2 system also it is seen that a smaller energy barrier is associated when methane is adsorbed on the Pt-site.

8.3.4 Pd_3Pt AND $PtPd_3$

The first step of the association reaction forming an electrostatically bound methane complex takes place at two active sites for the $PtPd_3$ system. For this system the reaction is carried out at singlet and triplet spin multiplicities as it play an important role in the stability of metal complexes. When Pt is the active site of the reaction, all the structures are stable at triplet state except for DA which is stable at singlet spin state. For the reaction carried on Pd as active site- the cluster and the metal complexes are found to have

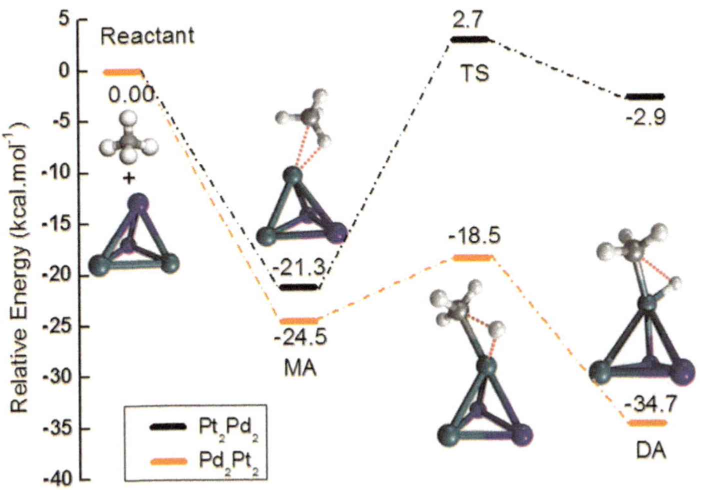

FIGURE 8.10 Potential energy curves for the activation of methane over bimetallic Pd_2Pt_2 system carried out for two active sites. The total energy of the reactants is taken as zero (Orange lines – Pt as the reactive site, black lines – Pd as the active site for carrying out the reaction).

attained stability in their triple state. The binding energy of methane to the Pt and Pd-sites are calculated as 20.14 kcal/mol^{-1} and 19.89 kcal.mol^{-1} respectively. The Pd-C/Pd-H bond lengths contract by 0.62 Å/0.42 Å and 0.97 Å/0.61 Å, respectively for the Pt and Pd-active sites of the cluster. Small amount of charge transfer is taking place as seen from Figures 8.11 and 8.12. The M-C-H bond angle increases along the reaction.

The reaction is exothermic when methane is bound to the Pt-site as seen from Figure 8.13, but endothermic when the reaction is carried out on the Pd-site. The reaction proceeds from the molecular adsorption to the DA of methane, via transition state, associated with an imaginary frequency of

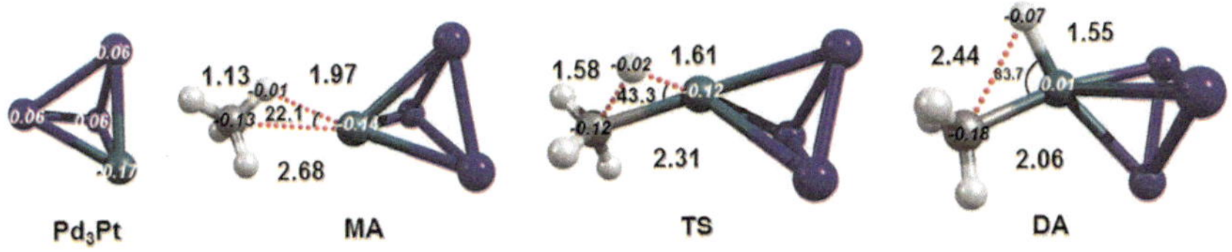

FIGURE 8.11 Optimized geometries of the transition and intermediate states involved in the activation of methane on $PdPt_3$-Pt as active site to carry the reaction, for selected structural parameters (distances in Å, bond angle in degrees) and Hirshfeld charges (shown in italic, in |e|).

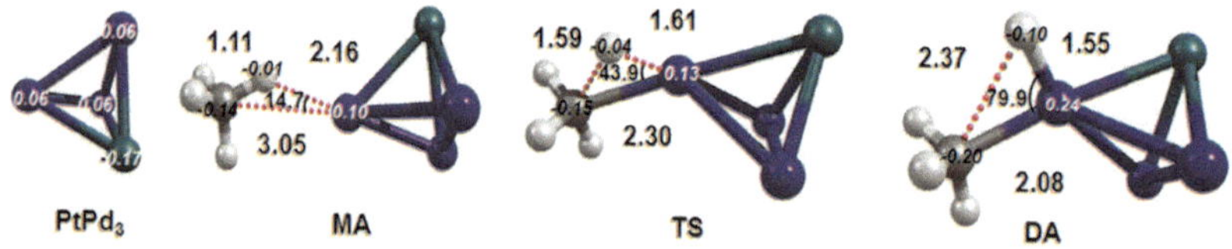

FIGURE 8.12 Optimized geometries of the transition and intermediate states involved in methane activation on Pt_3Pd system-Pd as active site, for selected structural parameters (distances in Å, bond angle in degrees) and Hirshfeld charges (shown in italic, in |e|).

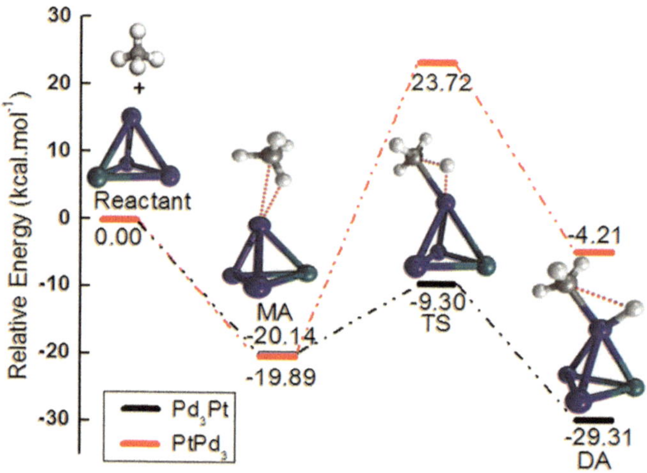

FIGURE 8.13 Potential energy diagram for the activation of methane over bimetallic $PtPd_3$ system carried out for two active sites. The total energy of the reactants is taken as zero. (Red lines – Pt as the reactive site, black lines – Pd as the active site for carrying out the reaction).

853.2i cm^{-1} for the Pt- active reaction site and 978.9i cm^{-1} for the Pd-site. Transition states involve barrier heights of 10.59 $kcal.mol^{-1}$ (active Pt-site) and 43.8 $kcal.mol^{-1}$ (active Pd-site), summarized in Table 8.1. For this system also, smaller energy barrier is linked to the Pt-site.

8.4 CONCLUSION

We have carried out a systematic comparative density functional study on the pristine Pt and Pd tetramer and their bimetallic counterparts at different spin multiplicities. The C-H bond breaking is the fundamental step of the conversion reaction of methane to value added products. The potential energy profile (including geometries and energies of reactant, intermediates, transition states and products) for methane C-H bond activation promoted by mono- and bimetallic Pd and Pt subnanoclusters investigated at the DFT level of theory to better understand the higher activity of the bimetallic system and preference of Pt-site over the Pd-site.

The study reveals that when methane binds with a higher affinity to a cluster, less energy is required to cross the TS barrier. Higher energy barrier for the activation of C-H bond in methane is associated with pristine as well as substituted Pd clusters (reaction carried on the Pd-site of the bimetallic Pt_mPd_n systems). Thus, for Pd-site of bimetallic clusters considered – the aforementioned reaction is mostly endothermic in nature, except on Pt_3Pd – which is weakly exothermic. Reaction for CH_4 dissociation on pure Pt and bimetallic cluster, where, Pt is considered as active site, are exothermic in nature. The lowest barrier height is obtained – $PdPt_3$ (Pt-site acting as the reactive site), followed by Pt_4, Pd_2Pt_2 and then Pd_3Pt. Thus the activity order is: $PdPt_3 > Pt_4 > Pd_2Pt_2 >$ $Pd_3Pt > Pt_2Pd_2$. Thus, Pt_4 and substituted Pt cluster (Pt-acting as active site for the reaction) are found to have higher activity for C-H bond breaking of methane.

ACKNOWLEDGMENTS

The author Pakiza Begum thanks the Department of Science and Technology, New Delhi, for INSPIRE Fellowship.

KEYWORDS

- **C-H cleavage**
- **density functional theory**
- **metal hydrogen bond formation**
- **methane**
- **mono and bimetallic subnanoclusters**

REFERENCES

1. Parkyns, N. D., Warburton, C. I., & Wilson, J. D. (1993). *Catal. Today 18*, 385–442.
2. Qiu, P., Lunsford, J. H., & Rosynek, M. P. (1997). *Catal. Lett. 48*, 11–15.
3. Xu, Y., & Lin, L. (1999). *Appl. Catal. A: Gen. 188*, 53–67.
4. Choudhary, V. R., Rajput, A. M., & Prabhakar, B. (1994). *Angew. Chem. Int. Ed. Engl. 33*, 2104–2106.
5. Gretz, E., Oliver, T. F., & Sen, A. *J. Am. Chem. Soc.* (1987). *109*, 8109–8111.
6. Vargaftik, M. N., Stolarov, I. P., & Moiseev, I. I. (1990). *J. Chem. Soc., Chem. Commun.* 1049–1050.
7. Brown, S. N., Myers, A. W., Fulton, J. R., & Mayer, J. M. (1998). *Organometallics* 17, 3364–3374.
8. Horn, R., & Schlögl, R. (2015). *Catal. Lett. 145*, 23–39.
9. Biswas, B., Sugimoto, M., & Sakaki, S. (2000). *Organometallics 19*, 3895–3908.
10. Choudhary, T. V., & Goodman, D. W. (2000). *J. Mol. Catal. A-Chem. 163*, 9–18.
11. Nizovtsev, A. S. (2013). *J. Comput. Chem. 34*, 1917–1924.
12. Sun, Q., Li, Z., Wang, M., Du, A., & Smith, S. C. (2012). *Chem. Phys. Lett. 550*, 41–46.
13. Labinger, J. A., & Bercaw, J. E. (2002). *Nature 417*, 507–514.
14. Bartholomew, C. H., & Farrauto, R. J. (2006). *Fundamentals of Industrial Catalytic Processes*, 2nd Edition, Wiley Interscience, NJ, Hoboken, p. 996.
15. Chen, M., Kumar, D., Yi, C.-W., & Goodman, D. W. (2005). *Science 310*, 291–293.
16. Begum, P., Gogoi, P., Mishra, B. K., & Deka, R. C. (2015). *Int. J. Quant. Chem. 115*, 837–845.
17. Misch, L. M., Kurzman, J. A., Derk, A. R., Kim, Y.-I., Seshadri, R., Metiu, H., McFarland, E. W., & Stucky, G. D. (2011). *Chem. Mater. 23*, 5432–5439.
18. Shao, M.-H., Sasaki, K., & Adzic, R. R. (2006). *J. Am. Chem. Soc. 128*, 3526–3527.
19. Enache, D. I., Edwards, J. K., Landon, P., Solsona-Espriu, B., Carley, A. F., Herzing, A. A., Watanabe, M., Kiely, C. J., Knight, D. W., & Hutchings, G. J. (2006). *Science 311*, 362–365.
20. Palagin, D., & Doye, J. P. K. (2015). *Phys. Chem. Chem. Phys. 17*, 28010–28021.
21. Yang, Y., Saoud, K., Abdelsayed, V., Glaspell, G., Deevi, S., & El-Shall, M. S. (2006). *Catal. Commun. 7*, 281–284.
22. Begum, P., Bhattacharjee, D., Mishra, B. K., & Deka, R. C. (2014). *Theor. Chem. Acc. 133*, 1418–1428.
23. Nizovtsev, A. S. (2013). *J. Comput. Chem. 34*, 1917–1924.
24. Pavlov, M., Blomberg, M. R. A., & Siegbahn, P. E. M. (1997). *J. Phys. Chem. A 101*, 1567–1579.
25. Russell, J., Zapol, P., Král, P., & Curtiss, L. A. (2012). *Chem. Phys. Lett. 536*, 9–13.
26. Delley, B. (2000). *J. Chem. Phys. 113*, 7756–7764.
27. Delley, B. (1990). *J. Chem. Phys. 92*, 508–517.
28. Becke, A. D. (1988). *Phys. Rev. A 38*, 3098–3100.
29. Lee, C., Wang, W., & Parr, R. G. (1988). *Phys. Rev. B 37*, 785–789.
30. Batista, V. S., & Coker, D. F. (1996). *J. Chem. Phys. 105*, 4033–4054.
31. Zhou, J., Xiao, F., Wang, W. N., & Fan, K. N. (2007). *J. Mol. Struct. (THEOCHEM) 818*, 51–55.
32. Delley, B. *Int. J. Quant. Chem.* (1998). *69*, 423–433.

CHAPTER 9

THEORETICAL ANALYSIS: ELECTRONIC AND OPTICAL PROPERTIES OF SMALL CU-AG NANO ALLOY CLUSTERS

PRABHAT RANJAN,[1] TANMOY CHAKRABORTY,[2] and AJAY KUMAR[1]

[1]*Department of Mechatronics, Manipal University Jaipur, Jaipur, Rajasthan – 303007, India*

[2]*Department of Chemistry, Manipal University Jaipur, Jaipur, Rajasthan – 303007, India, E-mail: tanmoy.chakraborty@jaipur.manipal.edu; tanmoychem@gmail.com*

CONTENTS

ABSTRACT

Due to diverse applications in the field of science and engineering, study of bi-metallic nano alloy clusters is very much popular. Among such nano clusters, the compound formed between Cu-Ag has gained considerable interest because of their unique optical, electronic and magnetic properties. Density functional theory (DFT) is one of the most popular techniques of quantum mechanics to study the electronic properties of materials. Conceptual DFT-based descriptors are indispensable tools to correlate the experimental properties of nano compounds and composites. In this report, we have studied $CuAg_n$; (n = 1–8) nano alloy clusters within theoretical framework of DFT at B3LYP level. The calculated nano alloy clusters show interesting odd-even oscillation behaviors, indicating even number of clusters show higher stability as compare to their neighbor odd number clusters. A close agreement between experimental bond length and computed data also reflect in this analysis.

9.1 INTRODUCTION

Since last decade, nanomaterials and nanotechnology have emerged as a new branch in the research domain of science and technology [1]. Due to presence of a large number of quantum mechanical and electronic effects, nanoparticles possess various unique physico-chemical properties [2–4]. The classification of nanoparticles is done in terms of size range of 1–100 nm. That particular size range exists between the levels of atomic/ molecular and bulk material [5]. But, there are still some instances of nonlinear transition of certain physical properties, which may vary depending on their size, shape and composition [6, 7]. A large number of available experimental and theoretical scientific reports describe the effects of size and structure which change the optical, electronic, magnetic, chemical and other physical properties of nanoparticles [1, 3, 4]. A deep insight into the research of nanoparticles with well-defined size and structure may lead to some other alternatives for better performance [8]. The nanoparticles, due to its enormous applications in the areas of biological labeling, photochemistry, catalysis, information storage, magnetic device, optics, sensors, photonics, optoelectronics, nanoelectronics, etc., have got immense importance [1, 3, 9–11].

The noble metals can be extensively applied in several technological areas due to its superior catalytic, magnetic and electronic properties [12–18]. There are number of instances where positive conjoint effects of two or more noble metals on the above-mentioned properties have been explained [13, 19–20]. Now-a-days, different compositions of nano alloys are being utilized for advancement of methodologies and characterization techniques [13, 19, 21]. A thorough study of core-shell structure of nano compounds is very much popular as because its properties can be tuned through the proper control of other structural and chemical parameters. Group 11 metal (Cu, Ag and Au) clusters exhibit the filled inner d orbitals with having one unpaired electron in the valence s shell [22]. This electronic arrangement is responsible to reproduce the exactly similar shell effects [23–27] which are experimentally observed for the alkali metal clusters [28–30]. Among the nano clusters of Group 11 elements, the compound formed between Cu-Ag is very much popular due to its large scale applications. The exact position of silver within Copper-Silver core-shell structure is highly important. The location of sliver has a controlling effect on the optical properties of such particles as because optical properties are governed by plasmon resonance frequency of silver, which is also dependent on its structural environment [31]. It has been already established that copper-silver bi-metallic nano clusters, as catalysts can enhance the reaction efficiency and selectivity [32, 33]. Though, a number of experimental studies have been done on this particular type of compounds, a theoretical analysis invoking density functional theory (DFT) is still unexplored.

Since the last couple of years DFT has been dominant method for quantum mechanical computation of periodic systems. Due to its computational friendly nature, DFT is very much popular to study the many- body systems [8]. Super conductivity of metal based alloys [34], magnetic properties of nano alloy Clusters [35, 36] quantum fluid dynamics [37], molecular dynamics [38], nuclear physics [39, 40] can be extensively studied by DFT methodology. Recently, we have established the importance of DFT-based descriptors in the domain of drug designing process and nano-engineering materials [41–47]. The study of DFT has been broadly classified into three sub categories viz. theoretical, conceptual, and computational [48–51]. The conceptual DFT is highlighted following Parr's dictum "accurate calculation is not synonymous with useful interpretation. To calculate a molecule is not to understand it" [52].

In this venture, we have successfully studied some bi-metallic nano-clusters containing Cu and Ag, in terms of DFT-based global descriptors, namely hardness, electronegativity, HOMO-LUMO Gap, softness, electrophilicity index and dipole moment. An attempt has been made to correlate the properties of instant compounds with their computational counterparts.

9.2 COMPUTATIONAL DETAILS

In this study, we have made an analysis on the bi-metallic Nano Alloy clusters of Cu-Ag_n where n = 1–8. 3d modeling and structural optimization of all the compounds have been performed using Gaussian 03 software package [53] within DFT framework. For optimization purpose, Becke, three parameter, Lee-Yang-Parr (B3LYP) exchange correlation with basis set LanL2DZ has been adopted. The used computation methodology in this paper is based on the molecular orbital approach, using linear combination of atomic orbitals. Z-axis has been chosen for the spin polarization axis. In this process, the symmetrized fragment orbitals (SFOs) are combined with auxiliary core functions (CFs) to ensure orthogonalization on the (frozen) core orbitals (COs). The quadrupole moment of molecule is calculated in terms of analytical integration methodology.

Invoking Koopmans' approximation [54], we have calculated ionization energy (I) and electron affinity (A) of all the nano alloys using the following equations

$$I = -\varepsilon_{HOMO} \tag{1}$$

$$A = -\varepsilon_{LUMO} \tag{2}$$

Thereafter, using I and A, the conceptual DFT-based descriptors viz. electronegativity (χ), global hardness (η), molecular softness (S) and electrophilicity index (ω) have been computed. The equations used for such calculations are as given under:

$$\chi = -\mu = \frac{I + A}{2} \tag{3}$$

where, μ represents the chemical potential of the system.

$$\eta = \frac{I - A}{2} \tag{4}$$

$$S = \frac{1}{2\eta} \tag{5}$$

$$\omega = \frac{\mu^2}{2\eta} \tag{6}$$

9.3 RESULTS AND DISCUSSIONS

A detail theoretical analysis of $CuAg_n$ (n = 1–8) nano alloy clusters has been performed interms of electronic structure theory. The orbital energies in form of highest occupied molecular orbital (HOMO)– lowest unoccupied molecular orbital (LUMO) gap along with computed DFT-based descriptors for instant nano clusters namely molecular electronegativity, global hardness, global softness and global electrophilicity index have been reported in the Table 9.1. The molecular dipole moment in debye unit is also reported in the Table 9.1. The quadrupole moment of the same nano clusters along with different axes are represented in the Table 9.2. Computed data from Table 9.1 reveals that HOMO-LUMO gaps of the Cu-Ag nano clusters are maintaining direct relationship with their evaluated global hardness values. As the frontier orbital energy gap increases, their hardness value increases. This trend is expected on the basis of experimental point of consideration. As the molecule possesses the highest HOMO-LUMO gap, it will be the least prone to response against any external perturbation. From the Table 9.1, it is clear that $CuAg_5$ is the least reactive species whereas $CuAg_8$ will exhibit maximum

TABLE 9.1 Computed DFT-Based Descriptors of $CuAg_n$; n = 1–8 Nanoalloy Clusters

Species name	HUMO-LUMO Gap (eV)	χ (eV)	η (eV)	S (eV)	ω (eV)	Dipole moment (debye)
CuAg	3.003	4.250	1.502	0.333	2.125	0.241
$CuAg_2$	1.823	3.850	0.912	0.548	1.925	0.398
$CuAg_3$	1.904	3.918	0.952	0.525	1.959	3.0075
$CuAg_4$	1.632	3.564	0.816	0.612	1.783	0.000
$CuAg_5$	3.047	3.891	1.523	0.328	1.945	0.258
$CuAg_6$	1.768	3.442	0.884	0.565	1.721	0.225
$CuAg_7$	1.851	3.918	0.925	0.541	1.959	0.958
$CuAg_8$	1.523	3.754	0.762	0.656	1.877	1.097

TABLE 9.2 Computed Quadrupole Moment in (a.u) of $CuAg_n$; n = 1–8 Nanoalloy Clusters

Species name	Quad- xx	Quad- xy	Quad- xz	Quad-yy	Quad-yz	Quad-zz
CuAg	–34.124	0.000	0.000	–48.297	0.000	–26.398
$CuAg_2$	–54.179	0.000	0.000	–54.077	0.000	–52.055
$CuAg_3$	–66.191	0.000	0.000	–70.058	0.000	–74.549
$CuAg_4$	–78.670	0.000	0.000	–94.785	0.000	–94.507
$CuAg_5$	–105.508	0.002	0.000	–104.899	0.000	–114.399
$CuAg_6$	–134.542	0.000	0.000	–124.213	0.000	–127.726
$CuAg_7$	–139.402	–1.362	0.000	–151.424	0.000	–154.708
$CuAg_8$	–174.630	–0.000	0.000	–160.596	0.000	–166.403

response. Though there is no such available quantitative data of optical properties of aforesaid clusters, we can assume that there must be a direct qualitative relationship between optical properties of Cu-Ag nano clusters with their computed HOMO-LUMO gap. The assumption is based on the fact that optical properties of materials are interrelated with flow of electrons within the systems which in turn depend on the difference between the distance of valence and conduction band. There is a linear relationship between HOMO-LUMO gap with the difference in the energy of valence-conduction band [55]. On that basis, we may conclude that optical properties of instant bi-metallic Nano clusters increase with increase of their hardness values. Similarly softness data exhibits an inverse relationship towards the experimental optical properties. Although the same type of direct relationships are observed in case of computed electronegativity, electrophilicity index and dipole moment along with their HOMO- LUMO gap, but for $CuAg_5$ and $CuAg_6$ cluster an exception is marked. The linear correlation between HOMO-LUMO gap along with their computed softness is lucidly plotted in the Figure 9.1. The high value of correlation coefficient (R^2 = 0.976) validates our predicted model.

The quadrupole charge separation is represented in form of quadrupole moment according to Buckingham convention in the Table 9.2. The quadrupole moment values in different axes are represented in atomic unit (a.u).

As per the cluster physics, dissociation energy and second difference of total energy have a marked influence on the relative stability of a particular system. These two energies are highly sensitive quantities and they exhibit pronounced odd-even oscillation behavior for neutral and charged clusters, as a function of cluster size [56]. The similar type of alternation behavior is also exhibited by the HOMO-LUMO gap of any particular compound. It has already been reported that cluster with an even number of total atoms exhibits less reactivity and possesses larger HOMO-LUMO

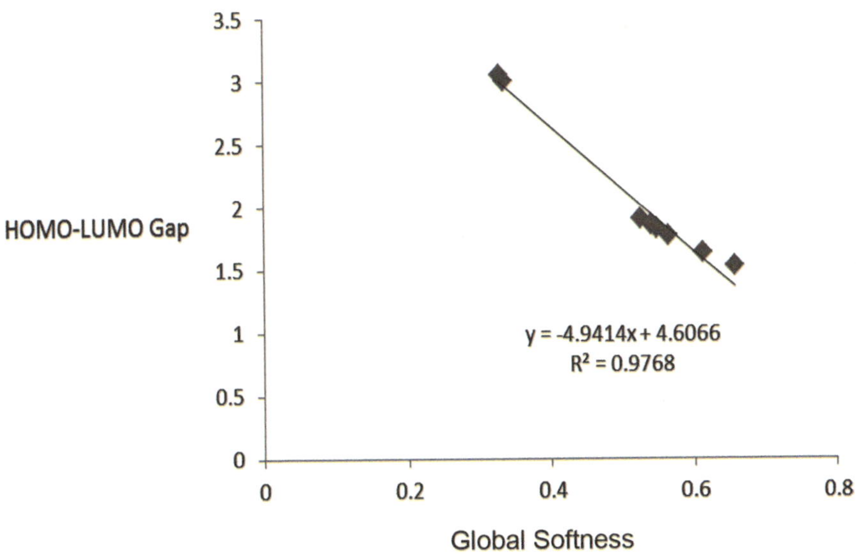

FIGURE 9.1 A Linear Correlation Plot between Global Softness Vs HOMO- LUMO Gap.

gap whereas the reverse trend is observed for the cluster containing a total odd number of the atoms [57–60]. The stability of the even number electronic cluster is actually an outcome of their closed electronic configuration which always produces extra stability. We have reported HOMO-LUMO gap as a function of cluster size in Figure 9.2. It is distinct from Figure 9.2 that the bi-metallic clusters containing even number of total atoms possess higher HOMO-LUMO gap as compared to the clusters having odd number of total atoms. So, our computed data supports the experimental odd-even alternation behavior of bi-metallic nano-clusters.

A comparative analysis has been made between experimental bond length [61–63] and our computed data of the species namely Ag_2, Cu_2 and CuAg. The same is reported in Table 9.3. A close agreement

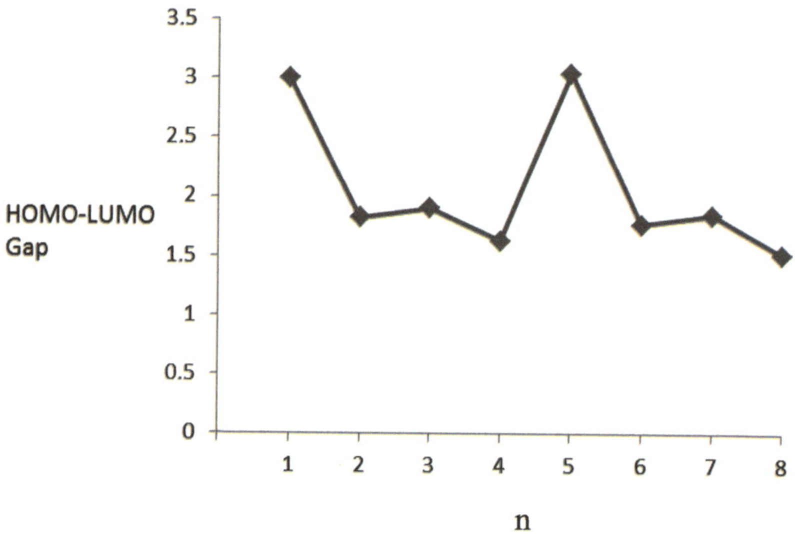

FIGURE 9.2 The size dependence of HOMO- LUMO Gap of $CuAg_n$ (n = 1–8) nano alloy clusters.

TABLE 9.3 The Calculated Bond Length (Å) of Ag_2, Cu_2, and CuAg Species

Species	Theoretical Bond Length	Experimental Bond Length
Ag_2	2.610	2.53[56]
Cu_2	2.259	2.22[57]
CuAg	2.432	2.37[58]

between experimental report and our computed bond length is reflected form the Table 9.3. It supports and validates our analysis.

9.4 CONCLUSION

In recent day, bi-metallic nano alloy clusters have got immense importance for its diverse nature of applications. A marked optical property is observed in case of nano cluster containing group 11 metals, namely Copper and Silver. In this paper, we have studied the system of $CuAg_n$; (n = 1–8) nano alloy clusters in terms of conceptual DFT-based descriptors namely global hardness, electronegativity, softness, electrophilicity index and dipole moment. In this analysis, it is observed that the computed HOMO-LUMO gap runs hand in hand along with it evaluated global hardness. As in absence of any quantitative benchmark, the optical property of Cu-Ag nano cluster has been assumed to be exactly equivalence of it HOMO-LUMO gap. Here our evaluated data reveals that optical property of these compounds maintains a direct relationship with hardness and inverse relationship with softness. This trend is expected from the consideration of other experimental facts. The high value of regression coefficient between hardness and HOMO-LUMO gap successfully supports our predicted model. The computed quadrupole moment data of above mentioned nano clusters also exhibit the quadrupole charge separation nicely. Our computed bond lengths for the species Cu_2, Ag_2, and CuAg are in very well agreement with the experimental data. In this study, the effect of even-odd oscillation behavior on HOMO-LUMO gap is also reflected.

KEYWORDS

- **bi-metallic nano alloy**
- **density functional theory**
- **electrophilicity index**
- **hardness**
- **softness**

REFERENCES

1. Zabet-Khosousi, A., & Dhirani, A.-A. (2008). "Charge transport in nanoparticle assemblies," *Chem. Rev., 108*, 4072–4124.
2. Daniel, M. C., & Astruc, D. (2004). "Gold nanoparticles: assembly, supramolecular chemistry, quantum-size-related properties, and applications toward biology, catalysis, and nanotechnology," *Chem. Rev., 104*, 293–346.
3. Ghosh, S. K., & Pal, T. (2007). "Interparticle coupling effect on the surface plasmon resonance of gold nanoparticles: from theory to applications," *Chem. Rev., 107*, 4797–4862.
4. Ghosh Chaudhuri, R. & Paria, S. (2012). "Core/shell nanoparticles: Classes, properties, synthesis Mechanisms, characterization, and applications," *Chem. Rev., 112*, 2373–2433.
5. Alivisatos, A. P. (1996). "Semiconductor Clusters, Nanocrystals, and Quantum Dots," *Science,* New series, *271*, 933–937.
6. Kastner, M. A. (1993). "Artificial Atoms," Phys. Today, *46* 24–31.
7. Haruta, M. (2002). "Catalysis of Gold Nanoparticles Deposited on Metal Oxides," *Cattech, 6*, 102–115.
8. Ismail, R. (2012). "Theoretical Studies of Free and Supported Nanoalloy Clusters," PhD Thesis, 20–38.
9. Roucoux, A., Schulz, J., & Patin, H. (2002). "Reduced transition metal colloids: A novel family of reusable catalysts?" *Chem. Rev., 102*, 3757–3778.
10. Munoz-Flores, B. M., Kharisov, B. I., Jimenez-Perez, V. M., Elizondo Martinez, P., & Lopez, S. T. (2011). "Recent advances in the synthesis and main applications of metallic nano-alloys," *Ind. Eng. Chem. Res., 50*, 7705–7721.
11. Murray, R. W. (2008). "Nanoelectrochemistry: Metal Nanoparticles, Nanoelectrodes, and Nanopores," *Chem. Rev., 108*, 2688–2720.
12. Teng, X., Wang, Q., Liu, P., Han, W., Frenkel, A. I., Wen, Marinkovic, N., Hanson, J. C., & Rodriguez, J. A. (2008). "Formation of Pd/Au Nanostructures from Pd Nanowires via Galvanic Replacement reaction," *J. Am. Chem. Soc., 130*, 1093–1101.
13. Ferrando, R., Jellinek, J., & Johnston, R. L. (2008). "Nanoalloys: from theory to applications of alloy clusters and nanoparticles," *Chem. Rev., 108*, 845–910.
14. Henglein, A. (1993). "Physicochemical properties of small metal particles in solution: "microelectrode" reactions, chemisorption, composite metal particles, and the atom-to-metal transition," *J. Phys. Chem., 97*, 5457–5471.
15. Davis, S. C., & Klabunde, K. J. (1982). "Unsupported small metal particles: Preparation, Reactivity, and Characterization," *Chem. Rev., 82*, 153–208.
16. Lewis, L. N. (1993). "Chemical catalysis by colloids and clusters," *Chem. Rev., 93*, 2693–2730.
17. Schmid, G. (1992). "Large clusters and colloids. Metals in the embryonic state," *Chem. Rev., 92*, 1709–1727.
18. Schon, G., & Simon, U. (1995). "A fascinating new field in colloid science: small ligand-stabilized metal clusters and possible application in microelectronics," *Colloid. Polym. Sci., 273*, 101–117.
19. Oderji, H. Y., & Ding, H. (2011). "Determination of melting mechanism of $Pd_{24}Pt_{14}$ nanoalloy by multiple histogram method via molecular dynamics simulations," *Chem. Phys., 388*, 23–30.

20. Liu, H. B., Pal, U., Medina, A., Maldonado, C., & Ascencio, J. A. (2005). "Structural incoherency and structure reversal in bimetallic Au-Pd nanoclusters," *Phys. Rev. B, 71*, 75403–75409.
21. Baletto, F., & Ferrando, R. (2005). "Structural properties of nanoclusters: Energetic, thermodynamic, and kinetic effects," *Rev. Mod. Phys.*, *77*, 371–423.
22. Alonso, J. A. (2000). "Electronic and atomic structure, and magnetism of transition-metal clusters," *Chem. Rev., 100*, 637–677.
23. Katakuse, I., Ichihara, T., Fujita, Y., Matsuo, T., Sakurai, T., & Matsuda, H. (1985). "Mass distributions of copper, silver and gold clusters and electronic shell structure: I," *Int. J. Mass Spectrom. Ion Processes*, *67*, 229–236.
24. Katakuse, I., Ichihara, T., Fujita, Y., Matsuo, T., Sakurai, T., & Matsuda, H. (1986). "Mass distributions of negative cluster ions of copper, silver and gold: I," *Int. J. Mass Spectrom. Ion Processes*, *74*, 33–41.
25. de Heer, W. A. (1993). "The physics of simple metal clusters: experimental aspects and simple models," *Rev. Mod. Phys.*, *65*, 611–676.
26. Gantefor, G., Gausa, M., Meiwes-Broer, K.-H., & Lutz, H. O. (1990). "Photoelectron spectroscopy of silver and palladium cluster anions; electron delocalization *versus* localization," *J. Chem. Soc., Faraday Trans.*, *86*, 2483–2488.
27. Leopold, D. G., Ho, J., & Lineberger, W. C. (1987). "Photoelectron spectroscopy of mass-selected metal cluster anions. I. Cu_n^-, n = 1–10," *J. Chem. Phys.*, *86*, 1715–1726.
28. Lattes, A., Rico, I., Savignac, A. de, & Ahmad-Zadeh Samii, A. (1987). "Formamide, a water substitute in micelles and microemulsions [xxx] structural analysis using a diels-alder reaction as a chemical probe," *Tetrahedron*, *43*, 1725–1735.
29. Chen, F., Xu, G. Q., & Hor, T. S. A. (2003). "Preparation and assembly of colloidal gold nanoparticles in CTAB-stabilized reverse microemulsion," *Mater. Lett.*, *57*, 3282–3286.
30. Taleb, A., Petit, C., & Pileni, M. P. (1998). "Optical properties of self-assembled 2D and 3D superlattices of silver nanoparticles," *J. Phys. Chem.* B, *102*, 2214–2220.
31. Langlois, C., Wang, Z. W., Pearmain, D., Ricolleau, C., & Li, Z. Y. (2010). HAADF-STEM imaging of CuAg core-shell nanoparticles," *J. Phys.: Conference Series*, *241*, 12043–12047.
32. Jankowaik, J. T. A., & Barteau, M. A. (2005). "Ethylene epoxidation over silver and copper-silver bimetallic catalysts: I. kinetic and selectivity," *J. Catal.*, *236*, 366–378.
33. Piccinin, S., Zafeiratos, S., Stampfl, C., Hansen, T. W., Hävecker, M., Teschner, D., et al. (2010). "Alloy catalyst in a reactive environment: The example of Ag-Cu particles for ethylene epoxidation," *Phys. Rev. Lett.*, *104*, 35503–1:35503–4.
34. Wacker, O. J., Kummel, R., & Gross, E. K. U. (1994). "Time-dependent density-functional theory for superconductors," *Phys. Rev. Lett.*, *73*, 2915–2918.
35. Illas, F., & Martin, R. L. (1998). "Magnetic coupling in ionic solids studied by density functional theory," *J. Chem. Phys.*, *108*, 2519–2527.
36. Gyorffy, B., Staunton, J., & Stocks, G. (1995). "In Fluctuations in Density Functional Theory: Random Metallic Alloys and Itinerant Paramagnets," Series B, *337*, 461–464, Proceedings of a NATO Advanced Study Institute on Density Functional Theory, August 16–27, 1993, Italy.
37. Kümmel, S., & Brack, M. (2001). "Quantum fluid-dynamics from density functional theory," *Phys. Rev.* A, *64*, 022506.

38. Car, R., & Parrinello, M. (1985). "Unified approach for molecular dynamics and density functional theory," *Phys. Rev. Lett., 55*, 2471–2474.
39. Koskinen, M., Lipas, P., & Manninen, M. (1995). "Shapes of light nuclei and metallic clusters," *Nucl. Phys. A, 591*, 421–434.
40. Schmid, R. N., Engel, E., & Dreizler, R. M. (1995). "Density functional approach to quantum hadrodynamics: Local exchange potential for nuclear structure calculations," *Phys. Rev. C, 52*, 164–169.
41. Chakraborty, T., & Ghosh, D. C. (2011). "Correlation of the drug activities of some anti-tubercular chalcone derivatives in Terms of the quantum mechanical reactivity descriptors," *Int. J. Chemoinf Chem. Eng., 1*, 53–65.
42. Chakraborty, T., & Ghosh, D. C. (2012). "Correlation of the drug activities and identification of the reactive sites in the structure of some anti-tubuculour juglone derivatives in terms of molecular orbital and the density functional descriptors," *Int. J. Chem. Model, 4*, 413–440.
43. Ranjan, P., Kumar, A., & Chakraborty, T. (2014). "Computational study of nanomaterials invoking DFT-based descriptors," Environmental Sustainability: Concepts, Principles, Evidences and Innovations, Excellent Publishing House, New Delhi, India, 239–242.
44. Ranjan, P., Kumar, A., & Chakraborty, T. (2014). "A theoretical analysis of bi-metallic $AgAu_n$ (n = 1–7) nano-alloy clusters invoking DFT-based descriptors," Research Methodology in Chemical Sciences: Experimental and Theoretical Approaches, Apple Academic Press, USA.
45. Ranjan, P., Dhail, S., Venigalla, S., Kumar, A., Ledwani, L., & Chakraborty, T. (2015). "A theoretical analysis of bi-metallic $(Cu\text{-}Ag)_{n=1-7}$ nano alloy clusters invoking DFT-based descriptors," *Mater. Sci. Pol.*, 33, 719-724.
46. Ranjan, P., Venigalla, S., Kumar, A., & Chakraborty, T. (2014). "Theoretical study of bi-metallic Ag_mAu_n (m + n = 2–8) nano alloy clusters in terms of DFT methodology," *New Front. Chem., 23*, 111–122.
47. Venigalla, S., Dhail, S., Ranjan, P., Jain, S., & Chakraborty, T. (2014). "Computational study about cytotoxicity of metal oxide nanoparticles invoking nano-QSAR technique," *New Front. Chem., 23*, 123–130.
48. Parr, R. G., & Yang, W. (1995). "Density-functional theory of the electronic structure of molecules," *Annu. Rev. Phy. Chem., 46*, 701–728.
49. Kohn, W., Becke, A. D., & Parr, R. G. (1996). "Density functional theory of electronic structure," *J. Phys. Chem., 100*, 12974–12980.
50. Liu, S., & Parr, R. G. (1997). "Second-order density-functional description of molecules and chemical changes," *J. Chem. Phys., 106*, 5578–5586.
51. T. Ziegler, (1991). "Approximate density functional theory as a practical tool in molecular energetics and dynamics," *Chem. Rev., 91*, 651–667.
52. Geerlings, P., & De Proft, F. (2002). "Chemical Reactivity as Described by Quantum Chemical Methods," *Int. J. Mol. Sci., 3*, 276–309.
53. Gaussian 03, Revision C.02, Frisch, M. J., Trucks, G. W., Schlegel, H. B., Scuseria, G. E., Robb, M. A., Cheeseman, J. R., et al., (2004). Gaussian, Inc., Wallingford CT.
54. Parr, R. G., & Yang, W. (1989). Density Functional Theory of Atoms and Molecules, Oxford, University Press, Oxford.

55. Xiao, H., Tahir-Kheli, J., & Goddard III, W. A. (2011). "Accurate Band Gaps for Semiconductors from Density Functional Theory," *J. Phys. Chem. Lett.*, *2*, 212–217.
56. Wang, H. Q., Kuang, X. Y. & Li, H. F. (2010). "Density functional study of structural and electronic properties of bimetallic copper-gold clusters: comparison with pure and doped gold clusters," Phys. Chem. *Chem. Phys.*, *12*, 5156–5165.
57. Ping, D. L. Yu, K. X., Peng, S., Ru, Z. Y., & Fang, L. Y. (2012). "A comparative study of geometries, stabilities and electronic properties between bimetallic AgnX (X = Au, Cu; n = 1–8) and pure silver clusters," *Chinese Phys B.*, *21*, 43601–43613.
58. Hakkinen, H., & Landman, U. (2000). "Gold clusters (AuN, $2 \leq N \leq 10$) and their anions," *Phys. Rev.*, *62*, 2287–2290.
59. Li, X. B., Wang, H. Y., Yang, X. D., Zhu, Z. H., & Tang, Y. J. (2007). "Size dependence of the structures and energetic and electronic properties of gold clusters," *J. Chem. Phys.*, *126*, 084505.
60. Jain, P. K. (2005). "A DFT-based study of the low-energy electronic structures and properties of small gold clusters," *Struct. Chem.*, *16*, 421–426.
61. Beutel, V., Kramer, H. G., Bhale, G. L., Kuhn, M., Weyers, K., & Demtroder, W. (1993). "High-resolution isotope selective laser spectroscopy of Ag_2 molecules", *J. Chem. Phys.*, *98*, 2699–2708.
62. Balbuena, P. B., Derosa, P. A., & Seminario, J. M. (1999). "Density functional theory of copper clusters", *J. Phys. Chem B*, *103*, 2830–2840.
63. Bishea, G. A., Marak, N., Morse, M. D. (1991). "Spectroscopic studies of jet-cooled CuAg", *J. Chem. Phys.*, *95*, 5618–5629.

CHAPTER 10

MULTISOLITONS IN SRR-BASED METAMATERIALS IN KLEIN-GORDON LATTICES

A. K. BANDYOPADHYAY,[1] BABUSONA SARKAR,[2] SANTANU DAS,[3] MOKLESA LASKAR,[4] and ANIRUDDHA GHOSAL[4]

[1]*Government College of Engineering and Ceramic Technology, W. B. University of Technology, 73, A. C. Banerjee Lane, Kolkata–700010, India, E-mail: asisbanerjee1000@gmail.com*

[2]*Department of Materials Science, Indian Association for the Cultivation of Science, Jadavpur, Kolkata–700032, India*

[3]*Department Materials Science and Engineering, University of North Texas, Denton, TX 76207, USA*

[4]*Institute of Radiophysics and Electronics, Calcutta University, Calcutta, India*

CONTENTS

ABSTRACT

Solitons are the solutions of nonlinear partial differential equations whose modal behavior can be described by a nonlinear Klein-Gordon (NLKG) equation that is derived through variational principle based on a generalized Hamiltonian. It is known that solitons gives robust solutions that can retain their shape to a longer extent without any distortion and hence also treated as particles. In the realm of their application in nonlinear optical fiber communications as well as in various nano-structured devices, NLKG equation seems to be a better option. It is a PDE with second order spatial and temporal terms that describe the nonlinear modes in the communication systems involving both ferroelectrics and metamaterials that are well-known nonlinear optical materials. Here, we derive NLKG for metamaterials as split-ring-resonators (SRR) for the applications in antenna. Various parameters involved in the Hamiltonian indicate interesting behavior against amplitude of magneto-inductive waves that also shows multi-soliton behavior in the metamaterials system.

10.1 INTRODUCTION

The fascinating world of optics deals with 'light' that is the ultimate means of sending information to and from the interior structure of materials. It packages data in a signal of zero mass and unmatched speed. However, light in a sense is 'one-handed' when interacting with atoms of conventional materials. This is because from the two field components of light, i.e., electric and magnetic, only the electric-field probes the atoms of a material, whereas the magnetic-field component of light is normally weak. Metamaterials, i.e., artificial materials with rationally designed properties, can allow both field components of light to be coupled to the 'metaatoms', enabling entirely new optical properties and exciting applications with such 'two-handed' light. Among the interesting properties is a 'negative refractive index'. The refractive index is one of the most fundamental characteristics of light–propagation in materials. With negative refractive index, metamaterials may lead to the development of a 'superlens' that is capable of imaging objects and fine structures, which are much smaller than the wavelength of light. Other exciting applications of metamaterials include 'antennae' with superior transmission properties, optical nanolithography and nanocircuits, and also 'metacoatings' that can make the objects invisible.

In the field of applied physics, a fertile ground was created with the tremendous surge of research activity on nonlinear optical materials and devices involving ferroelectrics and metamaterials. For such materials, the use of non-integrable Klein-Gordon (K-G) equation with second order spatial and temporal terms is already described [1–3] for modal behavior in various device applications including those in the nano-structured systems in terms of shape and velocity of the soliton (see later for other references). The main focus of this article is on the application of nonlinear K-G equation that could describe the modal behavior of soliton and multisoliton waves in metamaterials.

It may be pointed out that discrete solitons or discrete breathers (DBs) [also called (intrinsic localized mode (ILM)] are observed both in the integrable (viz. sine-Gordon equation) and non-integrable systems (viz. Klein-Gordon equation). However, the integrability imposes a criterion for obtaining DBs analytically that is possible only in the integrable systems, while for the non-integrable systems it is obtained by various numerical methods viz. (a) spectral collocation method, (b) finite-difference method, (c) finite element method, (d) Floquet analysis, etc. As evident from many numerical experiments, DBs mobility is achieved by an appropriate perturbation [4]. Within the ambit of integrable systems, although the above nonlinear differential equations, such as KdV equation, sine-Gordon equation and nonlinear Schrodinger equation (NLSE) give rise to soliton solutions, these equations are not discussed here in order to remain within our main focus. Next, let us discuss about metamaterials.

10.1.1 METAMATERIALS

The electrodynamics of substances with both negative dielectric constant (ε) and magnetic permeability (μ) is very interesting indeed. They are predicted to possess a negative index of refraction, and consequently they exhibit a variety of optical properties that are not found in positive indexed materials. About 46 years ago, Veselago [5] was the first to create a theoretical foundation of whether the refractive index of materials can possess negative values, and predicted a number of unusual optical properties associated with such negative index materials (NIM). A lucid account of transforming this theory into reality, i.e., metamaterials, was recently given by Litchinitser and Shalaev [6] and elaborated in many other important investigations [7–22]. The prefix 'meta' in Greek means 'beyond' or 'after'. This suggests that the material exhibits characteristics surpassing those in nature. In saying that metamaterials possess a negative index is a special case nowadays, if cloaking is considered, for example [7]. Also some anisotropic materials, such as calcite, exhibit negative refraction, as discussed by Boardman et al. [8].

Now, the question was: in what sort of materials can these properties be observed? This question was answered by Smith et al. [9] based on some theoretical work of Pendry et al. [10–12]. They fabricated a "metamaterial" (MM), i.e., an array of artificial structures consisting of metallic wires responsible for the negative permittivity and metallic "split ring resonator" (SRR) responsible for the negative permeability. The SRRs may have different shapes. The optical and electrical properties of a metamaterial can be harnessed by the proper use of SRRs that has application in the antennas. However, a very high accuracy is required for the fabrication of SRRs. Unlike natural materials, metamaterials also show relatively large magnetic response at THz frequency and hence their THz application assumes a lot of significance.

Thus, spatial periodic microstructure of metamaterials exhibits such significant macroscopic property as refractive index. With negative refraction, reversing light is a significant observation, and such materials have quite an interesting history in terms of various methods of fabrication and novel measurement techniques [13–22]. These metamaterials may show nonlinear features due to their field-dependence and bi-stability of states, as shown by many workers, notably by Litchinitser et al. [23], as nonlinearity could be an important feature in periodically discrete lattices. Apart from optical bistability, other nonlinear optical phenomena, such as (a) second harmonic generation (SHG), (b) optical parametric amplification, (c) modulation instability, and in many cases (d) soliton propagation, are also important considerations in such engineered materials [24–27]. Our focus is on the latter, i.e., on solitons and multisolitons in metamaterials system.

Another question was also raised: how to represent these types of materials in terms of a circuit model? From the theoretical standpoint, for application in antenna, if one makes a good approximation, the SRR assembly can be considered as equivalent to a nonlinear resistor inductor capacitor (RLC) circuit [28, 29] featuring a self-inductance L from the ring, a ring Ohmic resistance R to take care of dissipation, and a capacitance C from the split in the ring. Metamaterials are then formed as a periodic array of SRRs which are coupled by mutual inductance and arrayed in a material of dielectric constant (ε). Thus, these materials behave as capacitively loaded loops. It is known that these loops support wave-propagation. As the coupling is due to the induced voltages, these waves are referred to as magneto-inductive (MI) waves, as detailed in Ref. [15].

MI waves represent a vast area of active research in the field of (a) artificial delay lines and filters, (b) dielectric Bragg reflectors, (c) slow-wave structures in microwave tubes, (d) coupled cavities in accelerators, modulators, antenna array application, etc. Due to RLC circuit model, there exists a resonant frequency for this type of configuration. It is observed that these MI waves propagate within a band near the resonant frequency of the SRRs. Also, the magnetic permeability does not depend on the intensity of the electromagnetic field in the linear regime of MI wave propagation. Accordingly, the nonlinearity is incorporated in the system by embedding the SRRs in a Kerr-type of medium [28, 30], or by inserting certain nonlinear elements (e.g., diodes) [31–33].

Kourakis et al. [34] studied the self-modulation of the waves by NLSE that led to spontaneous energy localization via the generation of localized envelope structures (i.e., so called envelope solitons) and the dynamics of the nonlinear RLC circuit gave rise to a governing equation for SRRs in both space and time dimensions. There are important observations made on the appearance of multisolitons by controlling various parameters by a number of workers with numerical solutions [35, 36]. Lazarides and co-workers [28, 29] also studied classical DBs in metamaterial systems. After describing metamaterials, let us next discuss about discrete breathers.

10.1.2 DISCRETE BREATHERS IN K-G LATTICES

The historical significance of the term "breather" needs to be first elaborated. It can be created in translationally invariant nonlinear lattice models. It is in contrast to soliton solutions, which are moving wave-packets, i.e., nonlinear localized traveling waves that are robust and propagate without change in shape [37, 38]. Its temporal evolution has always been an area of intensive research. On the other hand, "breathers" are discrete solutions, periodic in time and localized in space, and whose frequencies extend outside the phonon spectrum, i.e., breathers are localized and time-periodic wave packets [39, 40].

The first discovered breather was in a sine-Gordon (integrable) system, which is also analytically tractable. There are various methods to characterize discrete breathers. Classical breathers can also be obtained in a non-integrable K-G system, as done numerically by well-known technique, such as spectral collocation method, which involves minimum errors in the analysis of different breather modes. Due to the localization, the length scale of such excitation assumes more significance that obviously drives us to the nano domain, whose importance in the field of solid state physics cannot be denied.

DBs or discrete solitons, also known as ILM, are nonlinear excitations that are produced by the nonlinearity and discreteness of the lattice. These excitations are characterized by their long time oscillations. These are highly localized pulses in space that are found in the discrete nonlinear model formulation. Unlike the plane wave like modes, DBs have no counterparts in the linear system, but exist only because of the system nonlinearity in a periodic lattice [1]. The frequencies of the DBs extend outside the phonon spectrum. They are formed as a self-consistent interaction or coupling between the mode and the system nonlinearity. In this way, DB modifies the local properties of the system at the DB peak, and the modified local properties of the system provide the environment for the DB to exist. As the continuum limit formulation cannot be applied to their study, the present formulation on discrete domains is appropriate to highly localized pulses having widths that are not large compared to the domain of interest. So, the question is about the appropriate length scale, which drives us to the nano-system. Thus, localization assumes more significance.

Localization is an important aspect for applications in a variety of devices under the broad field of solid state physics. It plays a crucial role in qualifying and quantifying a systems' operations, as in metamaterials. As said above, the extent of localization in the quantum regime assumes more significance for very small-structures, e.g., for nano devices. Now, the question comes: how do we get localization in a system or a lattice? Localization is evolved mainly either by 'disorder' in the lattice that first discovered by the pioneering work of Anderson [41], or by the systems' interplay of nonlinearity and discreteness [42], i.e., our attention is diverted towards DBs. The first one, i.e., Anderson localization, has been implemented in details in many types of devices. As nonlinearity arises in metamaterials by embedding them in Kerr-type medium [28, 30], they could also give rise to the localization.

Here, we shall mainly discuss about the localization due to nonlinearity by adding some nonlinear components in the governing equation and discreteness. After studying the existence of DBs in terms of discreteness and nonlinearity, the first question that comes to our mind: what is the importance of DBs in metamaterials? DBs seem to be quite versatile in managing localized energy, i.e., in targeted energy transfer (TET) or trigger mechanism [4]. Once DBs are formed, they can transport this energy efficiently by engaging the lattice in their motion, and moreover, under specific circumstances they can transfer this energy in selected lattices [1] (see references therein). Combining these facts from the model as well as some more general studies, in terms of an extrapolative argument, it can be said that DBs in nonlinear optical materials could in principle act as an able energy manager. Hence, the above explanations are given in brief to relate the localized waves of DBs in metamaterials.

Without going into the history, it can be said that there is a considerable amount of research activity on DBs since the paper of Sievers and Takeno [43] was published in 1988. Their existence has been theoretically proposed in several discrete many-body systems, and observed experimentally in different systems [44]. As DBs are time-periodic (spatially localized) excitations in a nonlinear spatially discrete lattice, a large volume of analytical and numerical studies has revealed the existence and properties of DBs in various nonlinear systems [40]. Detailed discussions of DBs have been reviewed extensively in the works of Sievers et al.

[40, 43, 44], Segev and coworkers [45]. Flach et al. [46, 47], and the subject was also studied with a lot of theoretical depth by Mackay and Aubry [48, 49] that was also followed by a presentation on "what we know about discrete quantum breathers" by Fleurov [50]. Here, another review by Flach and Gorbach [39] also needs to be mentioned, as it contains almost all the relevant references on DBs.

Furthermore, as the discreteness is found to trap the breathers, the moving breathers are non-existent in highly discrete nonlinear system that has been presented by Bang and Peyrard in the context of a K-G model [51]. These authors [52] did a numerical study on the 'exchange of energy and momentum' between the colliding breathers to describe an effective mechanism of 'energy localization' in K-G lattice that is arising out of the discreteness and non-integrability of the system. Here, the bright soliton solutions have been used for nonlinear dynamics of DNA molecules to demonstrate the generation of highly localized modes.

It is known that quasi-phase matching (QPM) is an important issue in a quadratic nonlinear photonic crystal (QNPC) or photonic band-gap materials with tunability. In an interesting work on QPM by Kobyakov et al. [53], the influence of an induced cubic nonlinearity on the amplitude and phase modulation was analytically studied to predict an efficient all-optical switching. Further, in the application front for a QNPC, a stable soliton solution was shown by Corney and Bang [54] for cubic nonlinearity and QNPC was found to support both dark and bright solitons even in the absence of quadratic nonlinearity, and also the 'modulation instability' in such systems was shown [55]. Trapani et al. [56] studied focusing and defocusing nonlinearities in the context of parametric wave mixing. A two-well potential has been used to derive kink solution of the nonlinear propagating waves in ferroelectrics in the context of a diatomic chain model [57–59]. The impact of this potential in case of a discrete system has been quite extensively studied by Comte [60].

In the context of discreteness, it should be mentioned that the Klein-Gordon (K-G) Hamiltonian with Landau two-well potential energy formulation can be discretized [61] and then go into the continuum limit [62] for nonlinear optical materials. The former manifests in the localization of sites and hence in the nonlinear regime, it is ripe for the formation of DBs – classical breathers and also quantum breathers on quantization of our Hamiltonian. It is pertinent to mention that there is a lot of research activity on both theoretical and experimental aspects of nonlinear optics in the realm of nanotechnology, i.e., on smaller length-scale of excitations. However, these investigations have been mainly carried out with the help of the NLSE.

In a recent work [1] it has been shown that nonlinear K-G equation can also be used in describing the modal behavior of metamaterials that shows breather pulses apart from dark and bright discrete breathers. Within the continuum limit dynamics, the nonlinear K-G equation with second order space and time derivatives has been used in a number of studies on domains and the motion of domain walls [2, 61–64], and also in some treatments of arrays of domains [1, 65, 66]. A particular facility in the treatment is that the K-G equation is a well-known equation of mathematical physics which exhibits a wide variety of interesting properties including soliton dynamics and has various applications to a large number of different physical systems.

The literature on soliton is so vast that it is very difficult to mention all the references. However, the Ref. [37, 38] are very useful for further references. As said earlier for soliton propagation in many optical systems, it is a common practice to use NLSE. However, we have recently shown that the NLSE can be derived through perturbation on K-G equation when progressive wave passes through nonlinear medium, such as lithium niobate ferroelectrics, where dispersion may take place, and discrete energy levels due to dipole-dipole interaction were estimated via 'hyper-geometric function' [3].

The dark and bright solitons in different systems are already known, particularly in the context of K-G equation. If the intrinsic field is only considered, then the modal dynamics for small oscillations could be characterized by the bound state in a limited range of frequency, revealed via 'associated Legendre polynomial'. The pairing and interplay between the dark and bright solitons occur. The disappearance of the bound state after a critical frequency gives rise to dark solitons in the unbound states that propagate through the domains. Above the upper boundary of the bound states, the estimated frequencies of dark solitons match with those experimentally found for 'acoustical memory', as revealed theoretically by Bandyopadhyay et al. [67] based on experimental work of McPherson et al. [68].

In nonlinear optical materials, such as ferroelectrics, the domain dynamics and the effect of impurities on such materials were lucidly presented by Phillpot and coworkers [69, 70] through density functional theory (DFT) approach and molecular dynamics simulations combined with thermodynamic calculations. Further on ferroelectrics, a set of ILMs or DBs of the domain array was also investigated in the context of K-G equation [2, 3]. This study was also extended from classical DBs [61–63] to quantum DBs (in terms of pinning transition for the effect of impurities on nonlinear optical properties) in ferroelectrics [64] with periodic Bloch function as well as on metamaterials using non-periodic boundary condition [65, 66].

Finally, some of the important applications in the nonlinear optical materials need to be mentioned. DBs have been investigated in several systems, viz. (a) solid state mixed-valence transition metal complexes [71], (b) quasi-one dimensional antiferromagnetic chains [72], (c) array of Josephson junctions [73], (d) oscillators in micromechanical system [74, 75], (e) optical waveguide systems [76], and (f) proteins, which is also an important molecules for biotechnology [77]. DBs modify system properties viz. lattice thermodynamics and introduce the possibility of non-dispersive energy transport [78, 79], because of their potential application for translatory motion along the lattice [80]. As mentioned earlier, these DBs are observed both in integrable and non-integrable systems. Next, let us briefly discus about the quantum counterpart of DBs that is a new field of research.

10.1.3 QUANTUM BREATHERS

Although we are not dealing here with quantum breathers, they deserve a brief mention, as they have implications in nano-devices. Also, it is quite pertinent to mention that, for the characterization of DBs or classical breathers [43–49, 61], the bulk system was the right tool, but when we are dealing with the smaller systems, (nano systems) we have to use a different tool of study, i.e., quantum physics, which brings us to the quantum breathers (QBs) [81, 82]. Once generated, QBs modify system properties such as lattice thermodynamics and introduce the possibility of non-dispersive energy transport, as generally described for DBs [78].

For QBs, it is important to consider detailed information on phonons and their bound state concept, which is sensitive to the degree of nonlinearity. The branching out of the QB state from the single-phonon continuum is quite noteworthy in nonlinear systems with charge defects [64]. Let us consider that the phonons in one sublattice may hop from one domain to another adjacent domain. This hopping might have some consequences with the change of nonlinearity thereby a relation could be worked out for the 'hopping strength' in metamaterials. It is determined by finding the phonon energy gap in the eigenspectrum by analyzing the QB state [83]. While the role of nonlinearity is still a debatable issue, at around room temperature there are very few phonons, and hence investigation is carried out only on two-phonon bound state (TPBS). It can be realized experimentally by a Raman scattering measurements similar to that done by Santori et al. [84].

It is known that the 'coupling' can be changed in the SRR systems through engineering of their geometry. Thus, detailed numerical analysis was done to explore whether the system shows any sensitivity on coupling by quantum calculations on metamaterials [66]. Now, let us discuss about K-G model, wherein such applications of QBs in metamaterials are in the evolution stage. This study is quite realistic in that it helps us to understand the localization due to nonlinearity in the quantum regime, which is essential for many small-structured nano-devices. Quantum localization in K-G lattice has been studied by many researchers in terms of four atoms, notably by Proville [85], dimer case for targeted energy transfer (TET) by Aubry et al. [86], delocalization and spreading behavior of wave-packets by Flach et al. [39, 83]. QBs are also characterized by various methods, as presented by Schulman in details [87].

The study of QB assumes special significance due to its possible application in the field of quantum computation. To name a few, Quach et al. [88, 89] studied reconfigurable metamaterials and Rakhmanov et al. [90] studied quantum Archimedean screw for superconducting electronics using metamaterials, and many others studied them for medicine to aeorospace applications. There are so many other applications, as given in Ref [65, 66] viz. (a) ladder array of Josephson junction for superconductors, (b) BEC in optical lattices, (c) optical waveguides, (d) micro-mechanical arrays, e) macromolecules like DNA, (f) SRR based metamaterials in antenna arrays,

(g) two-magnon bound states in antiferromagnets, and (h) two-phonon bound states (TPBS), i.e., QBs in metamaterials. As pointed out by Zheludev [22], the quantum-effect enabled systems via metamaterials route will bring a range of exciting applications in the future.

Further on quantum metamaterials, it needs to be elaborated that 'conventional' metamaterials consists of one of the most important frontiers of 'optical design', with applications in diverse fields. So far, these materials have been mainly 'classical' structures and interact with the classical properties of light. It is important to elaborate on the work of Quach et al. [89] who described a class of dynamic metamaterials, based on the quantum properties of coupled atom-cavity arrays, which are intrinsically lossless, reconfigurable, and operate fundamentally at the quantum level, i.e., by coupling controllable quantum systems into large structures, the concept of quantum metamaterials was introduced. This was used to create a reconfigurable quantum 'superlens' possessing a negative index gradient for single photon imaging by using the basic features of quantum superposition and entanglement of metamaterials properties, and this definitely opens up a new avenue for devices based on quantum science.

QBs could be studied by various means, such as 'temporal evolution of quanta', as different quanta first meet at a single point on the time-axis giving us a 'critical' time that falls in the THz range. It has to be noted that in K-G lattice, the levels of 'anharmonic potential energy' are non-equidistant that could have important implications in the applications. In various novel methods of preparation of metamaterials, the resonant frequency in the THz range has also been proposed to be an interesting field of study [91–94]. It is to be noted that although there is an extensive work going on in the area of metamaterials, this has been mainly in the domain of classical breathers [29, 63], including those on THz applications [93, 94]. Very few papers have been published on its quantum perspective, and hence motivating a study on metamaterials from its quantum point of view [64–66, 91].

Moreover, for many interesting device applications, it is also emphasized here that the 'phonon perspective' of viewing metamaterials is quite important, if we could control its temporal evolution. In such a case, we could make the fastest switch using metamaterials. For the benefit of the experimentalists, one has to theoretically study temporal evolution of the quanta and vary the relevant parameters eventually to optimize the results, and only then one could make the device applications, e.g., in making antenna arrays, modulation instability based gadgets, in making quantum metamaterials based superconductors and so on. Such engineered materials can also be used in the quantum computation, as also emphasized in Refs. [89] and [90]. The temporal evolution of the quanta at the correct value of coupling and dielectric permittivity could enable the required quantum coding for applications [91, 92].

Further, the proposed quantum metamaterials should allow the additional ways of controlling the 'propagation' of electromagnetic waves that are not possible by normal 'classical structures'. For the experimentalists, this should pave the way to design better metamaterials for the required applications, e.g., for having a better understanding on quantum coding in future for quantum computers. Indeed, as emphasized by Rakhmanov et al. [90], the 'coherent quantum dynamics' of 'qubits' determines the THZ optical properties in the system. Therefore, from the point of view of various theoretical and experimental investigations on THz and other applications in metamaterials, such a study assumes particular importance, as this is also a new area of research [65, 66, 91–94].

10.1.4 OTHER APPLICATIONS OF DBS AND SOLITONS

Both classical and quantum DBs have already been studied in details by Bandyopadhyay and coworkers [61–66]. The effect of coupling on the temporal evolution of the number of quanta for QBs in metamaterials has been studied in a K-G lattice under a non-periodic boundary condition [65, 66]. Let us also look at some other applications: The Fano resonance due to DBs has also been described on two channel ansatz in K-G lattice by a number of workers [95–98], albeit with different type of potential. Various parameters, such as dielectric permittivity, coupling, focusing-defocusing nonlinearity, are included in the Hamiltonian that is used to describe the nonlinear modes in SRR based metamaterials.

Thus, the Fano resonance due to DBs has also been explored in terms of these material parameters by Bandyopadhyay et al. [99]. This study was important for various applications, e.g., in the (a) biosensing technology, (b) spectral selectivity, (c) beam filtering, etc. Here, two different issues were tackled: (i) Fano resonance in metamaterials based on K-G lattice arises due to DBs, and (ii) defining the "breathers" that act like an "impurity" mode in a translationally invariant system leading to Fano resonance. Although this is a classical case, this translational invariance is not a necessary condition for the stability of DBs, albeit this is shown by Schulman in a quantum system [87].

Another important application, i.e., the rectification of current density (ratchet solitons), was also studied by Bandyopadhyay et al. [100] and Flach et al. [101] by the analysis of symmetry and its violation in such engineered materials. These studies were relevant for the applications in the (a) molecular motors and other non-equilibrium biological systems, (b) electric currents in superlattices, (c) voltages in coupled systems, i.e., in annular Josephson junctions revealing unidirectional fluxon motion that is driven by biharmonic microwaves, (d) magnetic flux cleaning in superconducting films, (e) Abrikosov vortex diodes, and (f) voltage rectifiers. An interesting review by Bliokh et al. [102] on unusual resonators also needs to be mentioned here. So far the description has been made in the realm of various possibilities of device applications in nonlinear optical systems, particularly metamaterials. At the core of it lies the solitons that are fundamentally important to our understanding of such devices. This gives us motivation to work on solitons on metamaterials, particularly on multisolitons in terms of various material parameters.

It is known that solitons can retain their shape to a longer extent without any distortion and hence treated as particles that are described by the amplitude of the MI waves forming the core part of our K-G model. If solitons are treated as particles, the main issue involves an understanding of the transfer of such solitons and their control. Now, as mentioned above, in case of metamaterials we are dealing with solitons which are basically MI waves in the coupled SRR loops. By adjusting the phase shift of the external ac driver, we could very precisely control the 'propagation direction' of these MI waves in such materials [100] that could throw light on the mechanism of 'quantum level switching'. This is based on flux exclusion superconducting quantum metamaterials, which have been recently investigated, and the combination of low losses and strong nonlinearities over the broad spectral range makes 'superconducting metamaterials' an ideal topic of research in the sub-optical range [103, 104].

Moreover, metamaterials can also be used for a diode type of application using ratchet behavior, as Carapella and Costabile [105] and Goldobin et al. [106] reported the ratchet effect for a fluxon trapped in an annular junction embedded in an inhomogeneous magnetic field and introduced the concept of fluxon based diode. These authors tried to control this kind of non-zero current (i.e., rectification) for such cases by only using spatial modulation, but the method adopted in Ref. [100] involves a study in a K-G lattice that could also allow for temporal control.

It is worth mentioning that Marques and co-workers [107–109] did an extensive study of different combinations of metamaterials; particularly on a 'new particle concept' with point electric dipoles for the design of metasurfaces with high frequency selectivity and planar configurations for one- and two-dimensional wave transmission in microwave circuits and filters [109]. The effect of cross-polarization was also considered in case of excitation of the multiple SRR or complimentary-SRR metasurfaces by a normally incident plane-wave [107, 109]. Moreover, some unexplained properties of the EM-wave propagation were explained by bianisotropy that is related to the magneto-electric coupling of the artificial constituents of the medium [109]. Although 1D system is discussed in this work, the bianisotropy may not appear to have an effect on the EM-wave propagation. However, the bianisotropy can also have an effect for the 1D structure, since polarization of impinging E and H field could modify the frequency response of the SRR elements; experimental results have been shown in microwave planar circuits in the implementation of devices such as filters and delay lines [107, 109]. In this context, this is not discussed here.

It is to be noted that unlike natural materials, these engineered materials show relatively large magnetic response at femtoseconds (fs) that manifests in the 'temporal evolution vs. number of quanta' curve, and hence their THz application assumes special significance [91, 92]. Singh et al. [93, 94] did an extensive work on THz application of metamaterials to highlight its importance. The flourishing fields of metamaterials clearly embody exotic applications for a new generation of novel photonic devices. From the early emergence of such materials, they have now gained a significant amount of global importance with considerable effort being made to extend their operation into the THz regimes and even optical window. However, as they stand, theoretical investigations show that losses, which are closely linked to the resonant behavior, will cause potential problems for all possible frequencies [110, 111]. In particular, such losses will kill any opportunity for a useful metamaterial operating around and above 30 THz. This is a challenging task that is being tackled by intensive research activity into this field. Notably, the behavior of the lifetime of quantum breathers against various parameters (dielectric constant, coupling, damping, etc.) that are included in our K-G model showed an interesting trend in the THz domain [65, 66, 91]. This gives us further motivation to work on K-G lattice for the multisolitons in metamaterials that is the main focus in our present article in terms of soliton profile [q(x′, t′)] at different values of non-dimensional time.

Here, the treatment focuses on a simple model system of a one-dimensional array of loosely coupled SRRs. The formulation is accomplished using a Taylor-expansion in the x-dimension. The magnetization of the metamaterial [1, 28, 29] is considered here to study the field behavior in SRRs. The dynamical equation thus obtained gives a reliable solution for device application as supported by the numerical solutions. The amplitude modulation of the soliton is not involved in the present analysis. Perturbation analysis is carried out for non-dimensional resistance (γ) and excited e.m.f (ε) in both non-dimensional space and time dimension. For different cases of the system, soliton solutions for the K-G equation for both bright and dark types are derived. Despite voluminous work done on SRRs based metamaterials, no derivation of classical dynamics from Hamiltonian of the system has been attempted to the best of our knowledge. Hence, an attempt is made in this direction in the present article.

The article is organized as follows: Section 10.2 describes the derivation of nonlinear K-G equation in an array of SRR loops in metamaterials through variational principle, with certain details on the theoretical basis of numerical experiment. In Section 10.3, results and discussion are presented with 2D figures for nonlinear modes, i.e., MI waves, against various material parameters, such as coupling, damping, focussing and defocussing nonlinearity, eventually to show multisolitonic behavior. Section 10.4 gives the conclusions.

10.2 THEORETICAL DEVELOPMENT

First of all, it should be noted that although Lazarides et al. [28, 29] worked on metamaterials through NLSE, in our recent work [1] we have observed that K-G type equation is a richer variety for dynamical study in terms of showing an additional breather pulse, while NLSE does not show such pulses. Moreover, on envelope NLSE, Boardman et al. [111] discussed some interesting aspects of diffraction management and modulation instability in metamaterials and found inaccuracy in the earlier investigations [28]. Without going into this debate, it can be said that we usually deal with the Hamiltonian given by Lazarides et al., but we derived K-G equation for metamaterials through variational principle and that is how the present case is different. However, as per Ref. [61, 62], the Hamiltonian developed by us for ferroelectrics can also take care of certain issues involved in metamaterials, if we suitably add a magnetic component in it. Our recent paper on the temporal evolution of quanta for quantum breathers in metamaterials in a K-G lattice under a non-periodic boundary condition also supports the same view [65, 66].

We consider a model of weakly coupled SRRs similar to that treated in Ref. [28] for the discrete limit. A typical SRR unit essentially consists of a wire ring with a cut made to form a slit. The ring operates as an inductor-resistor and the slit as a capacitor so that the SRR is essentially a RLC resonant circuit. Adding Kerr nonlinear dielectric filling to the SRRs' capacitor slits makes them a nonlinear response to applied electromagnetic (EM) field that arises due to the field dependence of the Kerr permittivity (ε). The SRR interacts as an RLC circuit with an externally applied EM field, giving a characteristic frequency response as determined by its inductive, capacitive and resistive components. In the limit, considered here, of weak nonlinearity and very small resistance, the resonant frequency of the free standing SRR is approximately that of its linear limit, i.e., $\omega_l = \frac{1}{\sqrt{LC_l}}$. In the presence of a time-dependent magnetic field of the form: $H = H_0\cos(\omega t)$ applied perpendicular to the plane of the SRR loop, and an induced electromotive force (emf) proportional to the area (S) of the loop is generated within the loop. This emf is given by: $emf = \mu_0 \omega S H_0 \sin(\omega t)$.

In addition to interactions with external fields, the SRR can have self-interactions and interactions with other SRRs. A time varying current within a single SRR leads to a self-inductive interaction, and two or more neighboring SRRs interact with one another through mutual inductive couplings. For the periodic arrays of SRRs that is considered here, it was show in Ref. [28] that only nearest neighbor SRR couplings are important.

The metamaterials under study are one-dimensional, discrete, periodic arrays of identical nonlinear SRRs. The loops of all of the SRRs lie in a common plane. The EM modes of interest have magnetic components perpendicular to the common plane of the SRRs with electric components transverse to the SRR slits. Only the magnetic component excites an emf in the SRRs, resulting in an oscillating current in each SRR loop and the development of an oscillating voltage difference across the slits. Neighboring SRRs of the linear array have their centers separated by a distance d such that ($x_n = nd$) for an integer gives the coordinates of the SRR along the x-axis. Each SRR has a self-inductance (L) and a weak mutual inductance (M) between its nearest neighbor loops. However, for more general (device-oriented) optimized models, a greater range of L and M values must be taken into account in practical applications [1, 28].

For a detailed description of the absolute dimension, ring/slot width, materials employed, it is useful to see Zharova et al. [112] and Zharov et al. [113], where two-dimensional transmission of EM waves was also considered. In these investigations, a finite slab of composite structure consisting of the cubic lattice of periodic arrays of metallic wires and SRRs was used, and the single-ring geometry of the lattice of cylindrical SRRs was chosen for simplicity. Also, it was assumed that the size of the unit-cell (d) of the structure was much smaller than the wavelength of the propagating EM field. In Ref. [1], the effects of further than first neighbor coupling in the SRR system were also discussed. Kivshar and co-workers [114] made a presentation on 'tuning metamaterials in near field interaction', but it was primarily focused for a pair-resonator based model having symmetric and antisymmetric resonances. However, for a 3-nearest neighbor based SRR system and assuming that at a particular frequency of the incident EM field, all of them resonate simultaneously. We are not dealing here with a particular resonance, as we are focused on the overall soliton behavior. Therefore, we consider the present assumption of nearest neighbor coupling for our system. Also, the bianisotropy has not been considered in our system, as done by Marques et al. [107, 109], as explained in Section 10.1. The resonators are shown in Figure 10.1.

Following Lazarides et al. [28, 29], the Hamiltonian of such a system involving the non-dimensional MI wave amplitude (q) and time (t) is given by:

$$\mathrm{H} = \sum_n [\frac{1}{2}\dot{q}_n^{\,2} - \lambda \dot{q}_n \dot{q}_{n+1} + V(q_n)] \tag{1}$$

where $\frac{dq_n}{dt'} = \dot{q}_n = i_n$ (i is the current in the SRR loop), λ = M/L (coupling parameter); the nonlinear on-site potential is:

$$V_n = \int_0^{q_n} f(q_n')dq_n' \tag{2}$$

And after truncation, this is expressed as:

$$f(q_n) \approx q_n - \left(\frac{\alpha}{3\varepsilon_l}\right) q_n^3 \tag{3}$$

Where, non-dimensional time $(t') = t\omega_l$. With a magnetic field H and magnetization σ of each SRR, the Hamiltonian is modified as:

$$\mathrm{H} = \sum_n [\frac{1}{2}\dot{q}_n^{\,2} - \lambda \dot{q}_n \dot{q}_{n+1} + V(q_n)] - \sigma H \tag{4}$$

Now, as B = μ_0(H+σ), we can write

$$\mathrm{H} = \sum_n [\frac{1}{2}\dot{q}_n^{\,2} - \lambda \dot{q}_n \dot{q}_{n+1} + V(q_n)] - \sigma(\frac{B}{\mu_0} - \sigma) \tag{5}$$

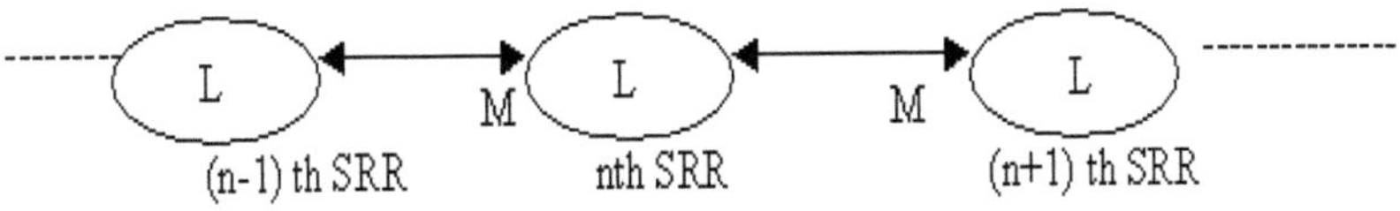

FIGURE 10.1 Three adjacent split-ring-resonators having self-inductance L and mutual inductance M.

Here, the magnetization (σ) of metamaterial, is given by [115]:

$$\sigma = \frac{n_m}{2c}\pi a^2 i \frac{B}{|B|} = \eta i \frac{B}{|B|} = \eta \dot{q} \frac{B}{|B|} \tag{6}$$

where $\eta = \frac{n_m}{2c}\pi a^2$ is a constant for a particular system, c is the velocity of light, a is the SRR radius, i is the current in SRR, n_m is the concentration of SRRs. Therefore, after putting the value of σ, the Hamiltonian takes the form:

$$\mathrm{H} = \sum_n [\frac{1}{2}(1-\eta^2)\dot{q}_n^{\,2} - \lambda \dot{q}_n \dot{q}_{n+1} - \frac{\eta \dot{q}_n}{\mu_0}|B| + V(q_n)] \tag{7}$$

To apply the variational principle, the Langrangian for the system with general external field interactions is given by:

$$L = \sum_n [\frac{1}{2}(1-\eta^2)\dot{q}_n^{\,2} - \lambda \dot{q}_n \dot{q}_{n+1} - \frac{\eta \dot{q}_n}{\mu_0}|B| - V(q_n)] \tag{8}$$

From the above Lagrangian, we get:

$$\frac{\partial L}{\partial \dot{q}_n} = (1-\eta^2)\dot{q}_n - \lambda \dot{q}_{n+1} - \lambda \dot{q}_{n-1} - \frac{\eta}{\mu_0}|B| \tag{9}$$

And,

$$\frac{\partial L}{\partial q_n} = f(q_n) = q_n - \frac{\alpha}{3\varepsilon_l} q_n^3 \tag{10}$$

From the variational principle, by using Euler-Lagrange equation of motion, we get:

$$\frac{d}{dt}\left(\frac{\partial L}{\partial \dot{q}_n}\right) - \frac{\partial L}{\partial q_n} = 0 \tag{11}$$

After a few mathematical steps, the governing equation in the x-direction is derived here to show the K-G dynamical equation as:

$$\frac{\partial^2 q}{\partial t'^2} - ab\frac{\partial^2 q}{\partial x^2} + b\left[q - \frac{\alpha}{3\varepsilon_l}q^3\right] - b\Lambda\Omega\sin(\Omega\tau) + \gamma\frac{\partial q}{\partial t'} = 0 \qquad (12)$$

$a = \lambda'/(1+2\lambda')$ and $b = 1/(1+2\lambda')$; λ' being an interaction constant or coupling within the SRR system in a K-G lattice. Here, is the amplitude of MI wave; is phase shift, Ω is a non-dimensional frequency factor, t' is the non-dimensional time, ε_l is the linear part of the dielectric permittivity that appears to be an important parameter to guide the soliton behavior, $\alpha = +1$ (-1) corresponds to a self-focusing (self-defocusing) in a nonlinear Kerr medium, and finally a damping term (γ) is included [1].

In order to perform stability analysis as done in nonlinear systems, we have differentiated the Klein-Gordon equation for different cases following Refs. [61, 62] for the dynamics of $q(x',t')$ with the appropriate transformation of variables. The dark-type soliton solution is given by:

$$q = \pm\tanh(\theta_1) \qquad (13)$$

where, θ_l is described as:

$$\theta_1 = \frac{x' - \sqrt{k}Vt'}{\sqrt{\frac{2k}{b}\sqrt{1-V^2}}} \qquad (14)$$

and V is a component of the soliton velocity. Following the same method, the bright-type soliton solution is given by:

$$q = \pm\sqrt{2}\sec h(\theta_2) \qquad (15)$$

Where θ_2 is described as:

$$\theta_2 = \frac{x' - \sqrt{k}Vt'}{\sqrt{\frac{k}{b}\sqrt{V^2-1}}} \qquad (16)$$

Our numerical analysis is done for the solitons with the above Eqs. (13) and (15) as a function of various parameters that were part of the above K-G equation (12).

10.3 RESULTS AND DISCUSSION

By using Euler-Lagrange dynamical equation of motion, a nonlinear K-G equation is derived based on the Hamiltonian for SRR based metamaterials. It is interesting to study the behavior of the system under various conditions at different values of the non-dimensional time, as per Eqs. (13) and (15). First, a Gaussian wave pattern was taken as an initial condition while solving the partial differential equation. The numerical simulation was done for two sets of values of damping parameter ($\gamma = 0.01$, and $\gamma = 0.05$); the first set corresponds to Figures 10.2–10.5 and the second one to Figures 10.6–10.9. In these two cases four sub-cases are shown for two values of coupling ($\lambda = 0.01$ and $\lambda = 0.02$) for two values of $\alpha = +1$ and -1 for focusing and defocusing nonlinearity respectively.

In this manner, the effect of damping could be evident from the above two sets of figures. For a given set, the effect of coupling could also be observed and finally for a given damping value and coupling, the effect of changing the focusing nonlinearity to a defocusing condition. Let us also take four different values of non-dimensional time for marking the respective curves as: $t' = 0$ which is termed as A, $t' = 10$ as B, $t' = 20$ as C and $t' = 30$ as D. The marking A to D is a matter of convenience for discussion.

For a case with zero field and no damping, i.e., Hamiltonian breathers, symmetric breathers were normally observed in 3D figures (not shown here). It is noted that in the simulation of 2D figures of polarization (P) with site index (n), i.e., distance, the peaks have been found to be symmetric Gaussian bands and even in 3D pictures, the same type of symmetric breather bands were observed. In the present case, with very low damping ($\gamma = 0.01$) and smaller coupling ($\lambda = 0.01$) with focusing nonlinearity ($\alpha = +1$) condition, different MI waves are shown in Figure 10.2, with bisolitons appearing at higher time.

However, in case of defocusing nonlinearity ($\alpha = -1$), for non-zero damping ($\gamma = 0.01$) with smaller coupling ($\lambda = 0.01$), the waves tend to

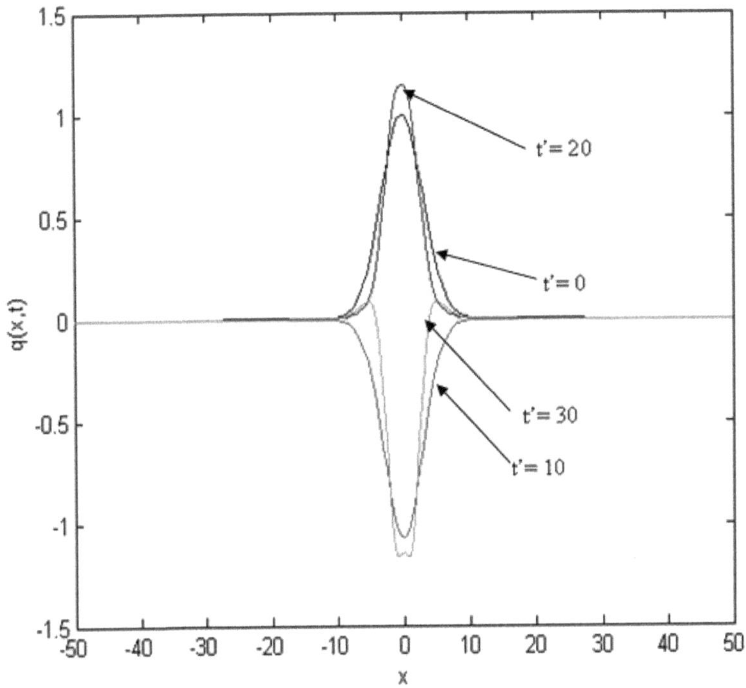

FIGURE 10.2 Schematic plot for MI solitonic waves for $\gamma = 0.01$, $\lambda = 0.01$, $\alpha = 1$.

show a bisolitonic behavior at higher time, as shown in Figure 10.3. For K-G lattice, at the highest time of our numerical study, it always shows bisolitonic behavior in defocusing nonlinearity. Theses bisolitons may arise due to cross-phase modulation mechanism and/or vector soliton formation mechanism, which is mainly attributed to corresponding eigenvalues of the equation, as proposed by a number of workers [35, 116].

Complex behavior is observed in all the figures under simulation that seems to be a promising condition for controlling the nature of the solitonic behavior in such materials. For example, B (i.e., nondimensional time of $t' = 10$) always follows the sequence of single and bisoliton in focusing and defocusing conditions respectively. One more important aspect observed for B is that it always shows dark solitonic behavior. On the other hand for D ($t' = 30$), bright and dark solitons are sequentially observed in focusing and defocusing conditions respectively up

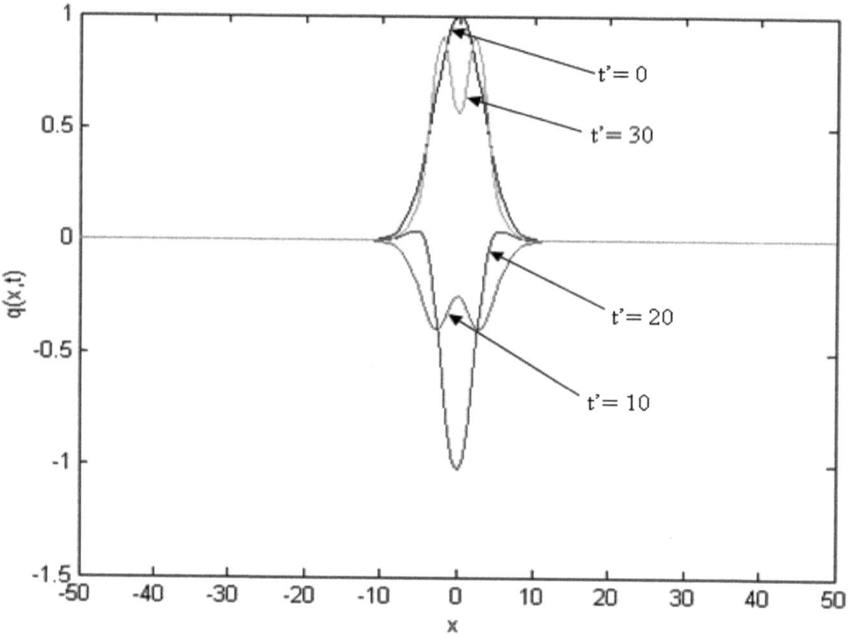

FIGURE 10.3 Schematic plot for MI solitonic waves for $\gamma = 0.01$, $\lambda = 0.01$, $\alpha = -1$.

to a certain extent and thereafter it shows dark solitons continuously. For C (t′ = 20), it just starts to show bisolitonic behavior after damping starts increasing in the system. The appearance of multisolitons by controlling various parameters has been shown by various authors in interesting investigations with numerical solutions [35, 36].

With still smaller damping ($\gamma = 0.01$), the coupling value increases from 0.01 to 0.02, these data are plotted in Figures 10.4 and 10.5 for focusing and defocusing conditions respectively. At higher nondimensional time of t′ = 30, there is a slight appearance of bisolitons in focusing nonlinearity, while in defocusing condition, there are bisolitons both t′ = 10 and t′ = 30; the former is inverted to the negative axis and the latter is inverted to positive axis. This shows the effect of focusing and defocusing conditions on the bisolitonic behaviors. In SRR based metamaterials, the above results are shown for 2-D features of solitons. Lazarides and coworkers [28, 29] have shown 3-D pictures of classical breathers, but that were based on NLSE formalism.

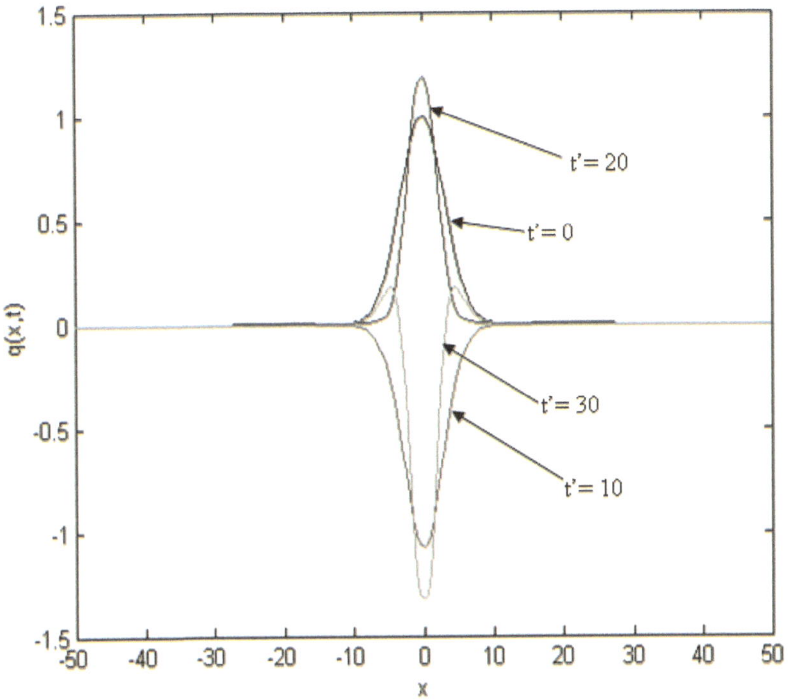

FIGURE 10.4 Schematic plot for MI solitonic waves for $\gamma = 0.01$, $\lambda = 0.02$, $\alpha = 1$.

In the second set of data, with higher damping ($\gamma = 0.05$), but with still smaller coupling ($\lambda = 0.01$), the computer simulation results for MI waves or modes are plotted in Figures 10.6 and 10.7 for focusing and defocusing conditions respectively. It is seen that at $t' = 20$, there is distinct appearance of bisolitons in focusing condition itself with sharper amplitude, while in defocusing condition, the bisolitons are not so clear thereby showing the effect of damping quite noticeable. In focusing condition (Figure 10.6), at $t' = 10$ and $t' = 30$, the curves are merged together in the negative axis and in defocusing condition (Figure 10.7), while the former show bisolitons with lower amplitude, the latter shows a very strong bisolitonic mode. So, the effect of damping is quite noticeable in this second set of data as well as the effect of focusing and defocusing nonlinearity on the bisolitonioc mode, as also observed above.

However, compared to the investigations of Lazarides et al. [28, 29] on 3D representations of solitonic data, here we have derived a nonlinear

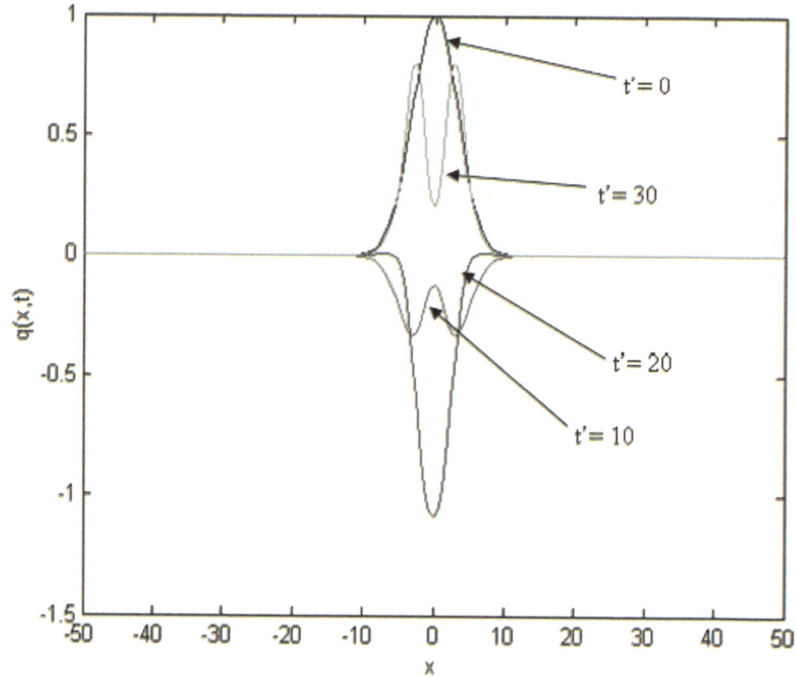

FIGURE 10.5 Schematic plot for MI solitonic waves for $\gamma = 0.01$, $\lambda = 0.02$, $\alpha = -1$.

K-G equation [1], where coupling between different SRR elements could be relative more important. By using spectral collocation method, it was interesting to see the overview of K-G breathers in 3-D pictures [66]. Here, the breathers were symmetric, while the effect of damping could be noticeable, where relatively high value of damping was applied. In metamaterials, the damping is generally not considered very high [1] so that breather oscillations could be considered stable. By comparing these features with those of other nonlinear optical materials, such as lithium niobate ferroelectrics, it is noted that the dissipation of the breathers with damping was relatively more pronounced in such systems [63] than that observed in metamaterials.

The energy losses expected in Eq. (12) depend on the system chosen for study and could be a consideration in such system design that is built for the observation of various modes. As per the considerations of losses in a number of different designs of SRR systems, the resistive losses,

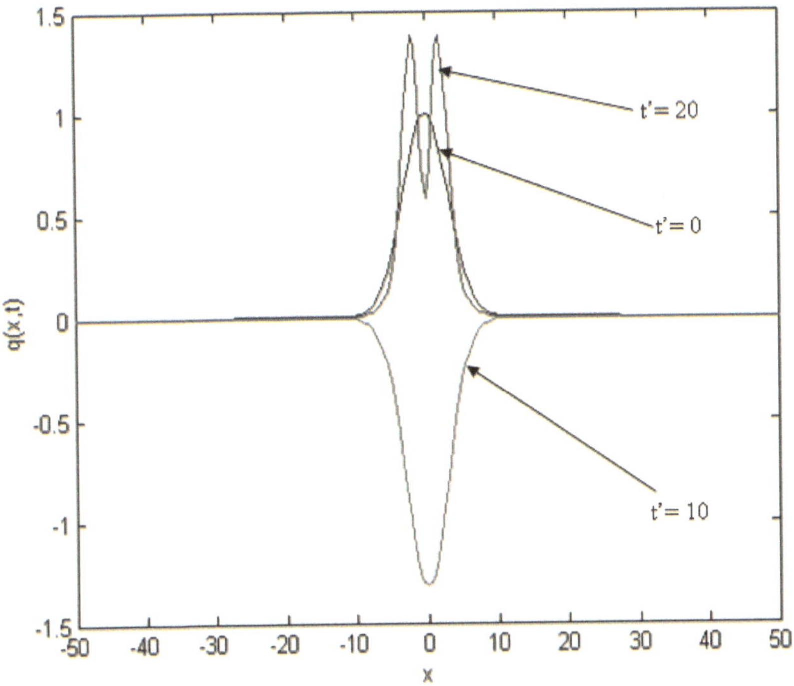

FIGURE 10.6 Schematic plot for MI solitonic waves for $\gamma = 0.05$, $\lambda = 0.01$, $\alpha = 1$.

in general, dominate radiative losses in metamaterials. This is due to the fact that these materials are designed to operate at wavelengths that are much greater than the radii of the SRR. The values of in the previous studies, including both the small radiative effects and the larger resistive effects, have ranged from 0.01 to 0.0016 and these are essentially small effects [29, 117]. It is noteworthy that despite the damping parameter is only taken into account as the resistance in the SRR system, the losses may come from various other material parameters, which have been extensively studied by various workers [118, 119] by impedance mismatch, loading effect, etc. However, these losses could be considered as damping in nature and hence these could be approximately taken to be embodied in the damping parameter (γ) in case of Klein-Gordon lattice. It has to be mentioned that Marques and co-workers [120] made an impressive study on reducing losses and dispersion effects on tunneling through waveguide system filled with metamaterials.

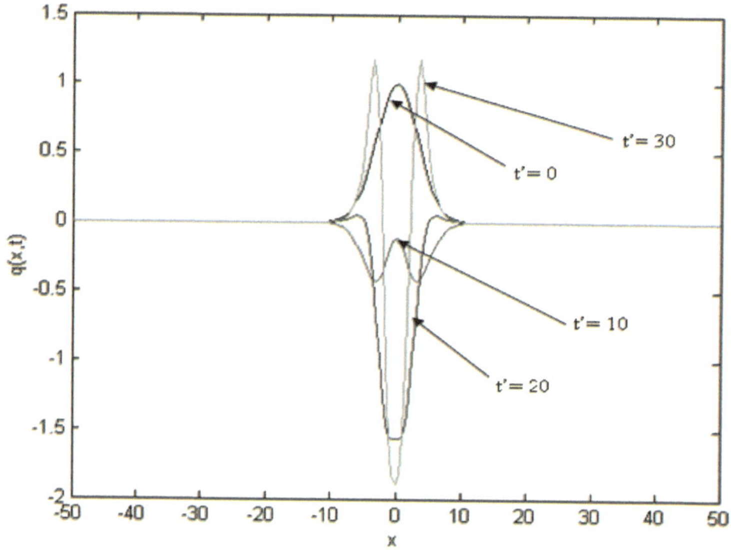

FIGURE 10.7 Schematic plot for MI solitonic waves for $\gamma = 0.05$, $\lambda = 0.01$, $\alpha = -1$.

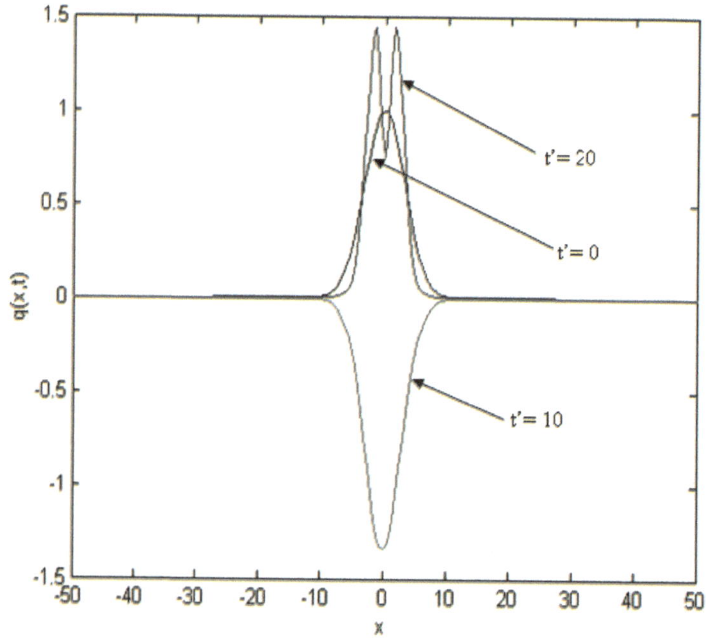

FIGURE 10.8 Schematic plot for MI solitonic waves for $\gamma = 0.05$, $\lambda = 0.02$, $\alpha = 1$.

With the damping value at this level, if we increase the coupling value ($\lambda = 0.02$), the data at focusing and defocusing conditions are plotted in Figures 10.8 and 10.9, respectively. It is seen that at t′ = 20, the strong bisolitonic behavior is still evident. The increase of coupling from 0.01 to 0.02 merely enhances the amplitude of the mode. For t′ = 10 and t′ = 30, the curves are again merged together in the focusing condition, while in the defocusing condition, both are showing strong bisolitonic modes with the latter having more stronger amplitude. All these features could be considered important for devices involving solitons. This seems to support our earlier quantum calculations [66, 91, 92] in that the tendency of hopping of phonons to the second excited state may be considered towards interpretation of some physical behavior of SRR based metamaterials for applications in antenna arrays.

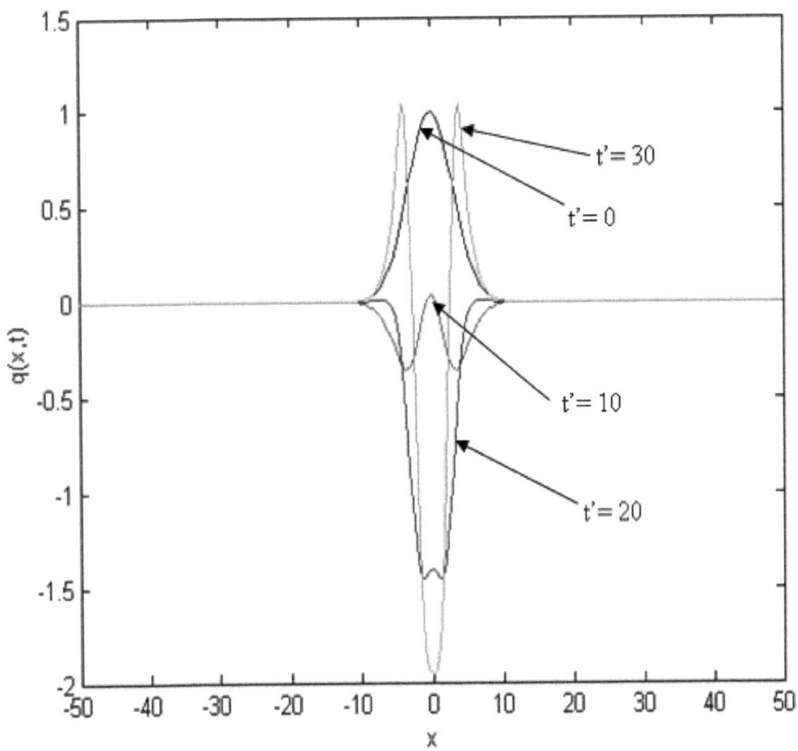

FIGURE 10.9 Schematic plot for MI solitonic waves for $\gamma = 0.05$, $\lambda = 0.02$, $\alpha = -1$.

This phonon-perspective is considered in terms of phonon hopping coefficient as well as the band gap between the unbound state and continuum phonons in the localized state [64–66]. To the experimentalists, this might give a sort of direction for the extent of change of design in terms of varying the spacing between the slits in the SRR rings and others that could influence the coupling in the system [66].

10.4 CONCLUSIONS

A special type of nonlinear partial differential equations (PDE) gives rise to a modal behavior that can be described by a nonlinear Klein-Gordon (NLKG) equation. This type of equation can be derived through variational principle based on a generalized Hamiltonian. Solitons are the solutions of nonlinear PDE and gives rise to robust solutions with applications in diverse field, such as nonlinear photonics and nano devices, when considered in a localized system with discreteness, in a variety of material systems (ferroelectrics, metamaterials, etc.). In the realm of their applications in nonlinear optical fiber communications as well as in various nano-structured devices, NLKG equation seems to be a better option. Here, we derive NLKG for metamaterials in terms of the modal behavior of SRR for the applications in antenna. Various parameters involved in the Hamiltonian, such as damping, coupling between two adjacent sites of split-ring resonators and focusing-defocussing nonlinearity play an important role in the overall solitonic behavior in the metamaterials system. After quite an exhaustive and informative introduction of the subject of metamaterials and different types of solitons, the mathematical deductions are made based on the variational principle that gives rise to solitons. The variation of the above parameters indicates interesting behavior about the amplitude of magneto-inductive waves that also shows multi-soliton behavior in the metamaterials system.

ACKNOWLEDGMENTS

The authors would like to thank Prof L. S. Schulman of Physics Department of Clarkson University (USA) and Prof. D. Frantzeskakis of Department of

Physics of University of Athens (Greece) for helpful comments on solitons and Prof. A. R. McGurn of Physics Department of Western Michigan University (USA) for valuable guidance.

KEYWORDS

- **Hamiltonian**
- **K-G equation**
- **metamaterials**
- **multi-solitons**
- **split-ring-resonators**
- **variational principle**

REFERENCES

1. Giri, P., Choudhary, K., Sengupta, A., Bandyopadhyay, A. K., & McGurn, A. R. (2011). *Phys. Rev. B. 84,* 155429–155437.
2. Bandyopadhyay, A. K., Ray, P. C., Vu-Quoc, L., & McGurn, A. R. (2010). *Phys. Rev. B 81,* 64104–64111.
3. Giri, P., Choudhary, K., Sengupta, A., Dey, A., Biswas, A., Ghosal, A., & Bandyopadhyay, A. K. (2012). *Phys. Rev. B. 86,* 184101–184108.
4. Chen, D., Aubry, S., & Tsironis, G. P. (1996). *Phys. Rev. Lett. 77,* 4776–4779.
5. Veselago, V. G. (1968). *Sov. Phys. Uspekhi. 10,* 509–514.
6. Litchinitser, N. M., & Shalaev, V. M. (2009). *J. Opt. Soc. Am. B. 26,* 161–169.
7. Alu, A., & Enghetta, N. (2007). *Opt. Exp. 15,* 3318–3332.
8. Boardman, A., King, A. N., & Velasco, L. (2005). *Electromagnetics. 25,* 365–389.
9. Smith, D. R., Padilla, W. J., Vier, D. C., Nemat-Nasser, S. C., & Schultz, S. (2000). *Phys. Rev. Lett. 84,* 4184–4187.
10. Pendry, J. B., Holden, A. J., Roberts, D. J., & Stewart, W. J. (1998). *J. Phys.: Condens. Matter. 10,* 4785–4809.
11. Pendry, J. B., & Smith, D. R. (2004). *Phys. Today. 57,* 37.
12. Pendry, J. B., Holden, A. J., Roberts, D. J., & Stewart, W. J. (1999). *IEEE Trans. Micr. Theory Tech. 47,* 2075.
13. Cheng, Q., Jiang, W. X., & Cui, T. J. (2010). *J. Phys. D: Applied Physics. 43,* 335406
14. Mei, Z. L., Bai, J., & Cui, T. J. (2010). *J. Phys. D: Applied Physics. 43,* 055404.
15. Syms, R. R. A., Shamonina, E., & Solymar, L. (2005). *Eur. Phys. J. B 46,* 301.
16. Tan, Y. S. & Seviour, R. (2009). *Europhys. Lett. 87,* 34005.

17. Li, M., Wen, Z., Fu, J., Fang, X., Dai, Y., Liu, R., Han, X., & Qiu, X. (2009). *J. Phys. D: Applied Physics 42,* 115420.
18. Di Falco, A., Ploschner, M., & Krauss, T. F. (2009). *New J. Phys. 12,* 113006.
19. Peng, L., & Mortensen, N. A. (2011). *New J. Phys. 13,* 053012.
20. Boardman, A. D. (2011). *J. Opt. 13,* 020401.
21. Plum, E. et al. (2011). *J. Opt. 13,* 055102.
22. Zheludev, N. I. (2011). *Opt. Photon. News*, March 31.
23. Litchinitser, N. M., Gavitov, I. R., & Maimistov, A. M. (2007). *Phys. Rev. Lett. 99,* 113902–113905.
24. Boardman, A. D., Egan, P., Velasco, L., & King, N. (2005). *J. Opt. Soc. Am. A 7,* S57–S67.
25. Gavitov, I. R., Indik, R., Litchinitser, N. M., Maimistov, A. I., Shalaev, V. M., & Soneson, J. E. (2006). *J. Opt. Soc. Am. B 23,* 535–542.
26. Popov, A. K., Slabko, V. V., & Shalaev, V. M. (2006). *Laser Phys. Lett. 3,* 293–297.
27. Popov, A. K., & Shalaev, V. M. (2006). *Opt. Lett. 31,* 2169–2171.
28. Lazarides, N., Eleftheriou, M., & Tsironis, G. P. (2006). *Phys. Rev. Lett. 97,* 157406–157409.
29. Eleftheriou, M., Lazarides, N., & Tsironis, G. P. (2008). *Phys. Rev. E 77,* 036608.
30. O'Brien, S., McPeake, D., Ramakrishna, S. A., & Pendry, J. B. (2004). *Phys. Rev. B 69,* 241101.
31. Lapine, M., Gorkunov, M., & Ringhofer, K. H. (2003). *Phys Rev. E 67,* 065601.
32. Shadrivov, I. V., Morrison, S. K., & Kivshar, Y. S. (2006). *Opt. Express 14,* 9344.
33. Shadrivov, I. V., Reznik, A. N., & Kivshar, Y. S. (2007). *Physica B 394,* 180.
34. Kourakis, I., Lazarides, N., & Tsironis, G. P. (2007). *Phys. Rev. E 75,* 067601.
35. Ablowitz, M. J., & Biondini, G. (1998). *Opt. Lett. 23,* 1668–1670.
36. Gabitov, I. R., Indik, R., Mollenauer, L., Shkarayev, M., Stepanov, M., & Lushnikov, P. M. (2007). *Opt. Lett. 32,* 605–607.
37. Dauxois, T., & Peyrard, M., "Physics of Solitons", (Cambridge Univ. Press, Cambridge, 2006) pp. 211.
38. Kivshar, Y. S., & Agarwal, G. P., "Optical Solitons: From fibers to photonic crystals" (Academic Press, London, 2002).
39. *Flach, S.,* & Gorbach, A. V. (2008). *Phys. Rep. 467,* 1–116.
40. Sato, M., Hubbard, B. E., & Sievers, A. J. (2006). *Rev. Mod. Phys. 78,* 137–157.
41. Anderson, P. W. (1958). *Phys. Rev. Lett. 109,* 1492.
42. Campbell, D. K., Flach, S., & Kivshar, Y. S. (2004). *Physics Today*, pp. 43.
43. Sievers, A. J., & Takeno, S. (1988). *Phys. Rev. Lett. 61,* 970–973.
44. Sievers, A. J., & Page, J. B. (1995). in "Dynamical Properties of Solids", Vol. *7,* Eds. Horton, G. K., & A. A. Maradudin, North-Holland, Amsterdam.
45. Fleischer, J. W., Segev, M., Elfremidis, N. K., & Christodoulides, D. N. (2003). *Nature 422,* 147–150.
46. Flach, S. (1995). *Phys Rev E 51,* 1503
47. Flach, S., & Willis, C. R. (1998). *Phys. Rep. 295,* 181.
48. Mackay, R. S., & Aubry, S. (1994). *Nonlinearity 7*, 1623.
49. Aubry, S. (1997). *Physica D 103,* 201–250.
50. Fleurov, V. (2003). *Chaos 13,* 676–682.
51. Bang, O., & Peyrard, M. (1995). *Physica D 81,* 9–22.

52. Bang, O., & Peyrard, M. (1996). *Phys. Rev. E 53,* 4143–4152.
53. Corney, J. F., & Bang, O. (2001). *Phys. Rev. E 64,* 047601–047603.
54. Kobyakov, A., Lederer, F., Bang, O., & Kivshar, Y. S. (1998). *Opt. Lett. 23,* 506–508.
55. Corney, J. F., & Bang, O. (2001). *Phys. Rev. Lett. 87,* 133901–133904.
56. Di Trapani, P., Bramati, A., Minardi, S., Chinaglia, W., Conti, C., Trillo, S., Kilius, J., & Valiulis, G. (2001). *Phys. Rev. Lett. 87,* 183902–183905.
57. Gonzalez, J. A., Guerrero, I., & Bellorin, A. (1996). *Phys. Rev. E 54,* 1265–1273.
58. Holyst, J. A. (1998). *Phys. Rev. E 57,* 4786–4788.
59. Benedek, G., Bussmann-Holder, A., & Bilz, H. B. (1987). *Phys. Rev. B 36,* 630–638.
60. Comte, J. C. (2002). *Phys. Rev. E 65,* 46619.
61. Bandyopadhyay, A. K., Ray, P. C., & Gopalan, V. (2006). *J. Phys.: Condens Matter 18,* 4093–4099.
62. Bandyopadhyay, A. K., Ray, P. C., & Gopalan, V. (2008). *Euro Phys. J. B 18,* 525–531.
63. Giri, P., Choudhary, K., Sengupta, A., Bandyopadhyay, A. K., & Ray, P. C. (2011). *J. Appl. Phys. 109,* 54105–54112.
64. Biswas, A., Choudhary, K., Bandyopadhyay, A. K., Bhattacharjee, A. K., & Mandal, D. (2011). *J. Appl. Phys. 110,* 24104–24111.
65. Mandal, S. J., Choudhary, K., Biswas, A., Bandyopadhyay, A. K., Bhattacharjee, A. K., & Mandal, D. (2011). *J. Appl. Phys. 110,* 124106–124116.
66. Mandal, B., Adhikari, S., Basu, R., Choudhary, K., Mandal, S. J., Biswas, A., Bandyopadhyay, A. K., Bhattacharjee, A. K., & Mandal, D. (2012). *Phys. Scr. 86,* 15601–15610.
67. Roy, R., Giri, P., Das, B., Choudhary, K., Ghosal, A., & Bandyopadhyay, A. K. (2014). *AIP-Advances 4,* 87101–87106.
68. McPherson, M. S., Ostrovskii, I., & Breazeale, M. A. (2002). *Phys. Rev. Lett. 89,* 115506–115509.
69. Lee, D., Behera, R. K., Wu, P., Xu, H., Li, Y. L., Sinnott, S. B., Phillpot, S. R., Chen, L. Q., & Gopalan, V. (2009). *Phys. Rev. B 80,* 060102–060105.
70. Xu, H., Lee, D., Sinnott, S. B., Gopalan, V., Dierolf, V., & Phillpot, S. R. (2009). *Phys. Rev. B 80,* 144104–144114.
71. Swanson, B. I., Brozik, J. A., Love, S. P., Strouse, G. F., Shreve, A. P., Bishop, A. R., Wang, W. Z., & Salkola, M. I. (1999). *Phys. Rev. Lett. 82,* 3288–3291.
72. Schwarz, U. T., English, L. Q., & Sievers, A. J. (1999). *Phys. Rev. Lett. 83,* 223–226.
73. Trıas, E., Mazo, J. J., & Orlando, T. P. (2000). *Phys. Rev. Lett. 84,* 741–744.
74. Sato, M., Hubbard, B. E., Sievers, A. J., Ilic, B., Czaplewski, D. A., & Graighead, H. G. (2003). *Phys. Rev. Lett. 90,* 044102–044105.
75. Dick, A. J., Balachandran, B., & Mote, C. D. (2007). *Nonlinear Dynamics 54,* 13–29.
76. Eisenberg, H. S., Silberberg, Y., Morandotti, R., Boyd, A. R., & Aitchison, J. S. (1998). *Phys. Rev. Lett. 81,* 3383–3386.
77. Edler, J., Pfister, R., Pouthier, V., Falvo, C., & Hamm, P. (2004). *Phys. Rev. Lett. 93,* 106405–106408.
78. Tsironis, G. P. (2003). Chaos *13,* 657–666.
79. Kopidakis, G., Aubry, S., & Tsironis, G. P. (2001). *Phys. Rev. Lett. 87,* 165501–165504.
80. Flach, S., & Kladko, K. (1999). *Physica D 127,* 61–72.

81. Scott, A. C., Eilbeck, J. C., & Gilhøj, H. (1994). *Physica D 78,* 194–213.
82. Pinto, R. A., Haque, M., & Flach, S. (2009). *Phys. Rev. A 79,* 052118–052125.
83. Nguenang, J. P., Pinto, R. A., & Flach, S. (2007). *Phys. Rev. B 75,* 214303–214308.
84. Santori, C., & Beausoleil, R. G. (2012). *Nature Photonics 6,* 10.
85. Proville, L. (2005). *Phys. Rev. B 71,* 104306–104311.
86. Maniadis, P., Kopidakis, G., & Aubry, S. (2004). *Physica D 188,* 153–177.
87. Schulman, L. S., Tolkunov, D., & Milhokova, E., *Chem. Phys. 322,* 55–74. (2006).
88. Quach, J., Makin, M. I., Su, C. H., Greentree, A. D., & Hollenberg, L. C. L. (2009). *Phys. Rev. A 80,* 063838.
89. Quach, J., Su, C. H., Martin, A. M., Greentree, A. D., & Hollenberg, L. C. L. (2011). *Opt. Exp. 19,* 11018.
90. Rakhmanov, A. L., Zagoskin, A. M., Savelev, S., & Nori, F. (2008). *Phys. Rev. B 77,* 144507–144513.
91. Mandal, B., Adhikari, S., Choudhary, K., Dey, A., Biswas, A., Bandyopadhyay, A. K., Bhattacharjee, A. K., & Mandal, D. (2012). *Quant. Phys. Lett. 1,* 59–68.
92. Bandyopadhyay, A. K., Das, S., & Biswas, A., "Computational and Experimental Chemistry: Development and Applications" (Apple Academic Press, Toronto and New Jersey, 2014) pp. 245–278.
93. Singh, R., Plum, E., Zhang, W., & Zheludev, N. I. (2010). *Opt. Exp. 18,* 13425.
94. Singh, R., Al-Naib, I. A. I., Koch, M., & Zhang, W. (2011). *Opt. Exp. 19,* 6312–6319.
95. Flach, S., Miroshchinko, A. E., Fleurov, V., & Fistul, M. V. (2003). *Phys. Rev. Lett. 90,* 84101–84104.
96. Miroshnichenko, A. E., Flach, S., & Kivshar, Y. S. (2010). *Rev. Mod. Phys. 82,* 2257–2298.
97. Vicencio, R. A., Brand, J., & Flach, S. (2007). *Phys. Rev. Lett. 98,* 184102–184105.
98. Kim, S. W., & Kim, S. (2001). *Phys. Rev. B 63,* 212301–212304.
99. Choudhary, K., Adhikari, S., Biswas, A., Ghosal, A., & Bandyopadhyay, A. K. (2012). *J. Opt. Soc. Am. B 29,* 2414–2419.
100. Biswas, P., Das, S., Sarkar, P., Choudhary, K., Baidya, P., Giri, P., Ghosal, A., & Bandyopadhyay, A. K. (2013). *J. Phys. D: Appl. Phys. 46,* 205102–205108.
101. Flach, S., Zolotaryuk, Y., Miroshnichenko, A. E., & Fistul, M. V. (2002). *Phys. Rev. Lett. 88,* 184101–184104.
102. Bliokh, K. Y., Bliokh, Y. P., Freilikher, V., Savelev, S., & Nori, F. (2008). *Rev. Mod. Phys. 80,* 1201–1213.
103. Savinov, V., Tsiatmas, A., Buckingham, A. R., Fedotov, V. A., de Groot, P. A. J., & Zheludev, N. I. (2012). *Scientific Reports 2*(450).
104. Lazarides, N., & Tsironis, G. P. (2007). *Appl. Phys. Lett. 90,* 163501–163503.
105. Carapella, G., & Costabile, G. (2001). *Phys. Rev. Lett. 87,* 077002–077005.
106. Goldobin, E., Sterck, A., & Koelle, D. (2001). *Phys. Rev. E 63,* 031111.
107. Marques, R., Medina, F., & Rafi-El-Idrissi, R. (2002). *Phys. Rev. B 65,* 144440.
108. Martin, F., Bonache, J., Falcone, F., Sorolla, M., & Marques, R. (2003). *Appl. Phys. Lett. 83,* 4652–4654.
109. Falcone, F., Lopetegi, T., Laso, M. A. G., Baena, J. P., Bonache, J., Beruete, M., Marques, R., Martin, F., & Sorolla, M. (2004). *Phys. Rev. Lett. 93,* 197401–197404.
110. Litchinitser, N. M., & Shalaev, V. M. (2009). *Nature Photon, 3,* 75–76.

111. Boardman, A. D., Egan, P., Mitchell-Thomas, R. C., Rapoport, Y. G., & King, N. J. (2008). *Proc. SPIE.* 7029, F1–F14.
112. Zharov, A. A., Shadrivov, I. V., & Kivshar, Y. S. (2003). *Phys. Rev. Lett. 91,* 37401–37404.
113. Zharova, N. A., Shadrivov, I. V., Zharov, A. A., & Kivshar, Y. S. (2005). *Opt. Exp. 13,* 1291.
114. Powell, D. A., Lapine, M., Gorkunov, M. V., Shadrivov, V., & Kivshar, Y. S. (2010). *Phys. Rev. B 82,* 155128.
115. Shadrivov, I. V., Morrison, S. K., & Kivshar, Y. S. (2006). *Opt. Exp. 14,* 9344.
116. Stratmann, M., Pagel, T., & Mitschke, F. (2005). *Phys. Rev. Lett. 95,* 143902.
117. Molina, M. I., Lazarides, N., & Tsironis, G. P. (2009). *Phys Rev E 80,* 46605.
118. Dolling, G., Enkrich, C., Wegener, M., Soukoulis, C. M., & Linden, S. (2006). *Opt. Lett. 31,* 1800.
119. Bossard, J. A., Yun, S., Werner, D. H., & Mayer, T. S. (2009). *Opt. Exp. 17,* 14771.
120. Baena, J. D., Jelinek, L., & Marques, R. (2005). *New J. Phys. 7,* 166.

CHAPTER 11

AB-INITIO TECHNIQUES FOR LIGHT MATTER INTERACTION AT THE NANOSCALE

JUAN SEBASTIAN TOTERO GONGORA and
ANDREA FRATALOCCHI

Primalight, Faculty of Electrical Engineering; Applied, Mathematics and Computational Science, King Abdullah University of Science and Technology (KAUST), Thuwal 23955-6900, Saudi Arabia

CONTENTS

11.1 INTRODUCTION

In recent years there has been a growing interest in the development of new computational methods to simulate and study realistic electromagnetic systems. Among different approaches, the Finite Difference-Time Domain (FDTD) algorithm is based on the solution of Maxwell's equations with no approximation and provides a reliable and flexible *ab-initio* framework to study realistic systems with arbitrary precision. However, one of the most challenging problems remains the numerical representation of dispersive active media, such as dyes, colloidal quantum dots. These materials play a crucial role in many different applications, ranging from plasmonics to nonlinear optics, laser physics, disordered photonics, photovoltaics and biological imaging. In this chapter we will describe a rigorous and efficient approach to model dispersive active materials in the FDTD framework. As is well known, a material is said to be dispersive when its optical properties, described in terms of a relative dielectric permittivity $\varepsilon_r(\omega)$, depend on the incident wavelength. While the representation of such materials is immediate in frequency-domain simulations, the inclusion in the FDTD algorithm requires a significant effort. In this work, we will introduce an optimized unconditionally stable second order algorithm, which is able to deal with any type of dispersion. In addition, we will model a fully three-dimensional quantum-mechanical active material with the aid of the Maxwell-Bloch (MB) formalism and we will show that the solution of MB equations can be simplified using concepts from group and graph theories.

This chapter is organized as follows: in Section 11.1 we introduce the total electric polarization, which describes the interaction between light and matter, and we discuss the main features of the FDTD algorithm. In Section 11.2, we describe the numerical approach to the description of dispersive materials in the FDTD domain. In Section 11.3, we introduce the MB approach to active media in the density-matrix formalism and we show how to use concepts from group theory to reduce the number of equations to be solved. Finally, we provide some examples of applications of the computational methods described above.

11.2 MAXWELL'S EQUATIONS AND TOTAL POLARIZATION

The equations describing the evolution of an electromagnetic field are Maxwell's equations

$$\begin{cases} \nabla \cdot \boldsymbol{E}(\boldsymbol{r},t) = 0, \\ \nabla \cdot \boldsymbol{B}(\boldsymbol{r},t) = 0, \\ \nabla \times \boldsymbol{E}(\boldsymbol{r},t) = -\partial_t \boldsymbol{B}(\boldsymbol{r},t), \\ \nabla \times \boldsymbol{B}(\boldsymbol{r},t) = \varepsilon_0 \mu_0 \partial_t \boldsymbol{E}(\boldsymbol{r},t), \end{cases} \tag{1}$$

where $\boldsymbol{E}(\boldsymbol{r}, t)$ and $\boldsymbol{B}(\boldsymbol{r}, t)$ are the electric and magnetic fields, respectively, and (ε_0, μ_0) the vacuum permittivity and permeability. Equation (1) describes the evolution of the electromagnetic field in free space and in the absence of any external charge or current. If one considers the propagation in a real material, conversely, Eq. (1) needs to be modified in order to include the medium response. This is obtained through the constitutive relations of the material and from the definition of the auxiliary displacement **D** and magnetization **H** fields, which for a non-magnetic medium read [23].

$$\begin{cases} \boldsymbol{D}(\boldsymbol{r},t) = \varepsilon_0 E(\boldsymbol{r},t) + \boldsymbol{P}_T(\boldsymbol{r},t), \\ \boldsymbol{H}(\boldsymbol{r},t) = \dfrac{1}{\mu_0} B(\boldsymbol{r},t), \end{cases} \tag{2}$$

where $\boldsymbol{P}_T(\boldsymbol{r}, t)$ is the total electric polarization of the material. In Eq. (2) the interaction between the electromagnetic field and the material is completely determined by the total polarization $\boldsymbol{P}_T$, whose expression depends on the nature of the interaction and on the model chosen to describe it. Without any loss of generality, we can divide the total polarization into its linear and non-linear contributions

$$\boldsymbol{P}_T(\boldsymbol{r},t) = \boldsymbol{P}_L(\boldsymbol{r},t) + \boldsymbol{P}_{NL}(\boldsymbol{r},t) \tag{3}$$

where the linear term expression is usually written as

$$\boldsymbol{P}_L(\boldsymbol{r}, t) = \varepsilon_0 [\varepsilon_r(\boldsymbol{r}, t) - 1] \tag{4}$$

The linear polarization term defines the relative dielectric permittivity $\varepsilon_r(\boldsymbol{r}, t)$ of the material, while the nonlinear polarization $\boldsymbol{P}_{NL}(\boldsymbol{r}, t)$ is related to higher order polarization terms (as in the case of nonlinear material response) and to the direct interaction between light and the atoms composing the medium. Following Eq. (2), Maxwell's equations in the presence of dispersive media are

$$\begin{cases} \nabla \cdot \boldsymbol{D}(\boldsymbol{r},t) = 0, \\ \nabla \cdot \boldsymbol{H}(\boldsymbol{r},t) = 0, \\ \nabla \times \boldsymbol{E}(\boldsymbol{r},t) = -\mu_0 \partial_t \boldsymbol{H}(\boldsymbol{r},t), \\ \nabla \times \boldsymbol{H}(\boldsymbol{r},t) = \partial_t \boldsymbol{D}(\boldsymbol{r},t) = \varepsilon_0 \partial_t \boldsymbol{E}(\boldsymbol{r},t) + \partial_t \boldsymbol{P}_T(\boldsymbol{r},t) \end{cases} \tag{5}$$

11.2.1 THE FDTD ALGORITHM

Researchers have developed many different numerical methods to solve directly Maxwell's equations (5). Among them, the finite differences in the time domain (FDTD) has gained increasing popularity. The FDTD algorithm, developed in the 1960s, constitutes an efficient and powerful *ab-initio* framework to study electrodynamical systems [34, 35]. In the FDTD framework Maxwell's equations are solved simultaneously in time and space and in the presence of any material distribution. This is achieved implementing a Yee grid to discretize the computational space, and a Leap-Frog scheme as the marching time algorithm (Figure 11.1).

The Yee grid has been specifically designed to overcome the computational difficulties arising from the presence of the curl $\nabla\times$ in Maxwell's equations. Rather than computing all the components of the fields in the

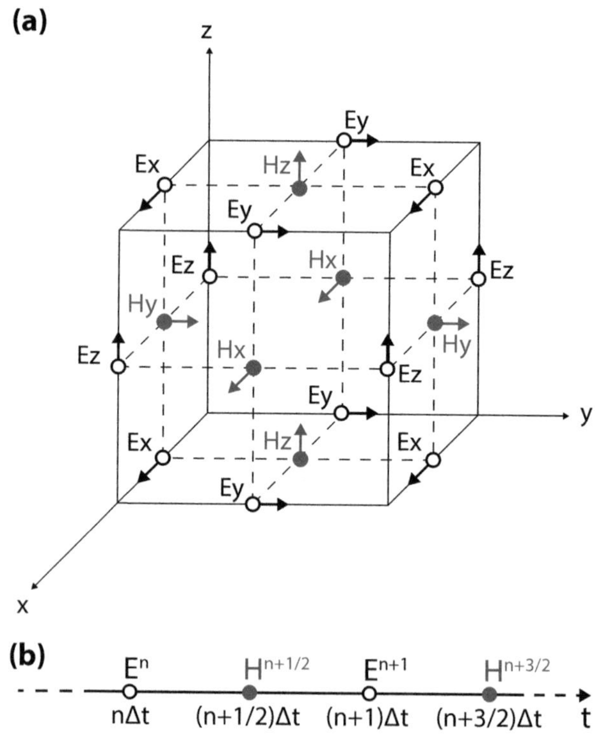

FIGURE 11.1 FDTD discretization schemes. (a) Yee grid. (b) Leap-frog marching algorithm.

same position, in fact, the electric and magnetic field components are positioned along facets or sides of the Yee cube, fulfilling the condition that each component is surrounded by its curl associates (Figure 11.1a). Time evolution, conversely, is derived within the Leap-Frog scheme, which is a second order, very robust time marching algorithm (Figure 11.1b). This allows for the precise knowledge of the electric and magnetic field in any point of space and time with arbitrary precision. The evolution in time of the electric and magnetic fields obeys to the standard Maxwell's update equations

$$\begin{cases} \partial_t H(\boldsymbol{r},t) = -\dfrac{1}{\mu_0} \nabla \times E(\boldsymbol{r},t), \\ \partial_t E(\boldsymbol{r},t) = \dfrac{1}{\varepsilon(\boldsymbol{r})} \nabla \times H(\boldsymbol{r},t) - \dfrac{1}{\varepsilon(\boldsymbol{r})} \partial_t \boldsymbol{P}_T(\boldsymbol{r},t). \end{cases} \tag{6}$$

where $\varepsilon(\boldsymbol{r}) = \varepsilon_0 \varepsilon_r(\boldsymbol{r})$ is the total dielectric permittivity of the material. The vector update Eq. (6) represent a system of six scalar differential equations, each ruling the evolution of a single component of the electromagnetic field. The discretization of the scalar equation on the Yee grid is based on a central difference expression of the derivatives as

$$\frac{\partial u}{\partial_x}(i\Delta x, j\Delta y, k\Delta z, n\Delta t) = \frac{u\left[\left(i+\dfrac{1}{2}\right)\Delta x, j\Delta y, k\Delta z, n\Delta t\right] - u\left[\left(i-\dfrac{1}{2}\right)\Delta x, j\Delta y, k\Delta z, n\Delta t\right]}{\Delta x} + O\left[(\Delta x)^2\right] \tag{7}$$

where Δx, Δy, Δz and Δt are the spatial and time discretization steps labeled by the discretization indices i, j, k, n. In the following we will adopt the convention to indicate the time dependence as a superscript and the spatial dependence as a subscript, i.e.,

$$E_x\big|_{i,j,k}^{n} = E_x(i\Delta x, j\Delta y, k\Delta z, n\Delta t) \tag{8}$$

As an example, the marching equation for the z-component of the electric field becomes

$$\partial_t E_x\big| = \frac{1}{\varepsilon_0}\left(\partial_y H_z - \partial_z H_y\right) \Rightarrow$$

$$\left.\frac{E_x\Big|^{n+\frac{1}{2}}_{i,j+\frac{1}{2},k+\frac{1}{2}} - E_x\Big|^{n-\frac{1}{2}}_{i,j+\frac{1}{2},k+\frac{1}{2}}}{\Delta t}\right| = \frac{1}{\varepsilon_{i,j+\frac{1}{2},k+\frac{1}{2}}}\left(\frac{H_z\Big|^{n}_{i,j+1,k+\frac{1}{2}} - H_z\Big|^{n}_{i,j,k+\frac{1}{2}}}{\Delta y} - \frac{H_y\Big|^{n}_{i,j+\frac{1}{2},k+1} - H_z\Big|^{n}_{i,j+\frac{1}{2},k+1}}{\Delta z}\right) \tag{9}$$

where the material properties of the material, described by the dielectric permittivity $\varepsilon(\boldsymbol{r})$, are discretized accordingly on the Yee grid.

The marching Eq. (6), in most cases, needs to be modified to include boundary conditions. The *Uniaxial Perfectly Matched Layer* (UPML), which is used to simulate open systems in the FDTD framework, represents a very robust implementation of absorbing boundary condition [2]. The UPML method is based on the definition of an absorbing material placed on the boundary of the computational domain. Such material can be introduced, in the frequency domain ω, by a complex coordinate stretching factor $S_{x,y,z}$ of the form

$$s_{x,y,z}(\omega) = \left(1 + \frac{i\sigma_{x,y,z}}{\omega}\right) \tag{10}$$

where $\sigma_{x,y,z}$ is a finite fictitious conductivity [34]. In the following, we assume a harmonic time dependence of the form $\boldsymbol{E}(\boldsymbol{r}, t) = \boldsymbol{E}(\boldsymbol{r}) \exp(jwt)$. The coordinate stretching can be included in Maxwell's equations as a modification of the dielectric permittivity $\varepsilon(\boldsymbol{r})$ of the form

$$\tilde{\varepsilon}(\boldsymbol{r}) = \frac{\mathcal{J}\varepsilon(\boldsymbol{r})\mathcal{J}^{\mathrm{T}}}{\det\mathcal{J}}, \quad \mathcal{J} = \begin{pmatrix} \frac{1}{s_x} & 0 & 0 \\ 0 & \frac{1}{s_y} & 0 \\ 0 & 0 & \frac{1}{s_z} \end{pmatrix} \tag{11}$$

Inserting the stretched $\tilde{\varepsilon}(\boldsymbol{r})$ into the marching equation for $\boldsymbol{E}(\boldsymbol{r}, \omega)$, and performing some trivial algebra, we obtain the final update equation, expressed in terms of a coupled system of equations, for an UPML enabled FDTD code

$$\begin{cases} \nabla \times H(\boldsymbol{r},\omega) = j\omega \begin{pmatrix} s_y & 0 & 0 \\ 0 & s_z & 0 \\ 0 & 0 & s_x \end{pmatrix} D(\boldsymbol{r},\omega), \\ \boldsymbol{D}(\boldsymbol{r},\omega) = \varepsilon_0 \varepsilon_\infty \begin{pmatrix} \frac{s_z}{s_x} & 0 & 0 \\ 0 & \frac{s_x}{s_y} & 0 \\ 0 & 0 & \frac{s_y}{s_z} \end{pmatrix} E(\boldsymbol{r},\omega), \end{cases} \tag{12}$$

which can be written in a more concise way as

$$\nabla \times \boldsymbol{H}(\boldsymbol{r},\omega) = j\omega \overline{\overline{s}} \cdot \boldsymbol{D}(\boldsymbol{r},\omega) \tag{13}$$

where we introduced the stretching tensor

$$\overline{\overline{s}} = \begin{pmatrix} \frac{s_y s_z}{s_x} & 0 & 0 \\ 0 & \frac{s_x s_z}{s_y} & 0 \\ 0 & 0 & \frac{s_y s_x}{s_z} \end{pmatrix} \tag{14}$$

11.3 DISPERSIVE MATERIALS IN THE FDTD FRAMEWORK

As we already mentioned, the electromagnetic properties of an inhomogeneous stationary material can be described, in the linear regime, in terms of the relative dielectric permittivity $\varepsilon(\boldsymbol{r}, \omega)$, expressed as a function of space and frequency. Depending on the frequency dependence of the dielectric permittivity, one can distinguish between *dispersive materials* (e.g., metals), whose refractive index depends on the incident wavelength, and *dielectric materials*, whose dielectric function can be considered constant across the frequency interval under examination. Introducing a dielectric material in the FDTD is simple, as it requires just to discretize the dielectric function $\varepsilon(\boldsymbol{r})$ over the Yee grid and pick-up the right value each time the electric and magnetic fields are updated. In single-wavelength simulations this approach can be extended to dispersive materials if one considers

$$\varepsilon_{\infty}(\boldsymbol{r}) = \varepsilon(\boldsymbol{r}, \omega = \omega_0) \tag{15}$$

where ω_0 is the frequency of interest. However, one of the advantages of the FDTD algorithm is its intrinsic ability to deal with broadband simulations. Dealing with fully dispersive material in this framework is a complex task, which requires the development of an efficient and robust algorithm.

11.3.1 ANALYTICAL REPRESENTATION OF DISPERSIVE MATERIALS

The standard procedure involved in the simulation of dispersive materials requires a nonlinear analytical representation of the dielectric function $\varepsilon(\omega)$ expressed in terms of a set of known functions. Historically, three models were available: the Debye relaxation, the Lorentz resonance and the Drude model of metals. These models are defined in terms of a set of parameters related to the physical phenomena occurring in the medium (e.g., intra-band transitions, infrared absorption). In each model the dielectric function is defined as a sum of complex poles or poles-pairs, whose analytical representation mimics the relevant features of the dielectric function of the medium (e.g., relaxation dynamics, resonances). Their expressions read [34]

$$\begin{aligned}
\varepsilon_{Debye}(\omega) &= \varepsilon_{\infty} + \sum_{k=1}^{N} \frac{\Delta\varepsilon_k}{1 + j\omega\tau_k}, \\
\varepsilon_{Lorentz}(\omega) &= \varepsilon_{\infty} + \sum_{k=1}^{N} \frac{\Delta\varepsilon_k \omega_p^2}{\omega_p^2 + 2j\,\omega\delta_k - \omega^2}, \\
\varepsilon_{Drude}(\omega) &= \varepsilon_{\infty} + \sum_{k=1}^{N} \frac{\omega_k^2}{\omega^2 - j\,\omega\gamma_k},
\end{aligned} \tag{16}$$

where ε_{∞} is the infinite frequency relative permittivity; $\Delta\varepsilon_k$ is the difference between the zero-frequency and the infinite frequency relative permittivities; τ_k is the pole relaxation time; ω_k is the frequency of the pole; δ_k is the damping frequency coefficient of the Lorentz oscillator; γ_k is the inverse of the relaxation time of a Drude pole.

These models, when considered singularly, are accurate only for a narrow frequency range for every set of parameters. In order to obtain a broadband model a sum of many terms is required in order to obtain an accurate representation of the dielectric function. More advanced techniques involve the introduction of mixed terms, as in the case of the Lorentz-Drude model, where Lorentz and Drude poles are coupled together, and whose applications are mainly in the analytical representation of metals. While these approaches provide a higher degree of accuracy, they usually require a large number of terms. In the case of an FDTD simulation, hence, a multi-pole formulation requires the computation and the storing of a large number of complex variables, significantly increasing the computational complexity [38].

Recently, a new analytical method, known as the Critical Point (CP) model, has been proposed to overcome these difficulties. While the CP model has been initially formulated in conjunction with the Drude model to describe Au dispersion in the 200 nm–1000 nm range [11], it has been shown that it allows for the description of a wide range of dispersive materials [17]. In the generalized CP model the dielectric function of the material is expressed as a sum of complex-conjugate poles of the form

$$\varepsilon_{CP}(\omega) = \varepsilon_\infty + \sum_{l=0}^{N} \left(\frac{c_l}{j\omega - d_l} + \frac{c_l^*}{j\omega - d_l^*} \right) \tag{17}$$

where c_l and d_l are the complex parameters which can be easily expressed in terms of the physical parameters of the Debye, Lorentz or Drude models [37] or extracted from the experimental dielectric function of the material using a nonlinear fit [39]. Using standard algebra, the expression of ε_{CP} can be simplified as

$$\begin{aligned} \varepsilon_{CP}(\omega) &= \varepsilon_\infty + \sum_{l=0}^{N} \left[\frac{c_l \left(j\omega - d_k^*\right) + c_l^* \left(j\omega - d_k\right)}{\left(j\omega - d_l\right)\left(j\omega - d_l^*\right)} \right], \\ &= \varepsilon_\infty + \sum_{l=0}^{N} \left[\frac{2\left(\Re\{c_l d_l^*\} - j\omega\Re\{c_l\}\right)}{-\omega^2 + |d_l|^2 - 2j\omega\Re\{d_l\}} \right], \end{aligned} \tag{18}$$

where $\Re\{a\}$ is the real part of the complex number a. Iterating and simplifying, the dielectric function $\varepsilon_{CP}(\omega)$ becomes

$$\varepsilon_{CP}(\omega) = \frac{\sum_l b_l (j\omega)^l}{\sum_l a_l (j\omega)^l} \tag{19}$$

where $0 < l < N$ and a_l, b_l are real parameters. This expression is easily implementable in an existing FDTD code and allows for a very fast and efficient numerical representation of dispersive materials.

11.3.2 NUMERICAL IMPLEMENTATION

Starting from Eq. (19), we developed a new implicit method, based on an unconditionally stable and implicit Crank-Nicholson algorithm. As in any FDTD technique, the starting point is given by Maxwell's update equations (6). Moving to the frequency domain ω, the update equation for $\boldsymbol{D}(\boldsymbol{r}, \omega)$ becomes

$$\begin{cases} j\omega \overline{\overline{s}} \cdot \boldsymbol{D} = \nabla \times \boldsymbol{H}, \\ \boldsymbol{D}(\boldsymbol{r},\omega) = \varepsilon_0 \varepsilon(\omega) \begin{pmatrix} \frac{s_z}{s_x} & 0 & 0 \\ 0 & \frac{s_x}{s_y} & 0 \\ 0 & 0 & \frac{s_y}{s_z} \end{pmatrix} \boldsymbol{E}(\boldsymbol{r},\omega), \end{cases} \tag{20}$$

where $\overline{\overline{s}}$ is the UPML tensor defined in Eq. (14). If we define a new vector $\boldsymbol{q}(\boldsymbol{r}, \omega)$ as

$$\boldsymbol{q}(\boldsymbol{r},\omega) = \varepsilon_0 \varepsilon_\infty \begin{pmatrix} \frac{s_z}{s_x} & 0 & 0 \\ 0 & \frac{s_x}{s_y} & 0 \\ 0 & 0 & \frac{s_y}{s_z} \end{pmatrix} \boldsymbol{E}(\boldsymbol{r},\omega) \tag{21}$$

here the displacement vector $\boldsymbol{D}(\boldsymbol{r}, \omega)$ can be expressed as

$$\boldsymbol{D}(\boldsymbol{r},\omega)=\frac{\varepsilon(\omega)}{\varepsilon_\infty}\boldsymbol{q}(\boldsymbol{r},\omega) \tag{22}$$

In the FDTD algorithm we have full knowledge of the fields at the previous timestep. As we will see, however, before computing the new value of the displacement vector $\boldsymbol{D}$ we will have to evolve the auxiliary field $\boldsymbol{q}$, so it's preferable to express the vector $\boldsymbol{q}$ $(\boldsymbol{r}, \omega)$ in terms of the known displacement field $\boldsymbol{D}$ $(\boldsymbol{r}, \omega)$ as

$$\boldsymbol{q}(\boldsymbol{r},\omega)=\frac{\varepsilon_\infty}{\varepsilon(\omega)}\boldsymbol{D}(\boldsymbol{r},\omega) \tag{23}$$

where in the last step we assumed that the material is causal, i.e., $\varepsilon(\omega) \neq 0$. Inserting Eq. (19) into (23) we obtain

$$\boldsymbol{q}(\boldsymbol{r},\omega)=\frac{\sum_l a_l (j\omega)^l}{\sum_l b_l (j\omega)^l}\boldsymbol{D}(\boldsymbol{r},\omega) \tag{24}$$

This expression is valid in the frequency domain, and we should rewrite it in the time domain in order to use it directly in the FDTD algorithm. The formulation (24), moreover, allows for an easy transfer to the time domain: considering that $(j\omega)^l \rightarrow \partial_t^l$, Eq. (24) can be rewritten as

$$\sum_{l=0}^{N} b_l \partial_t^l \boldsymbol{q}(\boldsymbol{r},t)=\sum_{l=0}^{N} a_l \partial_t^l \boldsymbol{D}(\boldsymbol{r},t) \tag{25}$$

Equation (25) is a N-th order differential equation representing the time evolution of the vector $\boldsymbol{q}$ $(\boldsymbol{r}, t)$. It's easy to see, at this point, that increasing the number of poles in the dielectric function representation increases the computational complexity significantly.

We can transform Eq. (25) into a system of N differential equations

$$\begin{cases} \boldsymbol{q}_0(\boldsymbol{r},t)=\boldsymbol{q}(\boldsymbol{r},t), \\ \quad\vdots \\ \boldsymbol{q}_l(\boldsymbol{r},t)=\partial_t \boldsymbol{q}_{l-1}(\boldsymbol{r},t), \\ \quad\vdots \\ b_N \partial_t \boldsymbol{q}_{N-1}(\boldsymbol{r},t)=-\sum_{m=0}^{N-1} b_m \boldsymbol{q}_m(\boldsymbol{r},t)+f(\boldsymbol{r},t), \end{cases} \tag{26}$$

where we defined the known term as

$$\boldsymbol{f}(\boldsymbol{r},t) = \sum_{l=0}^{N} a_l \partial_t^l \boldsymbol{D}(\boldsymbol{r},t) \tag{27}$$

The system (26), at this point, can be solved numerically using an implicit Crank-Nicholson algorithm, which is known to be unconditionally stable [29]. In the CN algorithm, the time discretization is based on a central differences scheme in time, with a marching step of the form

$$\frac{df}{dt}(n\Delta t) = F(f, n\Delta t) \Rightarrow \frac{f^{n+1} - f^n}{\Delta t} = \frac{F^{n+1} + F^n}{2} \equiv F^{n+\frac{1}{2}} \tag{28}$$

where Δt is the time discretization factor. The system (26) becomes then

$$\begin{cases} \boldsymbol{q}(\boldsymbol{r})_l^{n+\frac{1}{2}} = \dfrac{\boldsymbol{q}(\boldsymbol{r})_{l-1}^{n+1} - \boldsymbol{q}(\boldsymbol{r})_{l-1}^{n}}{\Delta t}, \\ b_N \left(\dfrac{\boldsymbol{q}(\boldsymbol{r})_{N-1}^{n+1} - \boldsymbol{q}(\boldsymbol{r})_{N-1}^{n}}{\Delta t} \right) = -\sum_{m=0}^{N-1} b_m \boldsymbol{q}(\boldsymbol{r})_m^{n+\frac{1}{2}} + \boldsymbol{f}(\boldsymbol{r})^{n+\frac{1}{2}}, \end{cases} \tag{29}$$

which finally provides the CN marching equations

$$\begin{cases} \dfrac{\boldsymbol{q}(\boldsymbol{r})_{l-1}^{n+1}}{\Delta t} - \dfrac{\boldsymbol{q}(\boldsymbol{r})_{l}^{n+1}}{2} = \dfrac{\boldsymbol{q}(\boldsymbol{r})_{l-1}^{n}}{\Delta t} + \dfrac{\boldsymbol{q}(\boldsymbol{r})_{l}^{n}}{2}, \\ \left[\left(\dfrac{b_N}{\Delta t} + \dfrac{b_{N-1}}{2} \right) \boldsymbol{q}(\boldsymbol{r})_{N-1}^{n+1} + \sum_{m=0}^{N-2} \left(\dfrac{b_m}{2} \right) \boldsymbol{q}(\boldsymbol{r})_m^{n+1} \right] \\ = \left[\left(\dfrac{b_N}{\Delta t} - \dfrac{b_{N-1}}{2} \right) \boldsymbol{q}(\boldsymbol{r})_{N-1}^{n} - \sum_{m=0}^{N-2} \left(\dfrac{b_m}{2} \right) \boldsymbol{q}(\boldsymbol{r})_m^{n} \right] + \boldsymbol{f}(\boldsymbol{r})^{n+\frac{1}{2}} \end{cases} \tag{30}$$

Before computing the evolution of the auxiliary field, we have to compute the known terms $\boldsymbol{f}(\boldsymbol{r})^{n+\frac{1}{2}}$, which are a function of the (known) displacement field $\boldsymbol{D}$ ($\boldsymbol{r}$, $n\Delta t$). By definition we have

$$\boldsymbol{f}(\boldsymbol{r})^{n+\frac{1}{2}} = \frac{\boldsymbol{f}(\boldsymbol{r})^{n+1} - \boldsymbol{f}(\boldsymbol{r})^{n}}{2} = \sum_{l=0}^{N} a_l \partial_t^l \boldsymbol{D}(\boldsymbol{r})^{n+\frac{1}{2}} \tag{31}$$

and defining a set of N variables $\{\boldsymbol{D}(\boldsymbol{r})_l, 0 \leq \mathrm{l} \leq N\}$, we can again rewrite the N-th order differential equation in a set of N first-order differential equations as

$$\begin{cases} \boldsymbol{D}(\boldsymbol{r})_0 = D(\boldsymbol{r}), \\ \vdots \\ \boldsymbol{D}(\boldsymbol{r})_l = \partial_t \boldsymbol{D}(\boldsymbol{r})_{l-1}, \\ \vdots \\ \boldsymbol{f}(\boldsymbol{r})^{n+\frac{1}{2}} = \sum_{l=0}^{N} a_l \boldsymbol{D}(\boldsymbol{r})_l^{n+\frac{1}{2}}, \end{cases} \tag{32}$$

which in the Crank-Nicholson algorithm becomes

$$\frac{\boldsymbol{D}(\boldsymbol{r})_l^{n^*+1} + \boldsymbol{D}(\boldsymbol{r})_l^{n^*}}{2} = \frac{\boldsymbol{D}(\boldsymbol{r})_{l-1}^{n^*+1} - \boldsymbol{D}(\boldsymbol{r})_{l-1}^{n^*}}{\Delta t} \Rightarrow \boldsymbol{D}(\boldsymbol{r})_l^{n^*+1} = \frac{2}{\Delta t}\left(\boldsymbol{D}(\boldsymbol{r})_{l-1}^{n^*+1} - \boldsymbol{D}(\boldsymbol{r})_{l-1}^{n^*}\right) - \boldsymbol{D}(\boldsymbol{r})_l^{n^*} \tag{33}$$

The resulting known term system will be

$$\begin{cases} \boldsymbol{D}(\boldsymbol{r})_0^{n^*} = \boldsymbol{D}(\boldsymbol{r})^{n^*} \\ \boldsymbol{D}(\boldsymbol{r})_l^{n^*+1} = \frac{2}{\Delta t}\left(\boldsymbol{D}(\boldsymbol{r})_{l-1}^{n^*+1} - \boldsymbol{D}(\boldsymbol{r})_{l-1}^{n^*}\right) - \boldsymbol{D}(\boldsymbol{r})_l^{n^*} \\ \boldsymbol{f}(\boldsymbol{r})^{n+\frac{1}{2}} = \sum_{l=0}^{N} a_l \boldsymbol{D}(\boldsymbol{r})_l^{n^* = n+\frac{1}{2}} \end{cases} \tag{34}$$

System (34) can be solved easily, and the resulting $\boldsymbol{f}(\boldsymbol{r})^{n+\frac{1}{2}}$ can be inserted directly into Eq. (30) at every timestep. In matrix form, Eq. (30) becomes

$$\begin{bmatrix} \frac{1}{\Delta t} & -\frac{1}{2} & 0 & \cdots & 0 \\ 0 & \frac{1}{\Delta t} & -\frac{1}{2} & \cdots & 0 \\ \vdots & 0 & \frac{1}{\Delta t} & \cdots & 0 \\ 0 & \vdots & \vdots & \ddots & -\frac{1}{2} \\ \frac{b_0}{2} & \frac{b_1}{2} & \cdots & \frac{b_{N-1}}{2} & \frac{b_N}{2} \end{bmatrix} \begin{bmatrix} \boldsymbol{q}_0^{n+1} \\ \vdots \\ \vdots \\ \vdots \\ \boldsymbol{q}_N^{n+1} \end{bmatrix} = \begin{bmatrix} \frac{\boldsymbol{q}_0^n}{\Delta t} + \frac{\boldsymbol{q}_1^n}{2} \\ \vdots \\ \vdots \\ \vdots \\ \left[\left(\frac{b_N}{\Delta t} - \frac{b_{N-1}}{2}\right)\boldsymbol{q}_{N-1}^n - \sum_{j=0}^{N-2}\left(\frac{b_j}{2}\right)\boldsymbol{q}_j^n\right] + \boldsymbol{f}^{n+\frac{1}{2}} \end{bmatrix} \tag{35}$$

whose solution vector $\left[q_0^{n+1} \cdots q_N^{n+1}\right]$ can be obtained inverting the coefficient matrix on the l.h.s. of (35). Due to its quasi-tridiagonal structure, the coefficient matrix can be efficiently inverted at every timestep using any general purpose computational method (e.g., LU decomposition) [29].

11.3.3 NUMERICAL EXAMPLE

In our simulations we use our own-made code NANOCPP, which has proved to be a highly-performing massively parallel FDTD simulator capable of scaling up to hundreds of thousands of processors [25]. NANOCPP is entirely written in C++, with meta-programming techniques and optimized classes designed to ensure high performance, stability and high accuracy. In order to describe a dispersive material, we provide as parameters the list of the coefficients in the numerator and denominator of (19). The standard procedure to obtain the list of coefficients involves the fitting of an experimental curve with a certain number of Drude/Lorentz poles. The resulting ε (ω), expressed as a sum of single poles, is then grouped in a single fraction using standard algebra packages. As an illustrative example, we performed a series of 60 simulations at different wavelengths to measure Ag dispersion. The computational box was set as a two-dimensional 2μm × 2μm square discretized on a 800×800 regular lattice with a 20 points thick UPML layer placed at the boundary of the computational domain. The electromagnetic source was set as a monochromatic plane wave of wavelength λ described using the TFSF formalism. The impinging wavelengths were uniformly distributed in the interval 150 nm $\leq \lambda \leq$ 750 nm and the TF region was placed at 200 nm distance from the border of the computational domain. A metallic Ag square was placed at the center of TFSF region, near its border (Figure 11.2a). The metal was simulated using a 4 poles analytic function, and the dispersion coefficients are shown in Table 11.1. The coefficients are compatible with data existing in the literature [39].

We extracted the real and imaginary parts of the refractive index $\tilde{n}(\omega)=n(\omega)+jk(\omega)$ from the attenuation inside the material and from the internal wavelength. The results, shown in Figure 11.2b, exhibit an excellent agreement beetween our simulated refractive index and actual experimental data.

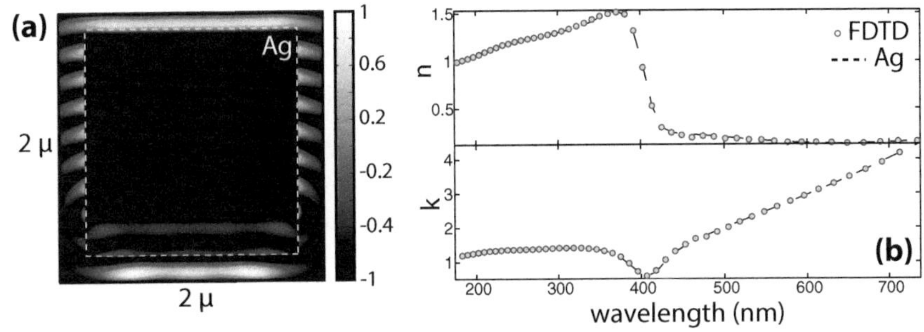

FIGURE 11.2 FDTD verification of dispersive materials. (a) Computational domain and field distribution for $\lambda = 350$ nm. (b) Real and imaginary parts of the complex refractive index as a function of the wavelength.

TABLE 11.1 Dispersion Coefficients for Ag in the 150 nm – 750 nm Range

Ag dispersion coefficients (150 nm – 750 nm)	
$b_0 = 1.6802599292925186 \times 10^{64}$	$a_0 = 0$
$b_1 = 1.8560640679695975 \times 10^{48}$	$a_1 = 8.69568126534369 \times 10^{45}$
$b_2 = 5.06708648108083 \times 10^{32}$	$a_2 = 9.622029823600002 \times 10^{31}$
$b_3 = 1.2294398178981344 \times 10^{16}$	$a_3 = 1.0445269 \times 10^{16}$
$b_4 = 1.1447$	$a_4 = 1$

11.4 MAXWELL-BLOCH EQUATIONS

11.4.1 GENERAL THEORY

Researchers have developed many methods to introduce a realistic model for gain and absorption in the FDTD simulations. The easiest approach is to introduce a material with a negative permittivity, as this would naturally produce gain. The presence of gain, however, highly reduces the stability of the simulations, and the computed fields tend to diverge even for short time intervals. A more advanced technique is based on the introduction of a set of Einstein rate equations, which describe the population dynamics of a *N*-level atomic system [31]. Such method, however, does not take into account any coherence effect and is not suitable for ultra-fast time scales [3]. In the following we will discuss an alternative theoretical approach in which Maxwell's equations are coupled with the atomic Bloch equations. The Maxwell-Bloch (MB) set of equations provides an explicit expression for the nonlinear polarization produced by the interaction between the photons and the electrons of the active material. Furthermore, the MB theory is fully derived from the quantum mechanical description of the atomic system, hence it allows for the simulation of realistic materials and direct comparison with experimental results event at the nanoscale.

In a quantum-mechanical formulation the state of the atomic system is described by a vector $|\psi\rangle$ of a complex Hilbert space. The natural basis to describe the system is provided by the complete set of energy eigenstates $\{\psi_n\}$, namely

$$\widehat{\mathcal{H}}|\psi_n\rangle = \varepsilon_n |\psi_n\rangle \tag{36}$$

being $\widehat{\mathcal{H}}$ the total Hamiltonian operator and $|\psi_n\rangle$ the eigen-state with energy ε_n.

The time evolution of the state vector is described by the Schrödinger equation

$$i\hbar\partial_t |\psi(\boldsymbol{r},t)\rangle = \widehat{\mathcal{H}}|\psi(\boldsymbol{r},t)\rangle \tag{37}$$

where $\hbar$ is the reduced Planck constant. If we assume a dipole-like interaction between the electric field and the electrons of the material, the total Hamiltonian can be written introducing a displacement operator $\widehat{\boldsymbol{Q}}$ as

$$\widehat{\mathcal{H}} = \widehat{\mathcal{H}}_0 + \widehat{\mathcal{H}}_{int} = \widehat{\mathcal{H}}_0 + e\boldsymbol{E}\cdot\widehat{\boldsymbol{Q}} \tag{38}$$

where the total Hamiltonian is split into an unperturbed term $\widehat{\mathcal{H}}_0$, representing the state of the atom when no electromagnetic field is present, and a dipole interaction term $\widehat{\mathcal{H}}_{int} = e\boldsymbol{E}\cdot\widehat{\boldsymbol{Q}}$ [5, 24]. This is the analog of the classical dipole interaction, where the interaction Hamiltonian was defined in terms of the positional displacement of the atom from its equilibrium position, namely $q = (r - r_0)$. In a quantum system, conversely, the displacement operator represents the displacement (i.e., transition) in the Hilbert space from the ground state to an excited state. Equation (38) produces a nonlinear polarization term P_{NL}, which depends on the expectation value of such displacement operator $\widehat{\boldsymbol{Q}}$, namely

$$\boldsymbol{P}_{NL} = -eN_a\left\langle\widehat{\boldsymbol{Q}}\right\rangle(t) \tag{39}$$

where e is the electric charge, N_a is the density of polarizable atoms and

$$\left\langle\widehat{\boldsymbol{Q}}\right\rangle(t) = \langle\psi(t)|\widehat{\boldsymbol{Q}}|\psi(t)\rangle \tag{40}$$

is the quantum expectation of the operator $\widehat{\boldsymbol{Q}}$ with respect to the time dependent state of the system $|\psi(t)\rangle$. Inserting the nonlinear polarization (39) into Maxwell's equation, we obtain

$$\begin{cases}\nabla\times\boldsymbol{E}(\boldsymbol{r},t) = -\mu_0\partial_t\boldsymbol{H}(\boldsymbol{r},t), \\ \nabla\times\boldsymbol{H}(\boldsymbol{r},t) = \partial_t\boldsymbol{D}(\boldsymbol{r},t) = \varepsilon_0\partial_t\boldsymbol{E}(\boldsymbol{r},t) - eN_a\partial_t\left\langle\widehat{\boldsymbol{Q}}\right\rangle(t)\end{cases} \tag{41}$$

Equation (41) describes the evolution of the electric and magnetic fields in the presence of a resonant system, whose quantum mechanical properties are associated to the evolution of the displacement operator $\widehat{\boldsymbol{Q}}$. In order to compute the nonlinear polarization term, hence, we need to study the evolution of the expectation value $\left\langle\widehat{\boldsymbol{Q}}\right\rangle$. Its evolution, as expressed by Eq. (40), is directly related to the dynamical evolution of the state vector, which is described by Eq. (37). While solving directly Eq. (37) would be the most direct approach, it is more convenient to study the evolution of the system using the density matrix formalism, which is based on the definition of the density of states operator $\hat{\boldsymbol{\rho}}(t)$ as

$$\hat{\boldsymbol{\rho}}(t) = |\psi(t)\rangle\langle\psi(t)| = \sum_l p_l(t)|\,\psi_l(t)\rangle\langle\,\psi_l(t)| \tag{42}$$

where $p_l(t)$ is the (normalized) probability associated to the eigenstate of the system $|\psi_l(t)\rangle$. Using this formalism, the expectation value of a generic operator $\widehat{\boldsymbol{O}}$ can be written in terms of the density matrix as

$$\left\langle\widehat{\boldsymbol{O}}\right\rangle(t) = \sum_l\langle\psi_l(t)|\widehat{\boldsymbol{O}}\hat{\boldsymbol{\rho}}(t)|\,\psi_l(t)\rangle = \mathrm{Tr}\left\{\widehat{\boldsymbol{O}}\hat{\boldsymbol{\rho}}(t)\right\} \tag{43}$$

The evolution of the density matrix operator is described by the Liouville equation of motion

$$i\hbar\partial_t\hat{\rho}(t)=\left[\hat{\mathcal{H}},\hat{\rho}(t)\right] \tag{44}$$

where $\left[\hat{A},\hat{B}\right]=\hat{A}\hat{B}-\hat{B}\hat{A}$ is the standard quantum commutator [28]. As we will see, solving Eq. (44) is a way less demanding operation than directly solving Eq. (37), provided an adequate reduction scheme is introduced. In our approach, we exploit the homomorphism between the Hilbert vector space and the *SU(N)* Lie Algebra, where N is the number of discrete levels, arbitrarily spaced, of the atomic system [18–21]. The transitions among the energy levels, which determine the dynamical evolution of the atomic system, can be excited by the vector components of the electromagnetic field, being $\hbar\omega_0$ the photon energy of the electromagnetic wave. The transitions are described in terms of transition-projectors, also known as level operators, of the form

$$\hat{P}_{mn}=|\psi_m\rangle\langle\psi_n| \tag{45}$$

and can rewrite the density matrix operator as

$$\left(\hat{\rho}(t)\right)_{mn}=\rho_{mn}(t)=\sum_j p_j\left(P_{nm}\mid\psi_j\rangle\langle\psi_j\mid P_{mn}\right)=\mathrm{Tr}\left\{P_{mn}\hat{\rho}(t)\right\}=\left\langle\hat{P}_{mn}(t)\right\rangle \tag{46}$$

where ρ_{mn} (t) are the N^2 time dependent components of the density matrix operator. In order to solve the system, then, we have to solve (41), coupled with (44) at each time-step. This problem is rather computationally demanding, as solving (44) involves the solution of N^2 complex-valued differential equations. However, one can simplify the equations investigating the symmetry properties of the transition-projectors. First of all, exploiting the orthonormality of the transition-projector basis, we can immediately write the $\hat{P}_{mn}$ commutator as

$$\left[\hat{P}_{ij},\hat{P}_{kl}\right]=|\psi_i\rangle\langle\psi_j|\psi_k\rangle\langle\psi_l|-|\psi_k\rangle\langle\psi_l|\psi_i\rangle\langle\psi_j|=\hat{P}_{il}\delta_{nk}-\hat{P}_{kj}\delta_{il} \tag{47}$$

The structure of the commutator unveils an important property of the $\hat{P}_{mn}$ set. In fact, if N is the total number of energy levels in the system, the full set of transition-projectors is constituted by N^2 elements, which under the commutator rule (47) generate the unitary group $U(N)$ [20]. The transition-projectors P_{mn}, furthermore, can be used to describe the $N^2 - 1$ generators of the $SU(N)$ Lie Algebra, which can be defined in terms of the normalized operators

$$\begin{cases} \hat{\boldsymbol{u}}_{lm} = \left(\hat{P}_{lm} - \hat{P}_{ml}\right), \\ \hat{\boldsymbol{v}}_{lm} = -j\left(\hat{P}_{lm} - \hat{P}_{ml}\right), \\ \hat{\boldsymbol{w}}_{l} = -\sqrt{\dfrac{2}{l(l+1)}}\left[\left(\sum_{j=1}^{l} \hat{P}_{jj}\right) - l\hat{P}_{(l+1),(l+1)}\right], \end{cases} \tag{48}$$

where $1 \leq l < m \leq N$ and $l \in [1, N-1]$ [7, 41]. In the following the generators of the $SU(N)$ algebra will be labeled as $\hat{\lambda}_k$ with $k \in [1, N^2 - 1]$. These operators, known as the Gell-Mann matrices or operators, obey the important symmetry properties

$$\begin{cases} \lambda_l = \lambda_l^{\dagger} \\ \mathrm{Tr}\left[\lambda_l\right] = 0 \\ \mathrm{Tr}\left[\lambda_l \lambda_m\right] = 2\,\delta_{lm} \end{cases}, \tag{49}$$

The Gell-Mann operators generate the Lie algebra of the $SU(N)$ group, as can be seen grouping them in a single vector $\boldsymbol{S}$, generally known as *coherence vector*, whose components are defined as

$$\boldsymbol{S} = \left(S_j\right)_{j \in \left[1, N^2-1\right]} = \left(\hat{\boldsymbol{u}}_{12}, \ldots; \hat{\boldsymbol{v}}_{12}, \ldots; \hat{\boldsymbol{w}}_1, \ldots, \hat{\boldsymbol{w}}_{N-1}\right) \tag{50}$$

and which are related by

$$\left[S_l, S_m\right] = \sum_{n=1}^{N^2-1} 2jf_{lmn} S_n \tag{51}$$

where f_{lmn} is the $SU(N)$ group completely antisymmetric structure constant [30]. Using the coherence vector components $S_l(t)$, it's possible to define the density matrix operator and the total Hamiltonian as

$$\begin{cases} \hat{\boldsymbol{\rho}}(t) = \dfrac{1}{N}\hat{\boldsymbol{I}} + \sum_{l=1}^{N^2-1} S_l(t)\hat{\lambda}_l, \\ \hat{\mathcal{H}}(t) = \dfrac{\hbar}{2}\left[\dfrac{2}{N}\left(\sum_{l=1}^{N}\omega_l\right)\hat{\boldsymbol{I}} + \sum_{l=1}^{N^2-1} h_l(t)\hat{\lambda}_l\right], \end{cases} \tag{52}$$

where

$$\begin{cases} S_l(t) = \mathrm{Tr}\left\{\hat{\rho}(t)\lambda_l\right\}, \\ h_l(t) = \frac{1}{\hbar}\mathrm{Tr}\left\{\widehat{\mathcal{H}}(t)\lambda_l\right\}. \end{cases} \tag{53}$$

This means that the evolution of the system, described by Eq. (44) as the evolution of N^2 complex density matrix elements, can be equivalently described in terms of the $(N^2 - 1)$ real components of the coherence vector $\boldsymbol{S}$. The time evolution of such components can be easily obtained from Eq. (44) substituting the expressions (52) for $\hat{\boldsymbol{\rho}}$ and $\widehat{\mathcal{H}}(t)$. We finally obtain

$$\partial_t S_l(t) = \sum_{m=1}^{N^2-1} \hat{\Gamma}_{lm} S_m \tag{54}$$

where the evolution coefficients Γ_{lm} are defined by

$$\hat{\Gamma}_{lm} = \frac{j}{2\hbar}\mathrm{Tr}\left\{\widehat{\mathcal{H}}[\hat{\lambda}_l, \hat{\lambda}_m]\right\} \tag{55}$$

which is the N-level generalization of the traditional Bloch torque equation for a two level system [5]. Equation (54) can be further extended to include relaxation times and coupling strengths, describing the dynamical properties of the energy levels, and the initial coherence vector $\boldsymbol{S}^{(0)} = \boldsymbol{S}(t = 0)$ containing the initial populations of the system. If we define a relaxation matrix γ as

$$\hat{\gamma}_{lm} = \frac{1}{T_{l\dagger}}\delta_{lm} \tag{56}$$

where T_l is the relaxation time of the l-th level, Eq. (54) becomes

$$\partial_t S_l(t) = \sum_{m=1}^{N^2-1} \hat{\Gamma}_{lm} S_m - \hat{\gamma}_{ll}\left(S_l - S_l^{(0)}\right) \tag{57}$$

which in the vector formalism becomes

$$\partial_t \boldsymbol{S}(t) = \hat{\boldsymbol{\Gamma}} \cdot \boldsymbol{S} - \hat{\gamma}\left(\boldsymbol{S} - \boldsymbol{S}^{(0)}\right) \tag{58}$$

As can be easily seen, Eq. (58) represents the full-vector evolution equation for the system, and it consists of $(N^2 - 1)$ real equations. Once the dynamics of the coherence vector is defined, we can compute the expectation value of $\langle\hat{Q}\rangle(t)$, which will be depending on the components of the vector $\boldsymbol{S}(t)$, and whose expression is in general depending on the number of levels and on the level structure describing the atomic system.

Up to this level, the dynamical equations are unconditionally deterministic. However, we can complete the quantum-mechanical treatment of light-matter interaction by introducing a numerical implementation of quantum noise [14]. This can be easily achieved introducing a fluctuating electric or magnetic field in the FDTD update equation, which simulates the presence of a thermal bath [32]. The statistical properties of the noise term represent a key element in the realistic description of a resonant material, and will be discussed extensively in the next subsection.

11.4.2 NUMERICAL IMPLEMENTATION OF A FOUR LEVEL SYSTEM

Before discussing the numerical implementation of the MB equations in the time domain, we need to specify the energy levels structure for the atomic system to be simulated. In our case, we consider a four-level system with three degenerate levels (Figure 11.3).

This structure allows for three different transitions from the level $|\psi_1\rangle$ to the upper levels, and the three levels degeneracy ensures that these transitions are mutually orthogonal. Having four energy levels, all our matrix operators will be 4×4 matrices, while the coherence vector will be a fifteen components real vector. Using the transition-projectors $\hat{P}_{lm}$, we can rewrite the total Hamiltonian $\hat{\mathcal{H}}$, defined in Eq. (38), as

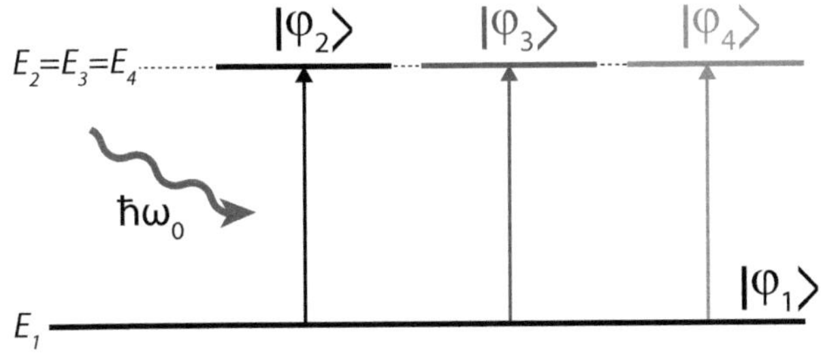

FIGURE 11.3 Energy diagram for a four levels system with three degenerate levels.

$$\begin{cases} \widehat{\mathcal{H}} = \widehat{\mathcal{H}}_0 + \widehat{\mathcal{H}}_{int} \\ \left(\widehat{\mathcal{H}}_0\right)_{ij} = \hbar\omega_0\left[\delta_{ij}\left(\delta_{i2}+\delta_{i3}+\delta_{i4}\right)\right] \\ \widehat{\mathcal{H}}_{int} = eE\cdot\widehat{\boldsymbol{Q}} = eE\cdot\left(\widehat{\boldsymbol{G}}+\widehat{\boldsymbol{G}}\ \right) \\ \left(\widehat{\boldsymbol{G}}\right)_{ij} = q_0\left[\delta_{i1}\left(\delta_{j2}\hat{\boldsymbol{x}}+\delta_{j3}\hat{\boldsymbol{y}}+\delta_{j4}\hat{\boldsymbol{z}}\right)\right] \end{cases} \tag{59}$$

where ω_0 is the energy difference between the ground level and the degenerate levels, δ_{ij} is the Kronecker delta, q_0 is the typical atomic length scale and $\left(\hat{\boldsymbol{x}},\hat{\boldsymbol{y}},\hat{\boldsymbol{z}}\right)$ are the Cartesian unit vectors. The evolution of the system is expressed by fifteen real equations of the form (58). The Gell-Mann matrices in the $N = 4$ case read [6]

$$\lambda_1 = \hat{\boldsymbol{u}}_{12} = \begin{pmatrix} 0&1&0&0\\1&0&0&0\\0&0&0&0\\0&0&0&0 \end{pmatrix} \quad \lambda_2 = \hat{\boldsymbol{u}}_{13} = \begin{pmatrix} 0&0&1&0\\0&0&0&0\\1&0&0&0\\0&0&0&0 \end{pmatrix} \quad \lambda_3 = \hat{\boldsymbol{u}}_{14} = \begin{pmatrix} 0&0&0&1\\0&0&0&0\\0&0&0&0\\1&0&0&0 \end{pmatrix}$$

$$\lambda_4 = \hat{\boldsymbol{u}}_{23} = \begin{pmatrix} 0&0&0&0\\0&0&1&0\\0&1&0&0\\0&0&0&0 \end{pmatrix} \quad \lambda_5 = \hat{\boldsymbol{u}}_{24} = \begin{pmatrix} 0&0&0&0\\0&0&0&1\\0&0&0&0\\0&1&0&0 \end{pmatrix} \quad \lambda_6 = \hat{\boldsymbol{u}}_{34} = \begin{pmatrix} 0&0&0&0\\0&0&0&0\\0&0&0&1\\0&0&1&0 \end{pmatrix}$$

$$\lambda_7 = \hat{\boldsymbol{v}}_{12} = \begin{pmatrix} 0&-j&0&0\\j&0&0&0\\0&0&0&0\\0&0&0&0 \end{pmatrix} \quad \lambda_8 = \hat{\boldsymbol{v}}_{13} = \begin{pmatrix} 0&0&-j&0\\0&0&0&0\\j&0&0&0\\0&0&0&0 \end{pmatrix} \quad \lambda_9 = \hat{\boldsymbol{v}}_{14} = \begin{pmatrix} 0&0&0&-j\\0&0&0&0\\0&0&0&0\\j&0&0&0 \end{pmatrix}$$

$$\lambda_{10} = \hat{\boldsymbol{v}}_{23} = \begin{pmatrix} 0&0&0&0\\0&0&-j&0\\0&j&0&0\\0&0&0&0 \end{pmatrix} \quad \lambda_{11} = \hat{\boldsymbol{v}}_{24} = \begin{pmatrix} 0&0&0&0\\0&0&0&-j\\0&0&0&0\\0&j&0&0 \end{pmatrix} \quad \lambda_{12} = \hat{\boldsymbol{v}}_{34} = \begin{pmatrix} 0&0&0&0\\0&0&0&0\\0&0&0&-j\\0&0&j&0 \end{pmatrix}$$

$$\lambda_{13} = \widehat{\boldsymbol{w}}_1 = \begin{pmatrix} 1&0&0&0\\0&-1&0&0\\0&0&0&0\\0&0&0&0 \end{pmatrix} \quad \lambda_{14} = \widehat{\boldsymbol{w}}_2 = \frac{1}{\sqrt{3}}\begin{pmatrix} 1&0&0&0\\0&1&0&0\\0&0&-2&0\\0&0&0&0 \end{pmatrix} \quad \lambda_{15} = \widehat{\boldsymbol{w}}_3 = \frac{1}{\sqrt{3}}\begin{pmatrix} 1&0&0&0\\0&1&0&0\\0&0&1&0\\0&0&0&-3 \end{pmatrix} \tag{60}$$

The resulting $\hat{\Gamma}$ matrix is a skew matrix with non-zero components

$$\begin{cases} \hat{\Gamma}_{1,2} = \hat{\Gamma}_{4,5} = \hat{\Gamma}_{9,10} = \omega_0 \\ \frac{1}{2}\hat{\Gamma}_{3,2} = \hat{\Gamma}_{7,4} = \hat{\Gamma}_{5,6} = \hat{\Gamma}_{12,9} = \hat{\Gamma}_{10,11} = \Omega_x \\ \hat{\Gamma}_{1,7} = \hat{\Gamma}_{2,6} = \hat{\Gamma}_{3,5} = \frac{1}{\sqrt{3}}\hat{\Gamma}_{8,5} = \hat{\Gamma}_{14,9} = \hat{\Gamma}_{10,13} = \Omega_y \\ \hat{\Gamma}_{1,12} = \hat{\Gamma}_{2,11} = \hat{\Gamma}_{3,10} = \hat{\Gamma}_{4,14} = \hat{\Gamma}_{5,13} = 2\sqrt{\frac{2}{3}}\hat{\Gamma}_{15,10} = \Omega_z \end{cases} \tag{61}$$

where we introduced the Rabi frequencies

$$\Omega_l = \frac{E_l \wp}{\hbar}, l = x, y, z \tag{62}$$

and $\wp = eq_0$ is the unperturbed atom dipole moment. In the *SU(4)* representation, the expectation value of the displacement operator becomes

$$\langle \hat{\boldsymbol{Q}} \rangle (t) = \mathrm{Tr}\left\{ \hat{\boldsymbol{\rho}}(t) \hat{\boldsymbol{Q}} \right\} = q_0 \left(S_1 \hat{\boldsymbol{x}} + S_4 \hat{\boldsymbol{y}} + S_9 \hat{\boldsymbol{z}} \right) \tag{63}$$

Equation (41), then, becomes

$$\begin{cases} \partial_t \boldsymbol{H}(\boldsymbol{r},t) = -\frac{1}{\mu_0} \nabla \times \boldsymbol{E}(\boldsymbol{r},t), \\ \partial_t \boldsymbol{E}(\boldsymbol{r},t) = \frac{1}{\varepsilon_0 \varepsilon_r} \left[\nabla \times \boldsymbol{H}(\boldsymbol{r},t) - eq_0 N_a \left(\partial_t S_1 \hat{\boldsymbol{x}} + \partial_t S_4 \hat{\boldsymbol{y}} + \partial_t S_9 \hat{\boldsymbol{z}} \right) \right] \end{cases} \tag{64}$$

which represent the update equations for the electric and magnetic field in the presence of an atomic resonant system in the MB formalism. These equations can be discretized using the usual FDTD procedure and the derivatives of the coherent vector components can be obtained directly from Eq. (58) at every time-step.

The system of equations corresponding to the vector equation (58) is composed by fifteen differential equations. The differential equations can be again solved using an implicit Crank-Nicholson algorithm as in the case of the dispersion auxiliary equations. In order to solve the system (58), again, we have to perform a numerical inversion of the coefficients matrix. While this is usually achieved implementing a *LU* decomposition of the coefficient matrix, exploiting concepts from graph theory allows for an additional speed-up of the matrix inversion [29]. The *LU* decomposition, in fact, is known to be inefficient in the presence of a highly sparse matrix, as in the case of our Γ_{lm} matrix. As we already mentioned, the Γ_{lm} matrix is a skew matrix, with only a few non-zero elements. The coefficient matrix of Eq. (58) is proportional to Γ_{lm} and inherits its sparse structure (Figure 11.4a). However, advanced graph theory algorithms allow for an efficient reordering of the matrix. In our numerical implementation, we opted for the reverse Cuthill-McKnee algorithm [9, 15], which produces a band matrix structure (Figure 11.4b).

Once the coherence vector is computed, we can implement the quantum noise generator simulating the spontaneous emission arising from the interaction with an external thermal bath. As we already mentioned, the quantum noise will be introduced in the electric field update equation in terms of a fluctuating random electric field of the form

$$\partial_t\left(\boldsymbol{E}(\boldsymbol{r},t)+\delta\boldsymbol{E}(\boldsymbol{r},t)\right)=\frac{1}{\varepsilon_0\varepsilon_r}\left[\nabla\times\boldsymbol{H}(\boldsymbol{r},t)-eq_0N_a\left(\partial_t S_1\hat{\boldsymbol{x}}+\partial_t S_4\hat{\boldsymbol{y}}+\partial_t S_9\hat{\boldsymbol{z}}\right)\right] \tag{65}$$

where $\delta\boldsymbol{E}(\boldsymbol{r},t)$ is the fluctuating random field. From a numerical point of view, the quantum noise was modeled using a pseudorandom Box-Muller

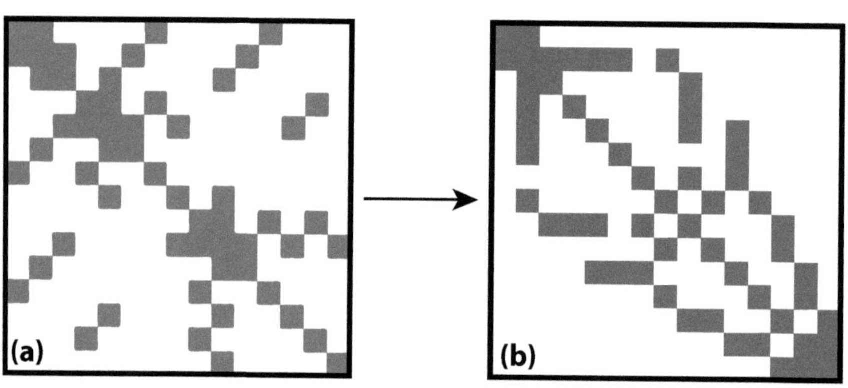

FIGURE 11.4 $\Gamma_{\rm lm}$ matrix reordering.

number generator [29]. The resulting statistical distribution of the noise term is a normal Gaussian distribution over the interval (0,1) of the form

$$\delta \boldsymbol{E}(\boldsymbol{r},t) = \sqrt{-2\xi_E \log(a)}\cos(2\pi b) \tag{66}$$

where ξ_E is the random electric field variance and (a,b) are random numbers following an uniform distribution in the interval (0, 1). This distribution ensures a white Gaussian noise time such as

$$\left\langle \delta\boldsymbol{E}(\boldsymbol{r},t)\delta\boldsymbol{E}(\boldsymbol{r},t')\right\rangle = \xi_E \delta\left(\frac{t-t'}{\Delta t}\right) \tag{67}$$

i.e., the fluctuating terms are delta-correlated in time. Finally, in order to express the quantum noise in terms of a realistic thermal bath, we relate the electric field variance ξ_E to the temperature of the external environment T_{ext}, imposing

$$\xi_E^2 = \frac{\pi^2 \varepsilon_0 \left(k_B T_{env}\right)^4}{15\left(c\hbar\right)^3} \tag{68}$$

where k_B is the Boltzmann constant and c is the speed of light [27].

11.4.3 EXAMPLES

As can be easily understood, the possibility to implement a realistic resonant medium opens a wide range of applications and investigations. As we mentioned, in our simulations we use our own-made FDTD code NANOCPP, which represents any dispersive active material in terms of the following parameters:

- $\{a_n, b_n\}$, representing the real coefficients used to describe the material dispersion using equation (19),
- N_a, representing the density of polarizable atoms, which is directly related to the incident pumping rate,
- q_0, representing the atomic length scale,
- ω_0, representing the transition frequency,
- τ_0 and τ_1, representing the upper and lower relaxation times.

11.4.3.1 Active Core-Shell Nanparticles

Starting from the pioneering work of G. Mie, spherical scatterers and resonators have proved to play a crucial role in many applications. This is mainly due to the high numbers of spherical modes, i.e., resonances, which can be excited from an impinging electromagnetic wave. These modes, moreover, can be computed exactly using analytical tools [4]. When a mode of the spherical structure is excited, the system behaves as a resonator, and is able to store or scatter high amounts of energy. Among different models and structures, core-shell micro and nanoparticles have had a wide diffusion in many fields, as in the case of biological imaging, medical diagnostics, nanolasers, and CQD photovoltaics [1, 12, 13, 22, 26, 36]. This is mainly due to the possibility to precisely control and engineer their resonances. Generally speaking, an active core-shell nanoparticle is a two layers structure: a dielectric or metal sphere of diameter d_1, constituting the core of the structure, surrounded by a dye shell of diameter d_2 (Figure 11.5a). The structure is illuminated by a plane wave of frequency ω_{inc}. The dye absorbs at the frequency ω_a close to the impinging frequency ω_{inc} and emits at a lower frequency ω_0 corresponding to one of the transitions of the four level atomic system. In our simulations, we assume that the population inversion is already reached, so we focus only on the emission frequency ω_0. If the structure is designed in such a way that one of its mode frequency is in proximity of the emission frequency of the dye, the emitted radiation can excite it and produce a standing radiation pattern (e.g., Mie whispering gallery modes) or a Surface Plasmon Polariton on the surface of the metallic nanoparticle constituting the core.

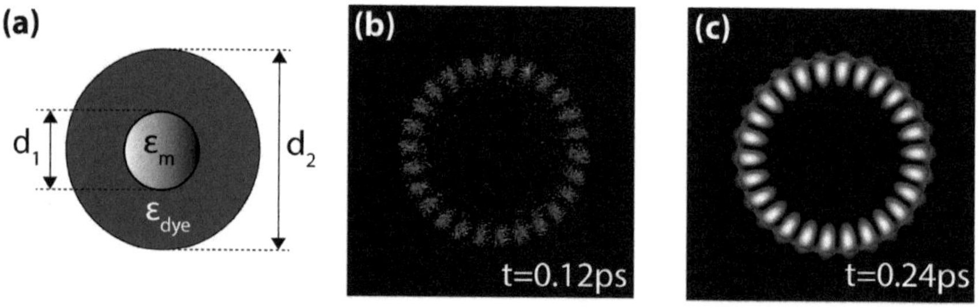

FIGURE 11.5 Active core-shell nanoparticle. (a) Scheme of the structure. (b-c) Emission regimes at different times.

Numerically, the dye is modeled as a dielectric medium with $\varepsilon_r = 2.62$, $q_0 = 0.1nm$, $T_0 = 100fs$, $T_1 = 10fs$ and $\omega_0 = 6.2788 \times 10^{15} \frac{rad}{s}$, which are parameters corresponding to a Rhodamine-B dye [33, 40]. The pumping rate is assumed to be constant and is set in terms of the polarizable atoms density $N_a = 10^{26}\ m^{-3}$. The structure was designed as metal core-shell: the core is a Ag sphere with diameter $d_1 = 300$ nm surrounded by a 200 nm thick Rhodamine layer ($d_2 = 500$ nm). As can be seen from Figure 11.6(b-c), the system moves from an incoherent spontaneous emission regime (Figure 11.3b) to a coherent emission regime. Varying the pumping rate, it's possible to study the dynamics of the emitting modes. In particular, at higher powers, multiple modes of the structure can be excited and many phenomena related to random lasers (e.g., frequency pulling, mode competition) can be observed [12, 13, 37, 38].

11.4.3.2 Organic Solar Cells Concentrator

As is well known, photovoltaics represent a field of great interest in our days. In recent years researchers have focused on designing, implementing and producing high-efficiency, low-cost solar cells. This can be obtained either reducing the cost of fabrication and maintenance, or increasing the conversion efficiency of the solar cell itself. While the production cost of semiconductors materials, as in the case of Si, has been significantly reduced, higher efficiencies can be obtained only introducing new materials or structures. In this context, the application of organic layers and heterostructures into the solar cell design is gaining increasing interest, as in the case of the Organic Solar cell Concentrators (OSC) [8]. A OSC consists of an organic dye layer deposited on top of a high-refractive index glass

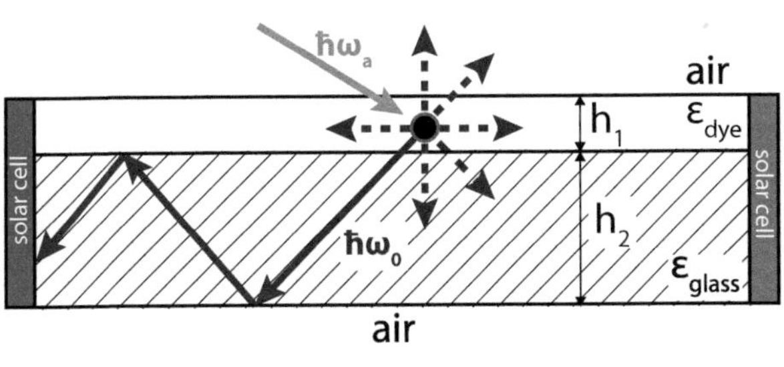

FIGURE 11.6 Scheme for an OSC enabled solar cell.

FIGURE 11.7 Field distribution at three different times for a FDTD simulation of OSC.

substrate (Figure 11.6). The organic dye, which can be even a multilayer structure, absorbs the incident sunlight and re-emits at lower frequency. A significant fraction (~80%) of the intrinsically omnidirectional spontaneous emission is captured by the glass substrate, which behaves as a slab waveguide. The emitted photons, then, can be captured and converted by a standard solar cell placed on the lateral sides of the glass substrate [15].

This device can be naturally simulated using FDTD, provided an opportune set of parameters is chosen for the organic dye. As an example, we simulated a small portion of OSC in order to show how the spontaneous emission of the dye molecules can excite one or more modes of the underlying glass waveguide. The structure is formed by a 100 μm × 10 μm high refractive index glass, with $\varepsilon_{glass} = 4$. A $1\mu m$ thick layer of organic dye was deposited on top of the slab. The dye parameters are $\omega_0 = 3.0381 \times 10^{15}$, $\varepsilon_r = 2.78$ and $N_a = 10^{10} m^{-3}$, which correspond to a AlQ3:DCJTB organic dye with doping concentration equal to 2% [10]. In Figure 11.7 we show the field distribution inside the OSC at three different times.

KEYWORDS

- **dispersive materials**
- **finite difference-time domain (FDTD)**
- **group theory**
- **Maxwell equations**
- **Maxwell-Bloch (MB)**
- **polarization**

REFERENCES

1. Adachi, M. M., et al., (2013). *Sci Rep-UK 3.*
2. Berenger, J. P., (1994). *J Comput Phys 114*(2), 185–200.
3. Bidegaray-Fesquet, B., et al., (2004). *Math Mod Meth Appl S 14*(12), 1785–1817.
4. Bohren, C. F. & Huffman, D. R. (1983). *Absorption and Scattering of Light by Small Particles*. New York, Wiley.
5. Boyd, R. W. (2008). *Nonlinear Optics*. Amsterdam; Boston, Academic Press.
6. Chen-Tsai, C. T. & Lee, Y. Y., (1965). *Chinese Journal of Physics 3*(1), 23.
7. Cornwell, J. F. (1984). Group Theory in Physics. London; Orlando, Academic Press.
8. Currie, M. J., et al., (2008). *Science 321*(5886), 226–228.
9. Cuthill, E., & McKee, J. (1969). Reducing the bandwidth of sparse symmetric matrices. Proceedings of the 1969 24th National Conference, ACM: 157–172.
10. Dalasinski, P., et al., (2004). *Opto-Electron Rev 12*(4), 429–434.
11. Etchegoin, P. G., et al., (2007). *J Chem Phys 127*(18).
12. Fratalocchi, A., et al., (2008). Opt Express 16(12), 8342–8349.
13. Fratalocchi, A., et al., (2008). *Physical Review A 78*(1).
14. Gardiner, C. W. & Zoller, P. Quantum noise: a handbook of Markovian and non-Markovian quantum stochastic methods with applications to quantum optics. Berlin; New York, Springer, 2004.
15. George, A. & Liu, J. W. H. *Computer Solution of Large Sparse Positive Definite Systems*. Englewood Cliffs, N.J., Prentice-Hall, 1981.
16. Goetzberger, A. & Greubel, W., (1977). *Appl Phys 14*(2), 123–139.
17. Han, M. H., et al., (2006). *IEEE Microw Wirel Co 16*(3), 119–121.
18. Hioe, F. T., (1983). *Physical Review A 28*(2), 879–886.
19. Hioe, F. T., (1985). *Physical Review A 32*(5), 2824–2836.
20. Hioe, F. T. & Eberly, J. H., (1981). *Phys Rev Lett 47*(12), 838–841.
21. Hioe, F. T. & Eberly, J. H., (1982). *Physical Review A 25*(4), 2168–2171.
22. Hirsch, L. R., et al., (2003). *P Natl Acad Sci USA 100*(23), 13549–13554.
23. Jackson, J. D. (1999). *Classical Electrodynamics*. New York, Wiley.
24. Jonsson, F. (2003). *Nonlinear* Optics, Kungl Tekniska Hogskolan.
25. Liu, C., et al., (2013). *Nat Photonics 7*(6), 474–479.
26. Liu, W., et al., (2012). *Acs Nano 6*(6), 5489–5497.
27. Loudon, R. (2000). *The Quantum Theory of Light*. Oxford; New York, Oxford University Press.
28. Messiah, A. (1999). *Quantum Mechanics*. Mineola, NY, Dover Publications.
29. Press, W. H. (2002). *Numerical Recipes in C++: The Art of Scientific Computing*. Cambridge, UK; New York, Cambridge University Press.
30. Puri, R. R. (2001). *Mathematical Methods of Quantum Optics*. Berlin; New York, Springer.
31. Siegman, A. E. (1986). Lasers. *Mill Valley*, Calif., University Science Books.
32. Slavcheva, G. M., et al., (2004). *IEEE J Sel Top Quant 10*(5), 1052–1062.
33. Sperber, P., et al., (1988). *Opt Quant Electron 20*(5), 395–431.
34. Taflove, A. & Hagness, S. C. (2005). *Computational Electrodynamics: The Finite-Difference Time-Domain Method*. Boston, Artech House.

35. Taflove, A., Johnson S. G. & Oskooi A (2013). *Advances in FDTD Computational Electrodynamics: Photonics and Nanotechnology,* Artech House Antennas and Propagation Library, USA
36. Tang, J. A. & Sargent, E. H., (2011). *Adv Mater 23*(1), 12–29.
37. Totero Gongora, J. S., Miroshnichenko, A. E., Kivshar, Y. S. & Fratalocchi, A. Energy equipartition and unidirectional emission in a spaser nanolaser. *Laser & Photonics Reviews* 10, 432–440 (2016).
38. Totero Gongora, J. S., Miroshnichenko, A. E., Kivshar, Y. S. & Fratalocchi, A. Anapole nanolasers for mode-locking and ultrafast pulse generation. *Nature Communications*, to appear (2017).
39. Udagedara, I., et al., (2009). *Opt Express 17*(23), 21179–21190.
40. Vial, A., (2007). *J Opt a-Pure Appl Op 9*(7), 745–748.
41. Vial, A., et al., (2011). *Appl Phys a-Mater 103*(3), 849–853.
42. Wuestner, S., et al., (2010). *Phys Rev Lett 105*(12).
43. Ziolkowski, R. W., et al., (1995). *Physical Review. A 52*(4). 3082–3094.

CHAPTER 12

SYNTHESIS AND CHARACTERIZATION OF MULTI-COMPONENT NANOCRYSTALLINE HIGH ENTROPY ALLOY

HEENA KHANCHANDANI,[1] JAIBEER SINGH,[1] PRIYANKA SHARMA,[1] RUPESH KUMAR,[1] ORNOV MAULIK,[1] NITISH KUMAR,[2] and VINOD KUMAR[3,4]

[1]*Department of Metallurgical and Materials Engineering, MNIT, Jaipur, India*

[2]*Centre for Nanotechnology, Central University of Jharkhand, Ranchi–835205, India*

[3]*Assistant Professor, Department of Metallurgical and Materials Engineering, MNIT, Jaipur–302017, India, Tel.: +91-141-2713457, E-mail: vkt.meta@mnit.ac.in*

[4]*Adjunct Faculty, Materials Research Centre, MNIT, Jaipur–302017, India*

CONTENTS

ABSTRACT

An AlMgFeCuCrNi based high entropy alloy was synthesized by mechanical alloying. Phase analysis at room temperature was investigated by using X-ray diffraction. It has been found that two phase solid-solution with body-centered cubic (BCC) and face-centered cubic (FCC) crystal structure forms in this alloy system. Damping capacity of powder samples was determined using dynamic mechanical analysis. Effect of sintering at different temperatures, such as 800°C, 850°C, and 900°C, on phase evolution and hardness was investigated.

12.1 INTRODUCTION

High entropy alloys are a new generation multi-component alloys and are quite different from traditional alloys, which are based on one or two elements. These multi-component alloys are solid solutions with equiatomic or near equiatomic compositions [1]. However, the first impressive report by Yeh et al. [2, 3] has shown that alloy systems with five or more metallic elements possess higher configurational entropy and therefore, favor the formation of solid solution phases, as opposed to the inferred complex structures consisting of many intermetallic compounds. The configurational entropy at equiatomic compositions for binary, ternary, quaternary, quinary and hexanary alloys is 5.8, 9.2, 11.6, 13.5, and 15.0, respectively. The formation of amorphous phase, which can be another competing phase in these multi-component systems, can be avoided by choosing elements carefully.

High entropy alloys are microcrystalline and their properties can be significantly enhanced if they can be synthesized in nanocrystalline form [4–7]. Various processing routes such as casting, sputtering, splat quenching, mechanical alloying (MA), etc. have been used to synthesize high entropy alloys in recent studies [8] and among them, MA is a widely used solid state processing route for the synthesis of homogeneous nanomaterials [8–11]. It results in a decrease in tendency of ordering and leads to extended solid solubility. Therefore, the present study is taken up to investigate the synthesis and characterization of nanocrystalline hexanary ($AlMgFeCuCrNi_{3.17}$) multi-component alloy using mechanical alloying.

12.2 EXPERIMENTAL DETAILS

The elemental powders of Al (98% purity), Mg (99% purity), Cr (99% purity), Cu (99.5% purity), Fe (99.5% purity), and Ni (99.9% purity) were mechanically alloyed using Fritsch Pulverisette-P6 high energy planetary ball mill with tungsten carbide balls. The ball-to-powder weight ratio was 10:1 and toluene was used as a process controlling agent in order to avoid excessive cold welding and also act as a reducing medium to avoid oxidation of the alloy (8). Rotational speed of disc was 300 rpm. In order to confirm the alloy formation and phase evolution during milling, powder samples were taken out after every 5 hr intervals (5, 10, 15, and 20 hr). After 20 hr of milling, the powder was completely taken out for further consolidation and characterization. After the successful synthesis of $AlMgFeCuCrNi_{3.17}$ alloy powders, the compaction was carried out in 12 mm diameter high speed steel die with a load 10 tons using simple hydraulic press (Kimaya Engineers – 15 tons manual heating and cooling press) at ~ 200°C. Three hot compacted samples were prepared followed by sintering in an electric resistance furnace at 800°C, 850°C, and 900°C, respectively, for 2 hr. The phases present in as milled and sintered samples were studied by X-ray diffraction (XRD) using X'Pert Pro Panalytical X-ray diffractometer with Cu Kα radiation. The crystalline nature of as milled powder was also studied by using transmission electron microscope (Tecnai G^2 20 FEI). For TEM studies, powder sample was kept in a beaker filled with ethanol which was further placed in an ultrasonic cleaner for about 50 min, then dispersed powder sample was allowed to settle down for 10 min. Powder sample was then spread on top surface of Cu grit. The hardness of the sintered samples was measured using micro hardness tester using 25 kg load. Dynamic mechanical analysis of powder sample was performed and temperature was varied from room temperature to 400°C at the rate of 5°C/min.

12.3 RESULTS AND DISCUSSION

12.3.1 PHASE ANALYSIS

Figure 12.1 illustrates the XRD pattern of $AlMgFeCuCrNi_{3.17}$ alloy as a function of milling time. It was evident from the Figure 12.1 that after 10 min of milling, peaks of all the elements present in the alloy system can be identified in the figure and as the milling proceeds, peaks of some elements start disappearing and width of Ni peak increases. Throughout the milling process, the decrease in intensity, broadening of the peak and its subsequent disappearance may result from the three factors: refined crystal size, high lattice strain and decreased crystallinity (10, 11). A careful observation of the XRD peaks in this alloy suggests some asymmetry in the peaks. Analysis of these patterns based on deconvolution of the peaks indicates the presence of FCC phase with a lattice parameter of 3.52 Å and BCC phase with a lattice parameter of 2.86 Å. Thus, the alloy is mainly composed of two phases, i.e., FCC and BCC structure with BCC as major

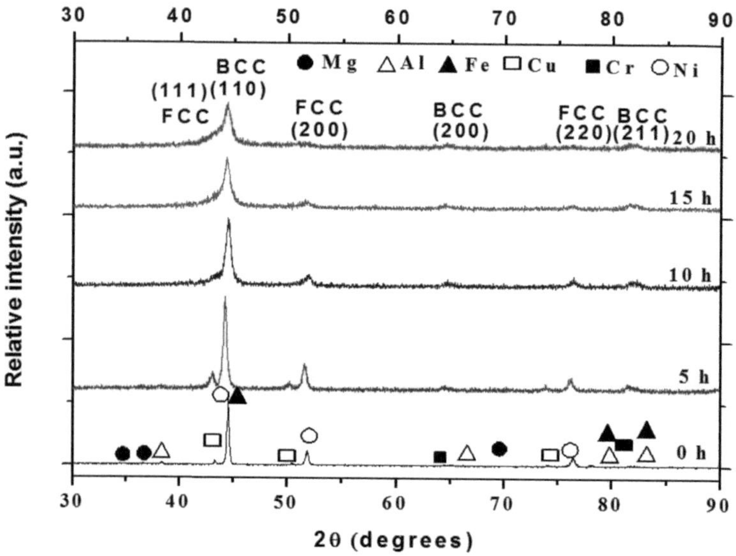

FIGURE 12.1 X-ray diffraction pattern of the $AlMgFeCuCrNi_{3.17}$ alloy as a function of milling time (0 to 20 hours).

phase in as milled condition. The crystallite size, calculated by Scherrer's formula after eliminating the instrumental and the strain contributions, after 20 hr of milling is 6.89 nm for FCC phase and 10.24 nm for BCC phase.

XRD pattern for sintered samples at three different temperatures, such as 800°C, 850°C and 900°C is shown in Figure 12.2(a) and shift in main XRD peak is illustrated in Figure 12.2(b). After sintering, ordered FCC structural phase was evolved as the major phase. The metastable state of the solid solution caused by MA converts to a more stable phase after annealing, thereby resulting in the phase transition from major BCC phase to FCC phase as a main phase. It was also observed that with an

increase in sintering temperature, peak shifts towards left side which indicates the development of tensile stresses in the sample. Based on above results, we can draw a conclusion that both the as-milled and the as-annealed $AlMgFeCuCrNi_{3.17}$ alloy samples mainly have a simple solid solution structure. The formation of amorphous phase has been avoided in this alloy system by choosing elements in such a way that the value of atomic size difference (δ) is 9.27 which is highly consistent with criteria for solid solution formation [12].

12.3.2 TEM ANALYSIS FOR POWDER

TEM bright field image and selected area electron diffraction pattern of powder particle shown in Figures 12.3 and 12.4 confirms the nanocrystalline nature of $AlMgFeCuCrNi_{3.17}$ high entropy alloy.

The analysis of rings in SAD pattern indicates that the phase has FCC + BCC structure. Bright field image suggests that further addition of alloying elements is permissible in the proposed high entropy alloy for the formation of solid solution.

12.3.3 MICRO HARDNESS TEST

Micro hardness test was done for all the three sintered samples. Hardness vs sintering temperature graph is shown in Figure 12.5. Hardness decreases with an increase in sintering temperature.

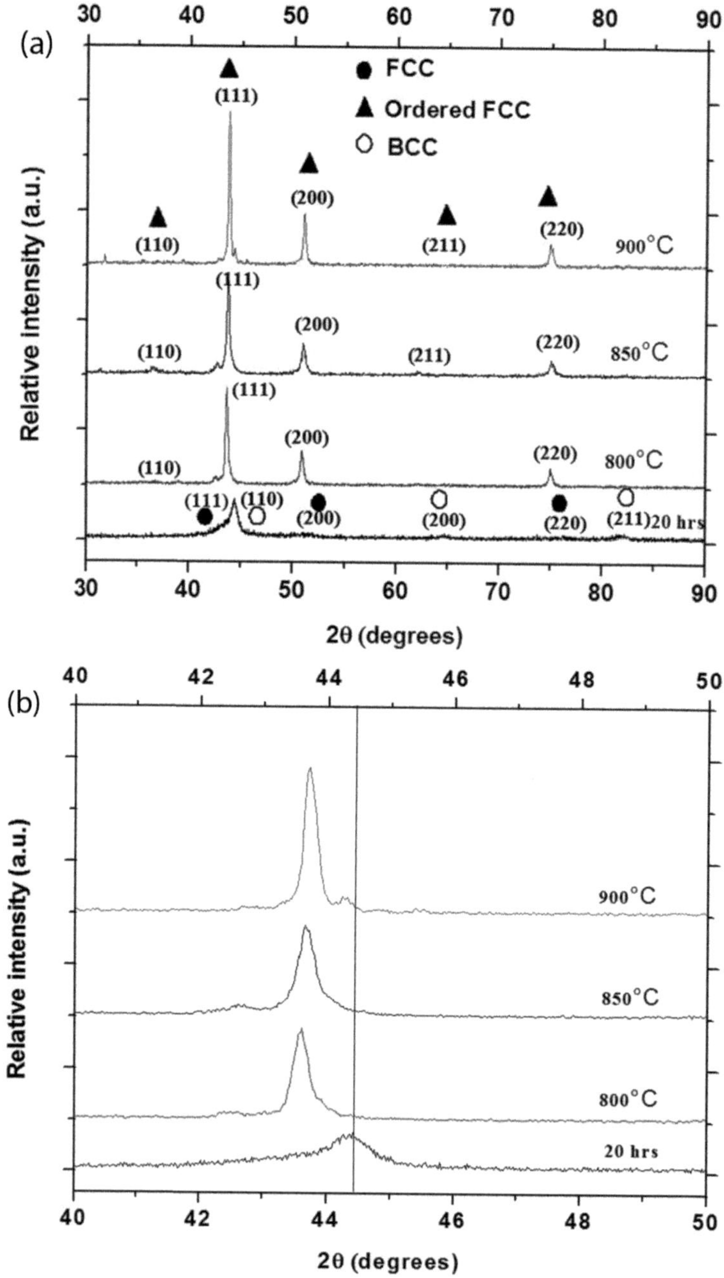

FIGURE 12.2 (a) X-ray diffraction pattern for samples sintered at different temperatures, (b) Shift in XRD peaks for sintered samples with respect to 20 h milled sample.

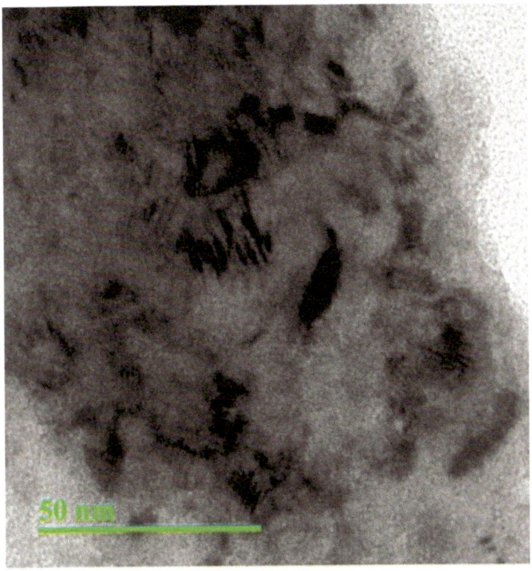

FIGURE 12.3 TEM bright field image of powder particles

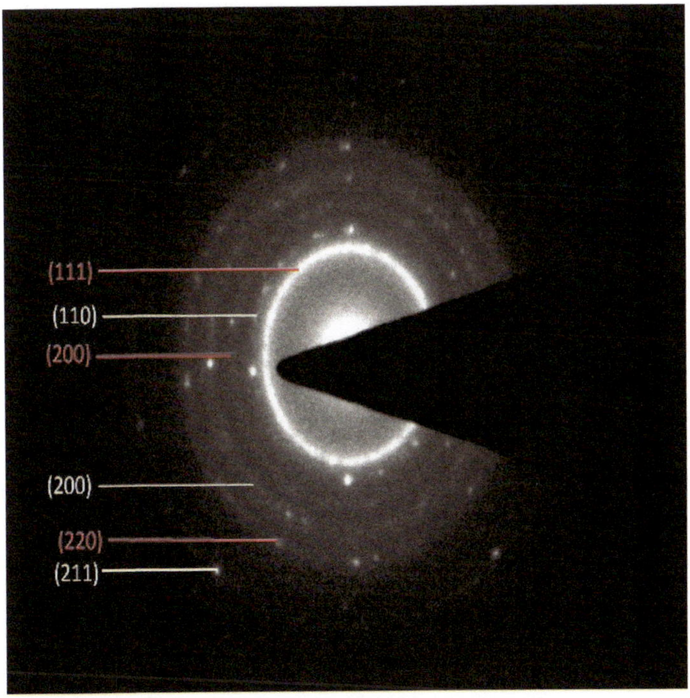

FIGURE 12.4 SAED pattern of powder particle of Figure 12.3.

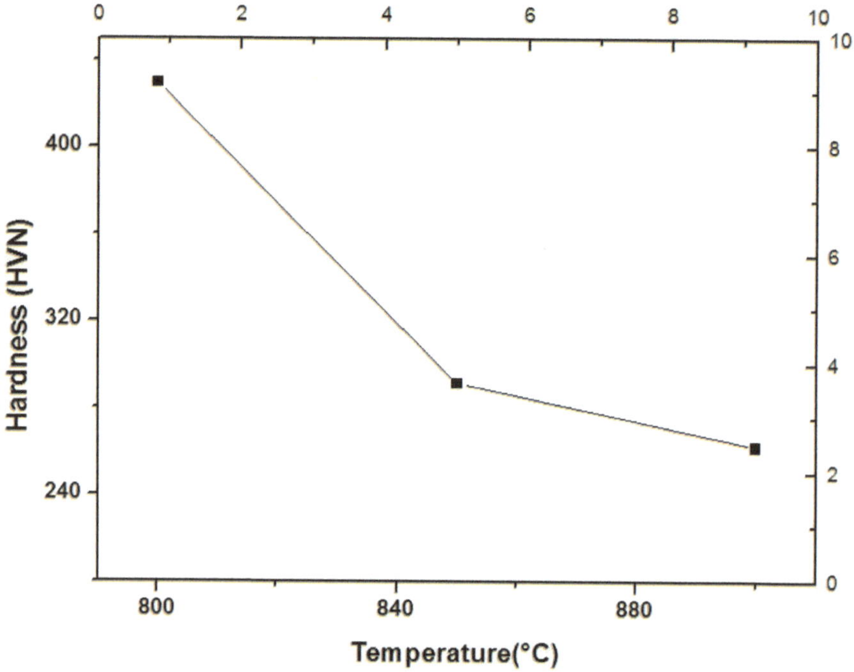

FIGURE 12.5 Variation of hardness as a function of sintering temperature.

Average value of hardness is 429.7 HVN for 800°C sintered sample, is 291.1 HVN for 850°C sintered sample and it is 262 HVN for 900°C sintered sample.

12.3.4 DYNAMIC MECHANICAL ANALYSIS OF POWDER

Dynamic response of as milled powder was characterized in terms of *tan δ*, loss modulus and storage modulus. *tan δ* vs. temperature graph is shown in Figure 12.6; loss modulus vs. temperature graph is shown in Figure 12.7 and storage modulus vs. temperature graph is shown in Figure 12.8.

tan δ is evaluated as:

$$\tan\delta = \frac{E''}{E'};$$

where E” is loss modulus and E’ is storage modulus.

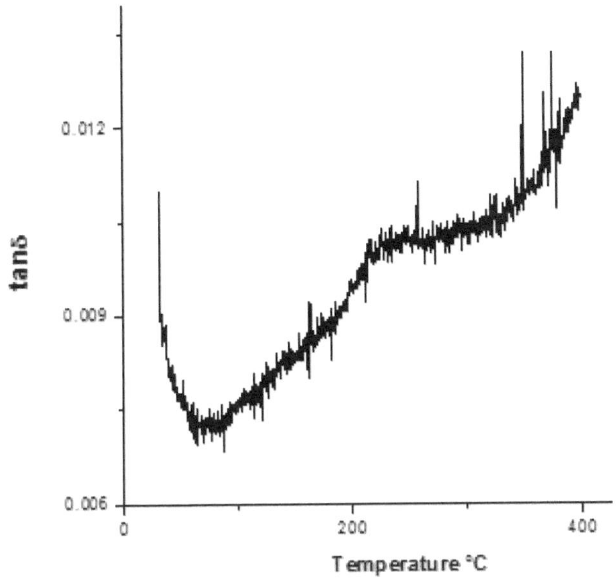

FIGURE 12.6 Variation of *tan δ* with respect to temperature.

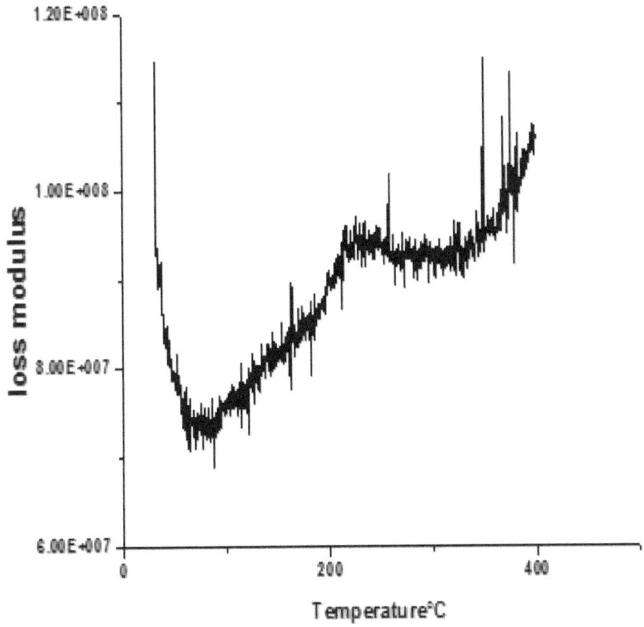

FIGURE 12.7 Variation of loss modulus w.r.t. temperature.

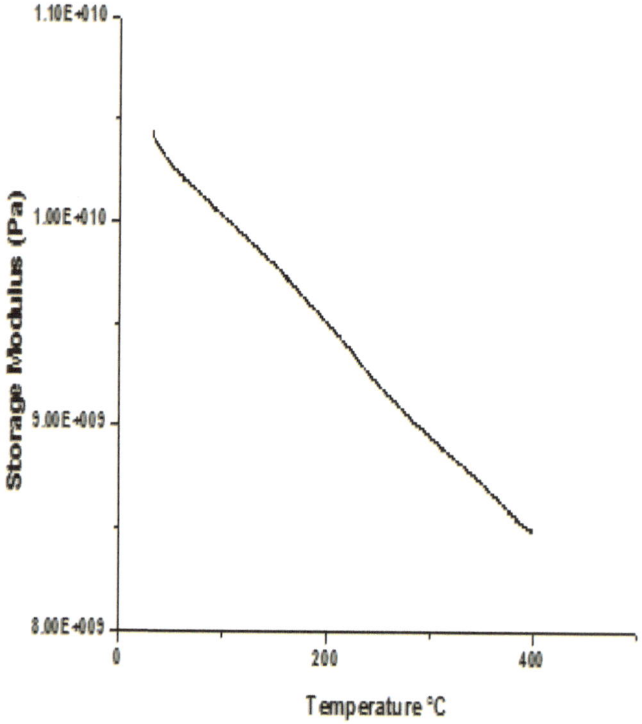

FIGURE 12.8 Variation of storage modulus w.r.t. temperature.

Loss modulus first decreases and then increases with temperature where as storage modulus continuously decreases with temperature. *tan* δ, which is a measure of damping, first decreases and then increases with temperature.

12.4 CONCLUSION

The nanocrystalline $AlMgFeCuCrNi_{3.17}$ alloy powder has been successfully synthesized by mechanical alloying. TEM study suggests that more alloying elements may be allowed to form solid solution. A major BCC structured solid solution with minor FCC phase was obtained after 20 hr of MA process which converts into major FCC phase after sintering at high temperature.

ACKNOWLEDGMENT

Authors thank DST-SERB Govt. of India project no. 178600036 for financial support.

KEYWORDS

- **dynamic mechanical analysis**
- **high entropy alloy**
- **mechanical alloying**
- **solid solution**
- **X-ray diffraction**

REFERENCES

1. Ranganathan, S. (2003). *Curr. Sci. 85*, 1404–1406.
2. Yeh, J. W., Chen, S. K., Gan, J. Y., Chin, T. S., Shun, T. T., Tsau, C. H., & Chang, S. Y., (2004). *Adv. Eng. Mater. 6*, 299–303.
3. Yeh, J. W., Chang, S. Y., Hong, Y. D., Chen, S. K., & Lin, S. J., (2007). *Mater. Chem. Phos. 103*, 41–46.
4. Tong, C. J., Chen, Y. L., Chen, S. K., Yeh, J. W., Shun, T. T., Tsau, C. H., Lin, S. J., & Chang, S. Y., (2005). *Metall. Mater. Trans. A 36*, 881–893.
5. Tong, C. J., Chen, M. R., Chen, S. K., Yeh, J. W., Shun, T. T., Lin, S. J., & Chang, S. Y., (2005). *Metall. Mater. Trans. A 36*, 1263–1271.
6. Hsu, Y. J., Chiang, W. C., Wu, J. K., (2005). *Mater. Chem. Phys. 92*, 112–117.
7. Chen, T. K., Shun, T. T., Yeh, J. W., & Wong, M. S., (2004). *Surf. Coat. Technol. 199*, 193–200.
8. Praveen, S., Murty, B. S., & Ravi Kottada, S., (2012). *Materials Science and Engineering A 534*, 83–89.
9. Gleiter, H., (1989). *Prog. Mater. Sci. 33*, 233.
10. Suryanarayana, C., (2001). *Prog. Mater. Sci. 46*, 1.
11. Varalakshmi, S., Kamaraj, M., & Murty, B. S., (2010). *Materials Science and Engineering A 527*, 1027–1030.
12. Sheng Guo, & Liu, C. T. (2011). Phase stability in high entropy alloys: Formation of solid-solution phase or amorphous phase, *Progress in Natural Science: Materials International 21*, 433–446.

INDEX

B

E

N

O

P

Q

R